VIDEO COMMUNICATIONS
THE WHOLE PICTURE

BY

JAMES R. WILCOX

WITH DAVID K. GIBSON

ILLUSTRATIONS BY

LAURA LONG AND MICHEL L. BAYARD

CMP**Books**
San Francisco

Published by CMP Books
An imprint of CMP Media, Inc.
600 Harrison St., San Francisco, CA 94107 USA
Web: www.cmpbooks.com Email: books@cmp.com

For individual orders, and for information on special discounts for quantity orders, please contact:
CMP Books Distribution Center, 6600 Silacci Way, Gilroy, CA 95020
Tel: 800-500-6875 or 408-848-3854 Fax: 408-848-5784 Email: bookorders@cmp.com

Distributed to the book trade in the US and Canada by:
Publishers Group West, 1700 Fourth St., Berkeley, CA 94710

Distributed in Canada by:
Jaguar Book Group, 100 Armstrong Avenue, Georgetown, Ontario, M6K 3E7 Canada

Library of Congress Cataloging-in-Publication Data
Wilcox, James R.
  Video communications : the whole picture / by James R. Wilcox, with David K. Gibson ; illustrations by Laura Long and Michel L. Bayard-- 4th ed.
     p. cm.
  Rev. ed. of: Videoconferencing & interactive multimedia. 2000.
  Includes bibliographical references and index.
  ISBN 1-57820-316-3 (alk. paper)

1. Videoconferencing. 2. Computer conferencing. 3. Interactive multimedia. 4. Telecommunication systems. I. Gibson, David K. II. Wilcox, James R. Videoconferencing & interactive multimedia. III. Title.

  HF5734.7.W56 2005
  384.55'6--dc22

                                        2005008087
Illustrations by Laura Long and Michel Bayard
Copy editing: James R. Wilcox
Text design and layout: Jon Lee and Associates

Transferred to Digital Printing 2008

ACKNOWLEDGMENTS

I get by with a little help from my friends.

The Beatles

I would like to acknowledge Toby Trowt-Bayard for her foresight and initiative in writing the pioneering first edition of this book, for inviting me to co-write the second edition, and for entrusting me with it. I hope that I have maintained, in the view of you, the reader, the high standards that Toby set.

I would like to thank Susie, for her faith, love, and tolerance. I would like to acknowledge Dick and Nancy, for inspiring and teaching me that our most worthwhile work is collaborative. I would like to thank Bailey and Evan, for their love and inspiration. I would like to thank Sister Joan Maiers, SNJM, and Bill Haworth for their profound effects on my writing and my life. I would like to acknowledge Dr. W. Edwards Deming and Les Paul for demonstrating what one can accomplish by passionately practicing what he loves and believes.

I would like to acknowledge Jessica Kersey at Polycom, Jean Rosauer at Tandberg, Michael Lentfer at Sprint, and Dan Hanrahan at VCON for their generous assistance. I would also like to acknowledge Jeff Hutkoff, PhC, for sharing his market insight, and Jim Luffred for sharing his knowledge of mobile communications. Jenny Child, at Comotiv, Allison Lattanze, at VSPAN, and Bob Friedler, at CompView were also helpful. Thanks to Stan Jones for providing refreshments and music, and keeping TheWorkingStiffs alive while I was writing.

The contributors to the standards organizations are the unsung heroes and pioneers who, along with the early adopters, deserve credit for removing barriers to worthwhile communications. It is a thrill to learn about and report on their work.

James R. Wilcox, CISSP

I would like to acknowledge my mentor Jim Wilcox for sharing his vision and insight into the world of video communications. The opportunity to contribute to the 4th Edition of this work that has been wrought through so much of Jim's tireless personal effort is one I will treasure throughout my life. More importantly, I want to acknowledge my friend Jim Wilcox whose consistent advice and persistent encouragement are exemplified in his passion for life, his incredible work ethic, and his genuine concern for his fellow man. A more real and "human" person has yet to cross my path.

I would like to thank my true love, Cyndi for her support and patience throughout our marriage and especially through the additional challenges experienced in the research, compilation, and delivery of this work. I would also like to thank my loving children Ryan, Aaron, Kaitlyn and Nathan for the food, hugs and kisses you delivered while "daddy is working and needs to be left alone." I would like to acknowledge my Mother and Father for teaching me that there is no goal that is beyond the grasp of those who are willing to work for it. I thank and acknowledge the grace and goodness of God to bless me with such a family.

Lastly, I would like to acknowledge all the scientists, developers, technologists and dreamers without whose crazy ideas we would still be sending smoke signals to one another. The "next big thing" that will reshape our world is probably scratched on a notepad or scrawled on a whiteboard in your office. Please go get it and make the future a reality for all of us!

David K. Gibson, CISSP

PREFACE

"The hinge of radical change almost always swings on new information systems."

James Champy,
Chairman CSC Consulting Group

This book is about video communication tools and the applications that are built upon them. Video communication represents an aggregation of technologies that includes information coding and compression, telecommunications and networking, broadcasting, and end-user presentational and data manipulation tools. The field of interactive video communication is not new, but it is rapidly evolving.

In this book, we seek to present clearly and concisely the complex subject of video communication. It is helpful, but not necessary, that the reader know a little about video communication and about networking. We start with the basics to build a foundation with which the reader can grasp the elementary issues. The intent is to spend equal time on technology, standards, and applications. In all cases, we want the reader to understand the business implications of interactive multimedia, and videoconferencing. Selling a solution, even if it is in an organization's best interest, requires that. It is our position that the purpose of technology, as fascinating as it is, is no more than to serve humanity's and nature's needs.

For those who are new to the topic, this book serves as an introduction to the various forms of video communication. For the technically advanced, it provides information for understanding and assessing new products, staying abreast of standards, and refining expertise. In any case, we encourage the reader to explore ways to implement video communication to enhance collaboration, facilitate competitive strategy, improve information delivery, and increase efficiency.

Interactive video communication strategies can be significant enablers for applications that span geographic and organizational boundaries. Such applications include group-to-group meetings, person-to-person desktop, videophone, one-to-many broadcasts, point-to-point and multipoint data collaboration, and variations thereof. A specific organizational or personal need is likely to warrant creativity in combining and implementing these applications. Some of the more successful application categories include telemedicine, distance learning, banking, legal and judicial, and joint engineering. Thorough needs-analysis, ingenuity, and senior management support will generally be necessary to ensure success.

For video communication to deliver on its promise, a strong interoperability model is required. With the adoption of numerous global standards, video communication applications can traverse a wide variety of networks including analog *POTS*, circuit-switched digital (ISDN, T-1), and IP-based networks. Interoperability is nearly universal, and the level of cooperation between industry leaders is unprecedented. This is a welcome development, given the industry's earlier tendency toward competing proprietary implementations. We admire those who profit by developing technologies that address human needs; we applaud even more those who adopt rather than impede better technologies that serve the public good.

A standards-based approach integrates tools from multiple sources. It is scalable, flexible, and based on best-of-breed components. Video communication system implementations that pursue this strategy are more likely to deliver an application-responsive, manageable, and cost-effective solution. Callers can leverage such systems regardless of what systems called parties use.

We believe that people will increasingly leverage the power of standards-based video communication and interactive multimedia to address important business processes that directly relate to organizational success and competitive advantage. Video communication facilitates teamwork. Many of the ways in which people have leveraged it to respond quickly to changing business conditions between remote regions of the earth were previously unimaginable. Systems based on such architecture can break through organizational barriers, and thereby provide links that bind trading partners in a flexible, personal, and highly efficient manner.

Although the focus in this book is on corporate enterprises, we emphasize that interactive video communication can help people complete ordinary tasks. Many valuable video communication applications provide breakthrough solutions for individuals and small organizations. Network choices abound such that anyone anywhere in the world can deploy video communication solutions.

Small or large, the principles that underlie video communication systems architecture remain the same. Standards-based systems are always better than closed ones. By promoting standards and supporting the efforts of groups such as the ITU-T, the IMTC, the IETF and the ISO, we overcome barriers to worthwhile communication. The promise of such standards is video communication, anywhere, any way, any time, with anyone.

We hope that you enjoy this book. We invite you to send questions or comments to: info@vcthewholepicture.com or visit us at www.vcthewholepicture.com.

James R. Wilcox, CISSP

CONTENTS

LIST OF FIGURES

PART ONE

VIDEO COMMUNICATION:
TERMS, CONCEPTS AND APPLICATIONS

1

AN INTRODUCTION TO VIDEO COMMUNICATIONS

Vision is the art of seeing things invisible.
Jonathan Swift

WHAT IS VIDEO COMMUNICATION?

What is *video communication?* To define the subject, we examine the (unfairly stigmatized) term that we used in the title of the first three editions of this book, *videoconferencing.* The term videoconferencing is rooted in two Latin words: videre, or "I see," and conferre, or "to bring together." Videoconferencing (or video communication) enables people to share visual information to overcome distance as a barrier to collaborative work.

The combination of the words videre and audio (the latter of which, in turn, stems from the Latin audire, or "to hear") gives us the word *video.* We define video as *a system that records and transmits visual information by conveying that information with electrical (or electronic) signals.* Although the term video, in its strictest sense, refers only to images, common vernacular reflects the common synchronization of audio with these images.

Video communication is real-time exchange of digitized video images and sounds (ideas and information) between conference participants at two or more separate sites. To meet our definition of video communication, an exchange must be interactive; it must be two-way. Transferred images may include pictures of participants, but may also include material as diverse as video clips, still pictures (e.g., product photos, diagrams, charts, graphics, data files, applications). Likewise, the audio portion may include discussions between meeting participants, but may include any form of digitized audio that is relevant to the discourse. Video and audio generally share the same communication path, but may also follow separate paths. Rather than anticipate every video communication possibility, we offer this concise definition: *interactive multimedia communication.*

Video communications bring together multiple persons or multiple groups into single multi-site meetings. They link two people through dissimilar computers, videophones (wired or unwired), or other communications-enabled devices. Video communications occur as point-to-point and multipoint events. Point-to-point videoconferences link participants in two sites; multipoint arrangements link more than two sites. The device that links three or more locations in a single conference is a multipoint control unit (MCU).

Initially, video communication consisted solely as group-system point-to-point or multipoint conferencing. However, the Internet and inexpensive (or free) software changed enabled *personal conferencing*. Early Internet-based video communication was little more than a novelty. However, business quality video communication was possible over the robust local area networks (LAN) that linked workstations to file- and application-servers; they also permitted connections to robust corporate wide area networks (WAN). With the networking infrastructure in place, video communication began transforming classic notions of *teamwork* and *workplace*. By merging interactive voice, video, and data onto a single network organizations foster rich collaboration.

At first, this collaboration could happen only within an organization or between organizations that standardized on hardware, software, and network architecture. However, widely-accepted standards and the Internet have overcome this limitation to make video communication nearly as easy as making a conventional telephone call. Standards provided the boost that interactive multimedia communications needed, and market growth resulted. A 2004 study by Wainhouse Research, an independent market research firm that specializes in the rich-media conferencing and communications fields, found that 31% of workers had participated in interactive video communication. Nearly three-quarters of respondents to a Wainhouse survey declared that videoconferencing use had increased in the last year; even more believed that it would increase in the coming year. Only about 30% cited expense as a major barrier to videoconferencing, and less than 18% found that integration or reliability issues were major barriers.

Companies that have historically maintained separate voice and data networks are now converging voice onto data networks. More than half of the decision-makers polled by Nemertes Research, a research firm that specializes in analyzing the business impact of technology, declared that their corporations want to converge networks specifically to include videoconferencing. More than a third intend to implement unified communications (UM) and nearly a third intend to implement collaborative tools. Nearly half of companies want carriers to offer managed services, nearly 20% are interested in hosted applications, and more than 10% want wireless integration services. The Nemertes Research study found that network convergence reduced annual local-loop costs by $9,600–$28,000 per site for large organizations, and by $4,800–$9,600 for midsize companies.

Service providers are trying to keep up. They anticipate investing nearly $5 billion in next-generation voice equipment between 2004 and 2007. Although voice over

Internet protocol (VoIP) softswitches, media gateways, and signaling gateways will consume the majority of that, by 2006, that investment will shift to complementary services (such as video communication).

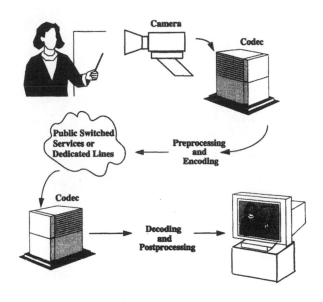

Figure 1-1. How Video Communication Systems Work.

With DSL and cable modems, video communication became possible, not just at work, but also at home. Now *wireless* broadband Internet access is enabling video communication from mobile devices. Cultural shifts take time but, as the explosive acceptance of the Internet demonstrates, these shifts eventually reach critical mass. We have learned to cope with the succession of new applications for our desktops and mobile devices, and we are now less resistant to adopting them. Graphical user interfaces (GUI) and on-line help utilities are making us increasingly proficient and increasingly able to adapt to new technologies. In this fertile environment, video communication seems destined to become a tool as fundamental as the telephone.

Video communication was not always an accessible, low-cost option for anyone who owned a computer. Until 1990, video communication was a tool for the business-elite. Moreover, until early 1994, when Intel introduced its ProShare family of personal conferencing products for Windows™ PCs, video communication was almost non- existent on the desktop. How video communication progressed from a tightly-engineered, customized boardroom implementation that cost as much as $1 million per site to *freebie* software is an interesting tale. It has been more revolutionary than evolutionary; it has been less a continuous progression of enhancements than a random string of breakthroughs.

Figure 1-2. Picturephone (AT&T)

THE HISTORY OF VIDEO COMMUNICATION

One of the first demonstrations of video communication in the U.S., Bell Labs demonstrated a crude video communication application between Washington DC and New York City in the 1920s. Others were engaging in experiments in Europe in the 1930s, where television technologies and systems were more mature than in the U.S. World War II intervened and, for almost two decades, the technology languished. In the late-1940s, Bell Labs resumed work on videoconferencing. After researching for a decade and a half, in 1964, it rolled out the world's first digital, interactive video-enabled telephone, Picturephone. Launched at the World's Fair in Flushing Meadow, New York, Picturephone weighed 26 pounds and sat on a nickel-plated pedestal with a base that measured about 10 inches in diameter.

Picturephone compressed a video signal; it squeezed it down for transmission over the equivalent of 100 telephone lines. Bell transmitted the audio signal separately. Displayed using less than one-half the resolution of a North American television pictures, the images were blurry, at 5.25 inches wide by 4.75 inches high, the screen was small. Nevertheless, it was a well-designed and engineered system.

Well ahead of its time, the Picturephone was a personal video communication tool. In the 1970s, attention shifted to group-oriented technologies. Nippon Electric Corporation (NEC) became the first company in the world to produce a group-oriented video communication system. British Telecom (BT), another pioneer, introduced its own version of a video communication system shortly thereafter.

BT persuaded various state-owned European telephone companies to conduct country-to-country video communication trials by creating facilities and availing them to the public. The efforts to render these public rooms operational paved the way for today's video communication interoperability standards.

Early video communication systems used television technology in its raw, analog form. The inability of networks to provide adequate bandwidth delayed progress in the area of digital transmission. It was not until the early 1980s, when low-cost, large-scale solid-state memory became available, that video communication advances began in earnest. The key was to store picture information during the processing sequence. Once *very large scale integration* (VLSI) technology appeared, it videoconferencing pioneers harnessed it with amazing results. The US *Defense Advanced Research Projects Agency* (DARPA), which wanted an inexpensive, small-profile system, commissioned one early video communication project. The resulting product, Widcom, used VLSI to aggregate large numbers of circuits on a single chip. The approach reduced the required bandwidth for image transmission to the digital equivalent of a telephone line. Unfortunately, transmitting a single image from this powerful system took too much time and Widcom, which launched in 1983, faded into obscurity for lack of a commercial application.

In the late 1970s and early 1980s, the installation of a corporate video communication network required a leap of faith. At that time, the *substitution paradigm*, or paying for video communication with travel costs avoided, served as the basis for most video communication business cases. Unfortunately, the cost of a system was $250,000—and that bought only the *codec.*

The codec is the heart of a videoconferencing system. A codec adapts audiovisual signals to the requirements of the digital networks that transport them. As the term implies, compression/decompression and coding/decoding are a codec's most important jobs. A codec must first turn analog signals into a continuous stream of zeros and ones (coding or digitizing), and then sort out the trivial bits from the ones that will be necessary for the session (compression).

Because video requires an uncompressed digitized telephone signal with a network transmission speed of at least 90 million bits per second (Mbps), it is easy to appreciate the job of a codec. Even after discarding the vertical and horizontal synchronization information that analog television signals require, an uncompressed digitized television signal requires at least 45 Mbps. Contrast this with a common modem's pseudo 56 Kbps (kilobits per second), and one comprehends that the uncompressed signal's bandwidth requirements are substantial. To squeeze full-motion video plus audio over such a line (having captured that signal using some type of standard camera and microphone) would require the discard of 99.99% of the information. Even with a common cable modem or DSL line for transmission, the compression burden would be enormous.

Today we can squeeze a sufficient videoconference over two *ISDN* (Integrate Services

Digital Network) B channels. We can squeeze a somewhat degraded version of the same signal over a plain old telephone service (POTS) line, the type of line that connects the vast majority of residential customers and many business telephone users to the public switched telephone network (PSTN). In 1978, the same degree of picture resolution and motion handling required six Mbps of bandwidth.

A coast-to-coast line that connected corporate video communication facilities rented annually for about $35,000. Furthermore, in addition to the $250,000 codec we mentioned earlier, a company in the early 1980s had to buy cameras, microphones, and speakers, and prepare a conference facility to offer acceptable acoustics and lighting. Consequently, early adopters of video communication built broadcasting studios at a cost of $500,000–$1,000,000 or more.

Figure 1-3. Custom Boardroom System c. 1993 (courtesy of AT&T).

Unless organizations conducted videoconferences all day, every day, the video communication projects of yesterday were rarely economically feasible. Implementation depended on senior management's belief in the strategic value of the technology. A few forward-thinking companies were up to the challenge. Early adopters of video communication include Aetna Life and Casualty, Atlantic Richfield, Boeing, Citibank, Chrysler, DEC, Eastman Kodak, Federal Express, Ford, Hewlett-Packard, J.C. Penney, Merrill Lynch, MONY, Proctor and Gamble, and Texas Instruments. Of the Fortune 500 companies listed in 1970, most who adopted video communication around that time have prevailed.

Nearly all these early adopters used group-oriented, room-based videoconferencing systems from Compression Labs, Incorporated (CLI) of San Jose, California. Founded in December 1976, CLI was initially dedicated to developing facsimile and video compression technologies. A CLI founder, Dr. Wen-hsiung Chen, will go down in history as one of the great pioneers of video compression. In 1963, Dr. Chen emigrated from Taiwan to the U.S. where he pursued his Ph.D. in mathematics.

While at the University of Southern California, he discovered a compression technique that became a fundamental part of CLI's first commercial system, the low bit-rate (compared to previous methods) VTS 1.5. The first VTS 1.5 reached the market in September 1982. It required a T-1 carrier (T-1 is the basis of the North American digital hierarchy, and is comprised of 24 channels, each of which is the equivalent of a common telephone line) for signal transmission, and provided very good picture quality considering the relatively limited bandwidth it required. CLI's system allowed companies to move from satellite-based videoconferencing to terrestrial transmission—the T in T-1 means terrestrial. As a result, the transmission costs associated with videoconferencing plummeted.

In 1984, CLI began to face competition from a Massachusetts company named PictureTel. A team of Massachusetts Institute of Technology (MIT) engineering students and their professor founded PictureTel (Danvers, Massachusetts). The group had invested a great deal of time researching low-bandwidth digital video signal processing. After a lot of trial and error, they developed a way to reduce the bandwidth requirement for transmitting an acceptable-quality videoconference to an amazing 224 Kbps—the equivalent of four telephone lines. Overnight, group-system videoconferencing became economical at low data rates.

PictureTel's product was the first software-based videoconferencing system to become commercially available. It became the cornerstone of a planned migration from a build-your-own product to an integrated system in which a single chassis included all necessary components. CLI responded with a similar product, the race began, and the two firms began grappling for control of an exploding market. While the two were busy working magic in their laboratories, a third company, VTEL (formerly VideoTelecom Corporation) challenged the competition. Founded in 1985, VTEL was the first company to offer a videoconferencing product that ran on a DOS-based PC. For the first time, customers could upgrade their videoconferencing systems by simply downloading new sets of manufacturer-provided floppy disks.

In 1980, only a few companies knew how to code and compress a video signal to affordable bandwidths. Manufacturers produced codecs at a rate of 100 per year. Today, software-only codecs are more powerful than the best 1980 model, which required a refrigerator-sized cabinet. These software-based codecs are now standard enhancements to personal computers. Because they are standards-based, they enable any-to-any communications, regardless of the platform on which they run.

FROM THE BOARDROOM TO THE DESKTOP

Video communication began with Picturephone, a personal conferencing product, and moved on to group-oriented systems. Economies of scale drove the migration. The exorbitant cost of early codecs, and of the bandwidth required to move a compressed audiovisual signal, required amortization across a group. The economics have changed but the market still lags.

The stigmatic legacy of boardroom videoconferencing lingers. Because early installations were costly, the first applications often reflected their boardroom environment. Housed in custom-designed and engineered studios, videoconferences were often formal events that only senior managers and executives attended.

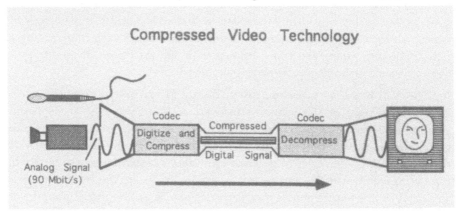

Figure 1-4. Compressed Video Technology (Michel Bayard).

These conference attendees were often so senior that a dedicated technician attended simply to operate the equipment. Videoconferencing was a tool for recovering productive time that would have been lost to travel for only the most in-demand people, who could not squander their time learning such technical complexities (it was more acceptable at that time for an executive to plead ignorance in the face of technology).

That paradigm has changed, and is changing. The ease-of-use challenges posed by early videoconferencing systems are disappearing as virtually every PC that is a video communication appliance (or can become one with a simple software download). Powerful processors make quick work of compression. The cost of a codec, cameras, microphones, and speakers are available at very low costs (if not included in the system), and broadband is ubiquitous. Competition has driven, and is driving, transmission costs increasingly. Fast processors, cheap bandwidth, and system operation simplicity all suggest a market that is poised to skyrocket. The market driver is demand for tools that allow people to lead just-in-time lives in a competitive global environment.

In this very competitive global economy, every competitor struggles to capture market share. Markets fragment and shift and niches repeatedly emerge for firms that act within increasingly short windows of opportunity. We are *all* resources or commodities in today's environment. Our productivity is vital to the survival of the enterprise for which we work. Furthermore, because there is so much to know and so much to do, we are forced to depend upon information technology for that which few of us can readily access from our brains. No matter how vast our foci become, the challenges intensify more quickly. Teamwork and collaboration, whether on a personal level or an enterprise level, have become mandatory.

Video communications enhance collaborative activities. However, since not every work environment is the same, each has adapted to its special needs. That is why another form of collaboration, that which leads to interoperability standards, has been key to the survival and prosperity of video communication. In December 1990, an international telecommunications standards organization, part of the United Nations, voted to accept a video communication interoperability standard. This standard allowed codecs from such manufacturers as CLI, PictureTel, VTEL and others to negotiate technically compatible sessions over communications channels of various bandwidths. The standards-making process did not end in 1990, it began. Today audiovisual standards exist for every application.

They share many common attributes. All are designed for digital networks, and all share the goal of a reduction of information to be transmitted. While the 1990 standard's target was group-system video, those ratified in 1996 were more inclusive; they clearly reflected that PC-based personal conferencing would be the next frontier.

PCs were an inevitable destination for video communication. In 1989, only VTEL was shipping a PC-based product and, at five feet tall and 200 pounds, few desks offered the necessary real estate. Three years later, several true desktop systems appeared but, with prices at $10,000–15,000, few organizations could afford them. In 1993, more than twenty desktop products entered the market. Most sold for $5,000–8,000 (including a computer), and several were in the $3,000 range. One of the companies that offered a product in 1993 was ShareVision.

ShareVision's ShareView product ran on an Apple Macintosh computer and transmitted video, voice, and computer data over ordinary telephone lines or POTS (plain old telephone service) lines. The package, which consisted of a combination video capture board and codec, and a network interface board, also included a color camera, a handset, a headset, and software. Initially, the system sold for $4,499, but the price soon dropped to less than $4,000. ShareView's major contribution was to enhance mere file transfer and image sharing with *application sharing*. A ShareView user could launch an application and share it with another party even though the second user did not have that application on their system. This feature made ShareVision one of the pioneers in the area of visually assisted collaborative tools. Throughout this book, we will refer to this area as *personal conferencing*. Many competitors have followed where ShareVision led.

In January 1994, almost exactly one year after the introduction of ShareVision, Intel introduced its ProShare family of personal conferencing (desktop) products. One ProShare family member, the Video System 200, allowed conferees to see each other in a quarter-screen window on a PC monitor while sharing a computer application. Other members of ProShare would forgo the face-to-face altogether, in favor of application sharing and voice communications. Intel owns the term *document conferencing*; in 1993, it bought the rights to use it from VTEL.

Critics of ProShare emphasized that it required the local telephone company to furnish ISDN BRI services, or the aggregation of two B channels into one 112/128 Kbps channel. The problem with ISDN was, and still is, availability. Although almost all telephone companies offer some form of high-bandwidth digital transmission service to business and residential consumers, not all offer ISDN. Intel can take much the credit for a surge in ISDN deployment by the Regional Bell Operating Companies (RBOC) and GTE; they worked cooperatively with them to demonstrate that customers wanted broadband connectivity (at a time when DSL and cable modems were little known or nonexistent). Remote access to enterprise network services and telecommuting (replacing the physical commute with a virtual one) led to many ISDN installations.

Besides complaining that ProShare required network services that were not always available, Intel's competitors also criticized the company for not embracing the standard that the International Telecommunications Union's Telecommunications Standardization Sector (CCITT/ITU-T) ratified in 1990. Without support for standards, ProShare could not interoperate with other systems.

Intel's response was rational: the standard in question, known as H.320, was overkill for the desktop, and too expensive to implement. Intel had managed to keep the ProShare Video System 200 priced at less than $1,500 for most installations, and less as time progressed. H.320 required expensive hardware codecs, but ProShare compressed with inexpensive software-only codecs. Intel eventually implemented the mandatory portions of the standard, but only after making their point. The desktop video communication market, and personal conferencing, would not happen until the ITU-T ratified a desktop standard. That standard also represented the demise of ProShare.

In 1996, the new personal conferencing standard, H.324, emerged. Its target was video communication over POTS. Even before H.324, another standard, aimed only at graphics conferencing, was ratified by the ITU-T. The formal name for this standard is Transmission Protocols for Multimedia Data, but it is better known by its ITU-T Recommendation number, T.120. T.120 is an umbrella standard or applications template. Among other things, it specifies how multipoint audiographics conferences can be set up and managed, and how binary file transfers should occur.

At the same time that the ITU's H.324 standard for videoconferencing over POTS was completed, yet another personal conferencing standard was completed. The H.323 standard set to define how videoconferencing systems communicate over local area networks (LANs) to integrate real-time voice, data, and video into networked devices such as PCs, workstations, and video-enabled telephones. Complementary standards include H.321, a standard for videoconferencing over fast-packet Asynchronous Transfer Mode (ATM) networks, H.322, a standard for videoconferencing over LANs that can guarantee the low latency (minimal time delays) required for moving frames of video in rapid succession. Clearly, the ITU, always the most prescient of organizations, envisioned desktop video communication as a very strong market. The standards that support the H.321, H.322, H.323, and H.324 umbrellas show a great deal of commonality. This commonality has led, and will continue to lead, to interoperability between systems that connect to diverse networks, including LANs, the Internet, telephone lines, and super-fast ATM-based networks.

THE MARKET FOR VIDEO COMMUNICATION

The developed world's omnipresent Internet fascination has both swelled the market for and changed the face of video communication. Internet-based video communication makes possible application sharing and whiteboarding, and store-and-forward video communications and combinations thereof. The time when videoconferencing was a series of technologies that were looking for applications, and when these applications were disconnected, has passed.

In his speech, "The Future of Collaboration on the Net," Dr. Norm Gaut (at the time, chair of PictureTel Corporation) stated that three drivers have been changing the entire IT industry: development of the Internet, the growing power of PC processors, and the deployment of new high-bandwidth networks. Dr. Gaut defined four waves of visual communication. Respectively, the first and second are dial-up room conferencing, and ISDN-based desktop conferencing. The third wave is video communication over LANs (including LAN-multicast in which multiple users receive a broadcast signal over a single *channel*). The last wave is videoconferencing over the Internet (we respectfully submit that the fifth wave is video communications over converged mobile devices). PCs have long-since integrated video communication-ready packages that include H.323 LAN-based video, H.324 codecs, video-ready modems, speakers, cameras, and microphones. These multimedia PCs are capable of video communication, Web surfing, DVD recording and playback, and interactive games. The technology works and it is ubiquitous; that does not mean that people are integrating it into their daily lives.

The challenge now is threefold: to reduce significantly the complexity of video communication, to improve the network so that the quality of the video communication experience compares with that of the telephone, and to change people's habits to include video communication as an alternative to conventional voice communication or travel. Rather than a complex process of pre-scheduling a call with an administrator or operator, one must be able to initiate video communication on the fly with anyone easily and spontaneously over the Internet. Internet-mediated video communications portend interactivity, multimedia, security, manageability, telephone functionality, and low cost. It is especially attractive in an environment of mergers and acquisitions, corporate downsizing, geographically-dispersed teams, global price competition, and the need to decrease response times.

The Internet is finally becoming as mature and reliable as the technologies that people would like it to support. Modern encryption routines such as IPSec and SSL provide interoperable standards for users to maintain confidentiality, to verify message integrity, and to control availability over insecure networks. Carriers are deploying Quality-of-Service (QoS) mechanisms, such as MPLS (Multi Protocol Label Switching) to ensure a level of service within a *best effort* network technology. Security and QoS technologies have often been complex and difficult to manage but, now that they are becoming transparent, companies are converging video, voice, and data on the fly, and in real-time over the Internet. As a result, they are managing fewer networks and finding it easier to communicate with employees, customers, and vendors.

Unfortunately, converging networks is not always easy, and PC-based IP telephone switches (often referred to as *softswitches*) do not completely alleviate the complexities long associated with programming PBXs (Private Branch Exchange, or conventional, connection-oriented phone switches). Converging networks still requires programming routers, network management software or devices, and security software and devices to interoperate. Nevertheless, corporations big and small are implementing voice over IP (VoIP) traffic over inter-branch T-1 data links; some are even leading the way by doing so over the Internet. Doing so enables them to provide better service by instantly linking incoming customer calls to CRM (customer relationship management) software. It enables them to cut costs by instantaneously linking all supply chain interactions to resource planning (ERP) software. It even enables healthcare providers to provide patients with access to personal healthcare records without waiting for a doctor or nurse, or worse, snail-mail. Integrated communications servers that support unified messaging (voice, video, and data, how and when you want it... wired or not), and presence are displacing conventional circuit-switched only, voice only, PBXs.

Historically, data (packet-switched) networks existed separately from voice and video (connection-oriented) networks and have been independently managed. Management packages that only recognize devices from a single manufacturer are of limited value outside homogenous networks (which are rare); such products should at least recognize, if not manage, devices from a variety of manufacturers (few such products recognize non-IP-enabled conventional PBXs). This is important in managing converged networks. Convergence requires expertise in conventional voice and video, and expertise in data networking.

The market has outgrown earlier distinctions between videoconferencing, interactive multimedia, and store-and-forward video communications. Whereas, in the past, video communications represented a technology that was looking for applications, applications now use a combination of video communications technologies. IP-based video communication is driving growth by making conventional videoconferencing a foundation on which to build applications for banking, government, distance learning, telemedicine, and various desktop applications. Of the 2003 videoconferencing market's paltry $550 million, IP-only endpoints comprised more than 23%. IP will comprise 85% of the 2007 videoconferencing end-points market. The real story is that, by 2009, the global market for web conferencing systems and services will reach $2.7 billion.

One of the fundamental reasons for the meteoric success of video communication today is reduced bandwidth demand. Because of dramatic improvements in the software that makes it possible, the *codec* (coder/decoder), requisite bandwidth has plummeted. In 1990, the H.261 codec made videoconferencing possible. However, business-quality conferencing commonly took place over 512 Kbps or even 768 Kbps connections (at a time when a 1.544 Mbps T-1 was considered *broadband*). In 1995, the H.263 codec reduced the bandwidth burden to 384 Kbps in most cases. In 2004, that burden declined by a third with the implementation of the H.264 codec.

Improvements in video communication standards and in hardware such as codecs, MCUs, cameras, and PCs are moving toward real-time DVD-quality video at 30 frames per second (FPS). USB and Firewire have simplified installation and reduced desktop and compact video communication costs. Infrared, Bluetooth, 802.x and now UWB and ZigBee are eliminating wires altogether. Such systems have eroded the conventional ISDN system end point market. ISDN-based videoconferencing comprises a large share of the business-quality videoconferencing market because it is still the safest bet. However, because it is circuit-oriented, it requires high per-minute costs. Moreover, it has long since been stigmatized as a legacy technology that IP has already supplanted. The perception is largely accurate; as WANs are becoming more robust, they are making packet-switched video communication more practical.

CONCLUSION

The Picturephone, was so far ahead of its time that it seemed a failure. Nevertheless, it was far more important than anyone might have comprehended. Its developers recognized that people naturally communicate not just aurally but also visually. Today's markets are proving that those Bell Labs pioneers did not labor in vain, and that their vision simply transcended their era. Successful technology applications are not possible without pioneers: scientists, engineers, marketers, and sellers. As important as their work is, however, it would all be for naught without visionary early adopters.

Eventually, successful applications achieve broad-based acceptance and move into the corporate mainstream. A typical conclusion to the cycle of research and development finds these applications delivered to and embraced by the casual consumer. With market growth comes acceptance, and with acceptance comes greater demand for ease-of-use, new capabilities, and lower cost. At some point, the beneficiaries take for granted the brilliance of every technological development.

The video communication industry is somewhere in the middle of that adoption continuum. It is a market that is becoming less constrained by technology than by human habits. The stigma of videoconferencing is that *next year* has always been the year it was *going to take off.* However, with the proliferation of broadband Internet, and now broadband wireless Internet to phones (if one can still call them that), the future of video communication is finally clear.

We dedicate the remainder of this book to fostering in the reader an understanding of video communication applications, the technologies upon which they depend, and the key decision points one must consider when making a product or service selection. Once mastered, we believe, more than ever, that most readers will extol the benefits of video communication, and will want to gain first-hand experience. It is our hope that they will make video communication a habit.

CHAPTER 1 REFERENCES

"How is IP Impacting the Current Videoconferencing Market?" *Frost and Sullivan.* 2004.

"Future of Conferencing Report on Video Conference Calls & Trends in H1 & H2 2003." *Telespan.* December 2004.

Green, Jeff. "Videoconferencing Market Trends." *Faulkner Information Services.* 2004

Barr, James G. "Web Conferencing Market Trends." *Faulkner Information Services.* 2004.

Doherty, Sean. "Digital Convergence, All-For-One System Offers High Performance At Reduced Cost." *Network Computing,* December 16, 2004, p 47.

Mcelligot, Tim and Vittore, Vince. "Quality and Security Hurdles Ahead." *Telephony.* November 8, 2004, p. 38.

"Vendors Vary Wildly on Real Costs for VOIP Rollouts; Meanwhile, Customers Grow Weary of Carriers' Convergence Inertia." *PR Newswire.* New York, November 16, 2004.

Bottger, Chris. "The Collaboration Revolution." *Successful Meetings.* October 2004, v53 i11 p32(1).

Wittmann, Art. "Is Video Broadcasting Coming to Your Network?" *Network Computing,* April 19, 1999, p152.

Anderson, J., Fralick, S.C., Hamilton, E., Tescher, A.G., and Widergren, R.D. "Codec squeezes color teleconferencing through digital phone lines," *Electronics* 57: 1984 pp. 113-115.

Barnard, Chester I. *The Functions of the Executive.* Harvard University Press, Cambridge, Massachusetts, 1979 p. 217.

Portway, Patrick and Lane, Carla, Ed.D. "Technical Guide to Teleconferencing and Distance Learning," *Applied Business teleCommunications.* San Ramon, California. 1992, p. 4.

2

VIDEO COMMUNICATION APPLICATIONS

Nothing in the world is single,
All things by a law divine
In one spirit meet and mingle
                              Percy Bysshe Shelley

VIDEO COMMUNICATION IN THE REAL WORLD

For decades, *next year* has promised to be *the year that videoconferencing takes off.* Its potential for fostering collaboration, conserving time, facilitating access, and increasing quality of life are clear.

Manufacturing companies leverage video communication for quality assurance and real-time process and equipment monitoring. Healthcare providers employ it to enhance patient care, and legal teams use it to streamline casework. Cultural acceptance of technology, diminishing network and equipment costs, and standardization now allow video communication to solve a wide range of business problems. Advances in group-system, application sharing, LAN, and even POTS standards have advanced interoperability such that it has transcended the confines of intra-organizational communications.

Notwithstanding a subset of the older population for whom modern communications and information technology may be an intimidating annoyance, society embraces technology. Proliferation and adoption of personal computers, telephones, and digital entertainment centers has happened.

The same is half-true for video communication. The means have become ubiquitous. Public rooms abound for providing high-end video communication. One can reserve a nearby Kinko's Copy Center videoconferencing room by dialing a toll-free number or by pointing a web browser. Once there, one will find a room-based videoconferencing system and accommodations that rival those in corporate conference rooms. At the desktop, video communication is even easier. Perhaps no computer ships today without video communication software, a modem or Ethernet jack, a microphone and speakers. Few, if any, computers stand completely disconnected from the Internet. Furthermore, while wireless networks are gearing up, mobile telephone users foreshadow end-to-end interactive video communications by augmenting their conversations with still pictures.

Businesses use personal conferencing to link employees, customers, suppliers, and strategic partners to save time and money. Legislators encourage video communication by rewarding organizations that adopt telecommuting initiatives and thereby reduce the number of vehicles on the roads. Only some rural residences today cannot gain broadband Internet access. Cable television service (CATV) providers and long-distance telecommunications providers (inter-exchange carriers, or IXCs) are frenetically competing in the local loop market that Local Exchange Carriers (LECs) historically considered their exclusive domain. Ample connectivity abounds for video communication from homes and businesses.

Any organization that relies upon geographically separated workers can benefit from video communication. Apparel makers use video communication to review concepts, artwork, type proofs, and sales and marketing collateral. Retailers use it to display merchandise and specify color or style, as well as recommend price and advertising messages. Pharmaceutical companies use video communication technology to develop new products. A common thread between chemical companies, railroads, and jewelers is that they all employ video communication.

Video communication saves travel expense. It cuts even more costs by improving productivity and communication. Many successful organizations find video communication indispensable for managing the invaluable time of strategic in-demand personnel. Some speed business cycles by uniting individuals who cannot meet in the same physical location. Many find that video communication allows for meetings that travel would preclude.

Video communication promises cultural implications. Live videoconferences provide an interactive alternative to staged, choreographed, one-way political conventions. The 1996 conventions laid the groundwork for this trend by sponsoring Worldwide Web sites and chat rooms on the Internet. Each party provided delegate groups at its respective convention with personal computers, and provided some with personal hand-held communications systems.

Before the 2004 Democratic National Convention officially started, more than 4,000 delegates made history by participating in a videoconference... without leaving their hotels. Polycom provided VSX 7000 units to each of the 22 hotels and the university that housed the 56 delegations. As a result, each delegation connected to a secure central host site using Polycom MGC 100 multi-point conferencing units for daily conferences with Democratic leaders, including former Vice President Al Gore, former Secretary of State Madeleine Albright, and New Mexico Governor Bill Richardson. "Polycom's technology helped us create an innovative environment to deliver information and better communicate with our delegates in real time... (through) a virtual Convention," said Rod O'Connor, DNCC CEO." Gore opened the series of daily, live video sessions.

With increasingly fast and inexpensive network bandwidth to offices and homes, video communication technology is becoming the basis for countless new products and services. A 500+ Kbps connection from a business or a home is usual, and it is becoming common from PDAs (personal digital assistants) that have mobile phone capabilities.

Figure 2-1. Videoconference with Set-top System (courtesy of Polycom).

Verizon launched such connectivity in about a dozen US markets in 2004, and Sprint followed by announcing nationwide deployment of similar networks for its own PDA and connection card (which insert into laptop computers) customers. A home user can view an MPEG2 video stream while browsing the Internet and making an Internet-based telephone call. Consider the following video communication applications:

- Executive, task force, and board member meetings
- Project management
- Desktop-to-desktop communications (wired or wireless)
- Merger and acquisition management
- Collaborative-engineering, product development, and product review
- Product Support
- Legal protocol (depositions, testimony, and remote arraignment)
- Surveillance
- Emergency Response
- Distance Learning & Training
- Large multi-site meetings
- Key personnel recruitment
- Telemedicine and health care provisioning
- Consulting
- Inter-company meetings with customers, suppliers, and partners
- New product announcements to restricted audiences

Although people generally associate video communication with business meetings, its applications are quite diverse and, ideally, creative. Following are some examples:

- To reunite families ravaged by the 2005 Tsunami disasters, VectorMAX Software donated video communications software and installed it onto IBM-donated laptops. Unicef, which said that many of the tens of thousands of children orphaned by the tsunami were unclaimed, and injured adult survivors lay unidentified in hospitals, put the computers in orphanages and in towns to help families find one another. The systems also served in cataloging the dead.

- The National Institutes of Health uses video communication to connect its 27 separate institutes and centers, as well as the 75 buildings on its campus.

- Video communication overcomes the time-consuming process of suiting-up to link workers in cleanrooms with co-workers outside. Moreover, online images do not introduce contamination, as do airborne-particulate-producing papers.

- Flight test engineers use video communication to share results with design engineers who are thousands of miles away. A trailing aircraft captures video that all parties instantly view together, while they are fresh, and allows engineers to schedule test flights more quickly.

- One dedicated violin master in Manhattan uses room-based video communication technology to instruct financially and physically challenged students in New Jersey, Minnesota, and Georgia.

- Advertising agencies use video communication public rooms to audition talent.

- Automobile manufacturers use video communication to co-locate engineering groups virtually, even if they are only minutes apart. The groups save time by sharing engineering drawings and 3-D models by concurrently viewing video of durability tests, vehicle crashes, and product demonstrations.

- Real estate firms leverage video communication for weekly reviews of listings and sales. Split-screen features enable sales managers to detail highlights of each listing, point out financial highlights, train salespeople, and show videos of properties while narrating to emphasize key points.

- Employers encourage workers to save time and alleviate traffic by using video communication systems to telecommute during major events such as the Olympics or political conventions.

- The Joint Warfare Center uses video communication between isolated Florida locations, and NASA provides in-service training for scientists, astronauts, program managers, and education specialists.

Senior Management Applications

Building a cohesive team in a global environment is a challenge. Through video communication, executives can include more team members in the strategic planning process, and create a virtual global campus. Downsized companies must achieve more work with less, and video communication is indispensable to executives who must simultaneously manage field and headquarters operations.

Investment firms use video communication for *crisis management* (e.g., a major swing in financial markets). Bear Sterns uses video to allow floor traders to receive up-to-the-minute financial reports from various news networks such as CNN and CNBC. The potential here for end-to-interactive multimedia via handsets is clear.

Executives and senior managers also use video communication to maintain rapport with important customers, strategic partners, suppliers, and other stakeholders. Video communication enables them to meet frequently with Wall Street analysts, legal experts, and consultants.

In executive applications, travel savings are usually a secondary objective. Typically, the real issues are time savings, productivity, and fostering of additional opportunities—areas in which one can express improvements in financial terms. Few, if any, will dispute the assertion that video communication can save creative potential that would otherwise be lost to excessive travel fatigue.

Executive applications are not cheap. Senior managers usually demand (and deserve) high-quality transmission, multiple large monitors, premium audio, and facilities that suit large audiences. However, the value that such systems can provide can often offset such costs with just one use.

PROJECT MANAGEMENT AND CONSULTING

Video communication is an ideal tool for managing projects over distance. It unites team members and subject-matter experts for prompt collective decision-making (or troubleshooting) in joint or distributed undertakings. Video communication improves the frequency and the efficiency of communications, and direct cost savings are often secondary to accomplishing the job.

In the mid-1980s, the aerospace industry discovered that inter-company video communication was valuable for managing contractors and subcontractors who cooperated on large projects. Many more companies are now using video for similar applications. Video communication works well in the management of construction projects. Experts at remote sites can inspect footage that is captured at the construction site, as can customers who are paying for the project.

Figure 2-2. Boardroom System (courtesy of Tandberg).

Air Products and Chemicals is one firm that has benefited from the use of video communication in project management. The company is a major international supplier of industrial gases and related equipment, chemicals, and environmental, and energy systems. It maintains project management control through a communications system that links its Allentown, Pennsylvania location with Bechtel in Pasadena, Texas.

Throughout one project, Bechtel provided engineering support to Air Products for a new plant that it was constructing at Air Products in Pasadena. The project team received project status updates through regularly scheduled video meetings. The Project Forum team met every other week to discuss project designs with engineers in Texas. The Project Steering Committee met every other week with the entire project team. Every six weeks, the team provided formal project review meetings for senior managers. Video communication allowed greater staff and senior management participation in the project; it even allowed controllers to join, for the first time, in discussions. Models, blueprints, and 35mm slides provided additional information about the project during meetings. The use of video communication saved more than $100,000 and eliminated more than 30% of the travel.

Figure 2-3. Dual Monitor System (courtesy of Tandberg).

FINANCIAL SERVICES

A Michigan based wholesale mortgage lender has improved the way it receives and processes loan applications from more than 4,400 loan originators and mortgage brokers nationwide. Its system allows officers to complete 8–10 mortgage applications each day rather than, as had previously been the case, only one. Video communication enables loan applicants, loan officers and underwriters to eliminate approval steps and avoid delays. In such meetings, underwriters input and make judgments about information, reconcile it, and verify it with the correct documentation. A company officer declared that the alternative would have been to hire 25,000 geographically dispersed loan underwriters.

Nationwide, more than 500 institutions installed more than 600 sites. The system reduced the typical time for processing a loan to five hours. The process begins when a loan originator faxes the applicant's completed loan forms to the mortgage center, calls the center to confirm receipt of the application, and requests the assignment of a loan underwriter. The underwriter calls the originator's office to arrange a videoconference and to designate a conference time. At the end of the videoconference, loan information is entered into the computer system for final processing and approvals.

MANAGEMENT OF MERGERS AND DECENTRALIZATIONS

An effective tool in encouraging post-merger communications between two disparate workgroups, video communication allows companies to establish an ad hoc management team and to keep that team communicating regardless of distance. A successful merger requires integration at all levels of a company; video communication is ideal in these situations because it promotes comprehensive team participation. The tool helps companies overcome the challenges of coordinating activities between two formerly unaffiliated companies and bridge both geographic and cultural chasms. It is particularly valuable for merging engineering, management information systems (MIS), legal, accounting, and human resources workgroups.

Visual communications tools are equally valuable in managing a decentralization project. When a department is split into multiple groups, video can foster a sense of teamwork, and facilitate regularly scheduled meetings.

JOINT-ENGINEERING AND PRODUCT DEVELOPMENT

Review and in-depth focus meetings often involve numerous people. Because video communication enhances face-to-face interaction with graphical information, the emergence of computer aided design (CAD) conferencing is little surprise.

Video communication helps reduce product time-to-market intervals. Delivering a high revenue-generating product ahead of schedule can make for a fast ROI. In one instance, Boeing simultaneously reduced the development time of its 767 by a year, generated revenues ahead of schedule, and cornered the market for a new generation of airplane. A manager at Boeing estimated that the accelerated development process resulted in profits that exceeded "every dollar ever invested" in video.

BancTec, a company that builds, markets and services financial document processing equipment for companies in the financial industry, uses video communication to shorten product development and delivery cycles. By linking the company's Texas headquarters to manufacturing facilities in Oklahoma, research and development teams develop rapid solutions to product quality issues. Remote visual inspection allows BancTec teams to implement corrective action quickly.

Video communication is beneficial for engineering and design applications even when coworkers are not in different states. For instance, NASA's Jet Propulsion Labs (JPL) uses desktop video communication units to connect departments throughout its 200 buildings—all of which fall within a twenty-mile radius. The systems enable JPL designers (even those in the same building) to collaborate on engineering drawings and allow employees to lecture and to "meet" with colleagues in Washington, D.C., Canada, and New Hampshire without the time and expense of travel.

Engineers, architects, and other technically-oriented professionals collaborate and make technical presentations during videoconferences. They commonly require systems that offer sophisticated audiographics—including high-resolution capabilities. They may also require the ability to store documents and images as electronic files for recall during a conference. Some need computer aided design (CAD) system interfaces. Such applications often require ceiling-mounted cameras to provide sufficient depth of field for capturing large documents and blueprints.

CUSTOMER SUPPORT

One manufacturer in the southeast integrates video into its product support offering. The company designs and manufactures packaging machinery that automates packing processes. Because an inoperable machine halts an entire production line, assisting customers with maintenance is a crucial part of the company's business.

Therefore, the company created an innovative system by which, if a machine fails, a customer simply places a video call to the manufacturer's technicians. A wireless remote camera with a 24-hour battery, and a RF system that transmits the video signal to an antenna over the factory floor, enables a technician to see the machine while talking to the customer. Using the system, which includes a wireless headset, high-resolution monitors, and 384 Kbps network links, a technician and a customer contact collaborate to quickly identify the problem and develop a solution.

Video communication allows this manufacturer to resolve 80% of customer service calls without sending technicians on-site. The company simultaneously limits the chances of stopping a production line, reduces the cost of service interruptions, maintains machines, and improves both the performance and lifespan of its equipment. It also gains strategic advantage over its competition. The customer leverages video communication to reduce costs and increase customer service.

A Los Angeles, California, company that sells computer chips and circuit boards uses video communication to share plans and finished products with a factory in Taiwan. BCA uses Panasonic's Vision Pro system to ensure that each product's design meets each customer's specific expectations. BCA eliminates surprises by allowing customers to scrutinize finished product over video before mass production begins; doing so without video would add significant time.

LEGAL AND JUDICIAL APPLICATIONS

Many states and counties use video for applications such as the remote arraignment of prisoners. San Bernardino County Municipal Court finds remote arraignment valuable in reducing or eliminating the cost, time, and security risks associated with transporting prisoners. It also frees Sheriff's deputies to enforce laws rather than drive buses. Moreover, the technology cuts arraignment time from hours to minutes, and enables it to record proceedings. When used to conduct remote arraignments, a video screen is often divided into quadrants. The prisoner appears in one quadrant, the district attorney in a second, the judge in a third, and the date, time, case number and information appear in the fourth quadrant for archival.

On the other side of the bench, clients are not always satisfied with local legal resources; they want the best-qualified attorney to represent them. Although communications by telephone, voice mail, email, and fax bridge much of this gap, they do not supplant face-to-face meetings. One law firm that serves large high-profile clients links 120 lawyers in six U.S. cities through personal video communication, which also simplifies taking depositions over distance. Legal Image Network Communications (LINC) provides attorneys with an opportunity, through video communication, to obtain depositions that would otherwise be prohibitively expensive. The visual component of a deposition is critical to attorneys who are skilled at reading body language and facial expressions. The service is available in more than twenty U.S. cities. Video communication also allows remote and sensitive witnesses (e.g., children or battered or mentally abused victims) to testify without being physically present in the courtroom. Many legal and judicial applications require large monitors, as well as VCRs, faxes, and flatbed scanners. Many legal applications rely on video communication public rooms to provide distant-end facilities.

By using video communication, attorneys collaboratively draft documents in real time, interactively discuss legal briefs, and conduct face-to-face meetings despite the thousands of miles that separate them.

SURVEILLANCE

Video surveillance enables remote location monitoring. Surveillance applications often incorporate lower-quality cameras and monitors. Applications include monitoring parking lots, corridors, and entrances. Some surveillance applications allow guards to press a button to unlock a door, and thereby remotely regulate admission to a building. Amtrak incorporates a unique video surveillance application to control and manage movable bridges in Atlantic City, New Jersey from a control center in Philadelphia, Pennsylvania. Four cameras, two on each side of a moveable bridge, relay pictures of the bridge to the control center. When a boat approaches the bridge, the control center performs a visual check for trains before remotely opening the bridge only long enough to let the boat pass. Before the use of surveillance technology, a bridge keeper was on location at all times. The installation pays for itself, and provides better visibility—the cameras provide four views of the bridge rather than the former bridge attendant's single view. Amtrak has since expanded the use of surveillance to Connecticut.

A Houston, Texas company offers continuous surveillance services to its customers without the high cost of security guards. Its system remains dormant until someone passes through a photoelectric or infrared beam or until someone manually presses an alert button. Then, the system notifies an operator at the company's control center. The operator may use the audio function of the system to advise the intruder to retreat, or may notify local police. The system allows operators to identify false alarms. Moreover, since the system captures intruders on videotape, it thoroughly serves the interests of justice.

Federal, state, and local law enforcement agencies used videoconferencing codecs to conduct security surveillance of both internal and external grounds during the 1996 Republican National Convention in San Diego. Stationed at the San Diego City Emergency Operations Center, the Joint Incident Command Post, and the San Diego Convention Center, law enforcement agencies simultaneously observed strategic areas, and surveyed the convention site and adjacent areas. Employing interactive, encryption-based codecs through a spread-spectrum microwave system, law enforcement agencies incorporated a helicopter-mounted camera and seventy land-based cameras. Authorities employed similar strategies to assist state and federal disaster field centers with emergency operations in south Sacramento and again in Southern California during the January 1994 Northridge earthquake.

EMERGENCY RESPONSE

The Federal Emergency Management Agency's (FEMA) video communication investment proved invaluable after terrorists attacked the Pentagon and the World Trade Center on September 11, 2001. FEMA used video communication to coordinate rescue efforts. A system linked FEMA to the White House and a wide range of federal agencies through secure dedicated links. Vice President Dick Cheney said the system was so important for coordinating the response that it was one of the reasons he initially chose to remain at the White House.

FEMA electronics maintenance manager, John Hempe, proclaimed, "We were online and communicating with the White House and other departments within minutes after the second aircraft struck in New York and before the hit in Washington." FEMA deployed a mobile videoconferencing unit in New York to relay twice-daily press briefings via satellite. Previously, it deployed video communication to improve communications between headquarters in Washington and the National Hurricane Center in Florida. John Hempe declared, "Being able to see the individuals you are interacting with adds a whole new dimension to meetings...our meetings are more lively and productive, as well as being far shorter than those conducted via telephone conference calls."

Figure 2-4. Emergency Response Application (courtesy of Intel).

Hempe said the time the system saves during emergency conferences, heightened communications, and a drop in costs speeded the adoption of videoconferencing. FEMA equipped each of its ten regional offices, and six other locations, with video systems comprised of a video camera, a television, a microphone, and either a PictureTel 760 or Polycom ViewStation linked by FEMA's private WAN or ISDN. Conferences run at speeds of up to 384 Kbps and provide video at a rate of 30 fps. FEMA uses PictureTel Prism and Montage conferencing servers running Microsoft Windows NT linked to an Oracle 8i database.

Duke Power Company, in North Carolina, leverages video to enable communication between emergency-response personnel. At two of its three nuclear power plants, Duke provides a network of desktop video systems that instantly allow face-to-face communication. At two plants (that generate more than half of the electricity it supplies) Duke relies on video communication to improve responsiveness and, therefore, its NRC ratings. Systems in each plant's technical support center link two other systems in each plant's operations support center. Duke uses desktop systems for voice and video dialers, programmable buttons, shared whiteboard, and tones to alert users to incoming calls.

Each system connects to the power company's fiber backbone network with an Incite Multimedia Hub that integrates voice, video, and data. Duke dedicates 384 Kbps of the network to videoconferencing, and equips each of the three sites with server software to manage the desktop (the two extras provide redundancy).

When an event occurs, each of the 800 members of the Emergency Response Organization reports to his station. In seconds, the video communication systems link the two support centers. A technician at the technical support center evaluates the problem and collaborates online with a user at the operations support center for a resolution. The TSC also calls the video communication unit at the Charlotte facility to inform outside agencies that decide whether to alert communities.

On July 16, 2004, video communication and the Freedom Calls Foundation enabled Army Specialist, Joshua Strickland to join his wife Dorothy, in Valdosta, Georgia, in celebrating his daughter Shelby's first birthday while he was stationed at Camp Cook, in Iraq. Camp Cook's Freedom Calls Foundation Facility, the first of its kind in Iraq, offers service members free videoconferencing so they can stay in touch with their families, and is made possible through private donations to the Freedom Calls Foundation. In addition, the foundation plans to install the videoconferencing capability at Army posts in the United States for families who do not have broadband Internet at home, and at military hospitals so troops can interact with their newborn children and new moms hours after birth.

John Harlow, executive director of the Freedom Calls Foundation, said, "Soldiers are now able to not only see, but participate in, milestone family events such as the birth of a new child, first communions, high school graduations, birthdays, weddings and anniversaries over the Freedom Calls Network."

The facility opened to 12,000 soldiers at Camp Cook in June 2004. Ed Bukstel, director of operations for the Freedom Calls Foundation, said, "They can talk just like they would over the dinner table." Harlow adds, "Although families may be separated by war, they need not be estranged by it."

DISTANCE LEARNING

Tens of thousands of students receive an education with the aid of video communication. Distance learning is not a replacement for, but an extension of, classroom learning. In distance learning, instructors' roles may not change, but their potential audiences increase. Video communication allows for a more collaborative classroom and for enhanced teaming outside of class. The success of distance learning correlates directly to a student's ability to assume responsibility for his own academic experience. Therefore, distance learning may be more promising for higher education than for primary education. Distance learning allows learning institutions to succeed regardless of location.

The move to use video technology in education started in the 1950s, when the City Colleges of Chicago began using television to deliver for-credit courses. In 1950, Iowa State University went on the air with WOI-TV. According to Dr. Carla Lane, a distance education consultant and an internationally recognized authority on distance learning, WOI-TV was the "World's first non-experimental, educationally owned television station." In 1963, the FCC allocated a portion of the microwave spectrum to Instructional Television Fixed Service. The California State University System was the first to apply for a license. In 1971, the British Open University began offering courses over television. As of 2005, it served more than 150,000 undergraduate and 30,000 postgraduate students, 10,000 of whom have disabilities. Nearly all of its students study part-time, and about 70% of undergraduate students work full-time.

In 1987, the Mind Extension University (ME/U) was formed. An education network that focuses on higher, for-credit courses, ME/U is a division of the Jones Cable network and is carried on a cable television channel. Instructors teach for-credit college courses from colleges and universities around the country. As with conventional courses, classes include tests and assignments. Students contact instructors by email, telephone, and videoconference. The ME/U has extended service to elementary and secondary school students.

An experiment in video-enabled distance learning shipped videoconferences over the Internet using *FrEdMail* (Free Educational Electronic Mail). The *Global Schoolhouse Project* linked students in Tennessee, California, Virginia and London, England and was part of a six-week joint-study curriculum about the environment. During the project, students read Vice President Al Gore's book, *Earth in the Balance*. At the end of the project, the students presented the information they had compiled to each other and, again, to government officials over the Internet. Funded in part by the National Science Foundation, the experiment was so popular that it became an ongoing experiment that linked students in Australia, Canada, Japan, Mexico, and Europe.

DISTANCE TRAINING

Most companies with more than 100 employees make use of computer-based training. As the term *distance learning* describes video-enabled instruction in an academic environment, the term *distance training* connotes video-enabled instruction in an organizational environment. Global competition and the increasing rate of change combine to make training workforces increasingly difficult for business, industry, and government institutions alike. Although traveling-shows can be an efficient way to train employees in remote cities, reaching all cities in this manner takes time. Video communication enables simultaneous, instantaneous, training no matter where people may be.

One of the nation's largest freight rail operators, Conrail, uses video communication to train a management team that is dispersed across five states. Training that used to take two days now takes two hours and includes valuable interactive discussion between instructors and employees. Wilson Learning (Eden Prairie, Minnesota) uses video communication equipment to offer its suite of corporate training courses (including sales and sales management, customer service, and supervisory skills) to nearly any point in the world.

The World Bank conducts *virtual seminars*. Through video communication on the Internet, the organization unites individuals who are addressing similar or related issues but are in such geographically separated locations as South Africa, Russia, and Egypt. The World Bank first installed room-based videoconferencing systems, but found desktop video to be less expensive, more convenient, and adaptable, and better suited for communications between small ad hoc groups.

PRODUCT ANNOUNCEMENTS

Many companies use videoconferences to introduce products to their sales forces before the products hit the market. In 1982, Champion Spark Plug used video communication to introduce a new product in what was, at the time, the largest business meeting ever held. Approximately 30,000 automotive parts sales representatives in 181 Holiday Inns throughout the U.S. participated in the event.

Since that time, video communications has become the preferred means of making product announcements, and the Internet has become the most common medium. Companies such as WebEx and PlaceWare (which Microsoft acquired) cast the die for such communication, and the market has since exploded.

Companies often conduct regular meetings in which remote territory sales managers participate by telephone and video communication. Subsequently, those companies introduce those products and services to resellers through live nationwide multipoint videoconferences in which participants at remote sites can interactively provide feedback and ask questions.

TELEMEDICINE

Telemedicine is one of the most widely-deployed and beneficial applications of video communication. The US military uses it to enhance health care for patients on remote bases where not all areas of medical specialty are represented.

Figure 2-5. Pediatrics Application (courtesy of Canvas).

Another popular application is the provision of health services at correctional facilities. Security issues significantly increase the cost of providing health care services to prison inmates, and persuading specialists to visit prisons can be challenging. Prisoners receive better care when specialists can support local physicians through video communication. Fitted scopes allow doctors to see a patient's inner ear or the back of a retina. Cameras can zoom in on a skin cancer, electronic stethoscopes can magnify a heartbeat, and machines can transmit digitized X-rays and lab results. Such off-the-shelf tools enable doctors to better diagnose and treat prisoners over distance more conveniently, and at less taxpayer expense.

As the abundance of aging baby-boomers stresses the already overburdened and expensive health care system, Telemedicine tools will allow physicians to remotely monitor and treat patients that would have otherwise been out of reach. This may decrease the number of days that patients must remain hospitalized, and permit visiting nurses to provide services to more patients than they could have personally visited. A visiting nurse can visually examine a group of patients before deciding which are in the greatest need of on-site care.

Videotaping doctor-patient consultations, which is a simple enhancement to video communication, can even decrease malpractice exposure because it allows review of a physician's diagnostic process. Many health care professionals prefer the process because a doctor on tape is a polite and courteous doctor.

CONCLUSION

Video communication may not suit every situation. Learning how to operate and maintain machinery may require an on-site instructor. Moreover, smaller audiences may make it difficult to cost-justify video communication infrastructure.

However, accepting that time spent commuting is generally time wasted, video communication promises increased productivity through better time management. Many people who drive to an office or a classroom could perform their work at home, at a remote office, or on the road with a computer and a network connection. With video communication, organizations need not limit their hiring choices by where people live; they can pick the best people no matter where they live. Schools and colleges can disregard time and distance limitations by delivering education through audioconferencing, audiographics, video, and computer-mediated communication.

While it is easy to adopt a predisposition that, "It may work for others but not for me," video communication boasts success in many diverse applications and environments. By shedding preconceptions and considering how to combine video communication with other processes and technologies such as broadcast media and store-and-forward video, one can find ways to remove barriers to worthwhile work. When one focuses on the needs, discards stereotypes, and applies the necessary technologies to solve communications challenges, the possibilities are practically endless.

Through the remainder of this section, we will detail instances in which organizations have leveraged video communication.

CHAPTER 2 REFERENCES

Miles, Donna. "Freedom Calls Offers Free Video Conferencing." *American Forces Press Service*. July 16, 2004.
http://www.defenselink.mil/news/Jul2004/n07162004_2004071601.html

Robb, Drew. "Videoconferencing Paid Off Sept 11 ." *Government Computer News*. September 24, 2001, Volume 20; Issue 29.

"Democratic Leaders, Speakers to Connect Daily with Delegates via Live Video Conferencing ." http://www.polycom.com/investor_relations/1,1434,pw-180-221-7552,00.html.

The Open University. http://www.open.ac.uk/about/ou/index.shtml

Gardner, W. David. "A maestro fiddles with videoconferencing," *Electronic Engineering Times*. August 26, 1996 n916 p. 26-27.

Hamblen, Matt. "Desktop video surge forecast," *Computerworld*. October 14, 1996 v30 n42 p. 85.

Hayes, Mary. "Face To Face Communication–Desktop videoconferencing helps utility respond to potential nuclear emergencies," *Information Week*. December 9, 1996, Issue 609.

Lazar, Jerry. "Feds get the picture on videoconferencing," *Federal Computer Week* August 26, 1996 v10 n25 p. 53-59.

Petropoulos, Gus, and Vaskelis, Frank. "College piggybacks video network on phone system," *Communications News* February 1996 v33 n2 pp. 24-25.

Pietrucha, Bill: "Videoconferencing Played Security Role at Republican Convention," *Newsbytes* August 20, 1996.

Riggs, Brian. "Team spirit guides WAN; Northrop Grumman picks SMDS for videoconferencing and document-conferencing apps," *LAN Times* July 22, 1996 v13 n16 p. 37-39.

"Soldiers Take Their Offices to the Battlefield," *Government Computer News* April 15, 1996.

3

CASE PROFILES

Sometimes you can tell a large story with a tiny subject.

Eliot Porter

AMERICA TRUE TEAM NAVIGATES COMMUNICATIONS CHALLENGE

As the crew of America True made a run at the America's Cup, video communication kept the team on course. The physical process of racing is just one aspect of what makes a successful America's Cup team; races require months of preparation. Complicating preparations for America True team members in San Francisco and Auckland was a vast body of water and some 6,513 miles. Polycom signed on as a corporate sponsor and exclusive videoconferencing supplier to America True, and helped it to navigate its communications challenge.

America True used Polycom ViewStation systems to communicate between team headquarters in San Francisco and the racing team in Auckland, New Zealand, where preliminary races began in fall 1999. The value of ViewStation quickly became apparent to America True CEO and Captain, Dawn Riley, the first female skipper of an America's Cup syndicate, and the first skipper of a co-ed America's Cup team. Riley used video communication regularly with the entire America's Cup team. Every Wednesday morning, she would host a staff meeting that included 25 people on two continents, corporate sponsors, and community supporters.

Riley claims that video communication allowed her to sail with the team in New Zealand, and still manage day-to-day operations in San Francisco. She proclaims, "When we're discussing pivotal subjects, such as sponsorships, upcoming events in both California and New Zealand, and sailing updates, it's critical that we are all in the same room, so to speak... That way, no details go unnoticed, we can keep everyone in the loop, and we stay on course to win the America's Cup."

America True held a video press conference on September 24, 1999 to allow the press an opportunity to meet the crew, to catch up on progress, and to tour the America True boat and the compound in Auckland. Phil Kaiko, America True's principal designer, used video communication to assist in faster decision-making on boat designs. He declares, "When I used videoconferencing about five years ago, the transmission was broken up and the camera angle wasn't wide enough to capture the whole room, so it was really a waste of time. Now, I'm seeing videoconferencing in a whole new light for enhancing our team productivity." Kaiko credits video communication for enabling the team to make timely decisions on designs and, subsequently, to get the boat in the water faster.

The America True team credits video communication for convenience, and for saving time and money. Without video communication, travel costs to accomplish the same tasks would have reached the tens of thousands of dollars. Said Chuck Riley, America True's vice chair, "We have been able to connect potential donors to the program with our Auckland activities and let them see the set-up there first hand; this would have been impossible without videoconferencing."

University of Tennessee Health Sciences

University of Tennessee Health Sciences Center uses video communications to reach remote patients, and save millions of dollars. The University of Tennessee Health Science Center employs 1,892 physicians who staff numerous area hospitals. The UT College of Medicine (UTCM) and community partners launched the MidSouth Telehealth Consortium to bridge the distances between its physicians and potential patients who live in isolated, underserved communities.

The telehealth system quickly demonstrated value for all partners through treatment cost savings, reduced treatment delays and improved health outcomes. The system quickly grew to include 67 telehealth connections that serve hospitals, health departments, prisons, pharmacies, distance education centers, district health offices, and a mobile dental clinic. The average rural telehealth patient lives 117 miles from Memphis, so by reducing two outpatient visits per person each year the system cut travel costs by $61.29 million per year.

The organization, staff and partnering agencies on the network have found value in a wide range of video communication applications, including specialty consultations, urgent care consults for trauma, continuing medical education for providers, healthcare education for the general public, staff training, administrative meetings, family consultations, and technical diagnostics for training and troubleshooting. UTCM also uses the telehealth model to aid an outreach center for applying technologies to meet community outreach health care needs, such as E-health, Minority Recruitment, Campus Faculty and Student Volunteer Efforts, and Health Disparities.

The UTHSC identified five priorities for improving the region's healthcare system:

1. Reduce delays in patient care

2. Improve patient safety

3. Improve prescription medication management

4. Provide resources to financially troubled rural hospitals

5. Improve continuing medical education support to rural providers

"The project partners led by the UTHSC decided upon a regional interactive video conference system utilizing dedicated T1 lines," said Dr. Karen Fox, the outreach center's Executive Director and Assistant Dean for Communication, who also serves as Assistant Dean for the College of Medicine.

Both efficiency and patient satisfaction benefit with Telehealth. The telehealth program efficiency has outpaced traditional medical clinics by 48% to 27%. These practices include an increase in efficiency across specialty fields such as dermatology, psychiatry, ENT pediatrics and adult, and neurology. Patient wait times are less than five minutes for more than 70% of patients using telehealth video communication, and 47% of patients agree that the quality of service of a video conference gives them more clinical attention than a regular office visit.

For those looking to start similar services, the UTHSC Outreach Center suggests:

1. Develop healthcare services in response to a true, significant, and ongoing community need.

2. Collaborate with project partners such as local and state health departments, local health education centers, regional patient advocacy groups, statewide clinician groups, and healthcare management organizations.

3. Conduct site visits and learn from others' experience.

4. Develop a system with inherent flexibility that changes as constituents' needs change.

The USHSC deployed a regional, interactive video communication system using dedicated T1 lines. The IP network utilizes Polycom's latest protocols (H.323) with 768 Kbps connectivity for point-to-point calls. The network includes three Polycom MGC's for bridging multiple video calls. This arrangement combines broadcast of training sessions to large audiences, with the advantages of interactive video.

PREVENTING NUCLEAR DESTRUCTION

The U.S. Department of Energy (DOE) and Russia's atomic energy ministry constructed a videoconferencing link between the two countries' energy departments during the year 2000 date rollover. U.S. Energy Secretary Bill Richardson and Russia's Minister of Atomic Energy Yevgeniy Adamov demonstrated the videoconferencing equipment that linked the DOE's Emergency Operations Center in Washington to a MinAtom Situation and Crisis Center that was dedicated October 2, 1999 in Moscow.

The videoconferencing system used a T1 line and off-the-shelf switches, codecs, and LCD video projectors. The DOE used internally developed Unix-based routing software for the internal routing of video signals, and videoconferencing allowed officials in Moscow and Washington to view graphics and stored pictures simultaneously to address issues that arose. Russian experts were in the DOE's emergency center and U.S. experts were in the MinAtom center to monitor and to provide advice on year 2000-related issues regarding Russia's electricity grid and nuclear power plants. Additionally, U.S. experts resided in Ukraine to monitor the site of the 1986 explosion and fire at the Chernobyl nuclear power station.

The DOE began working on the videoconferencing link in March after Richardson and Adamov signed the U.S.-Russian Joint Commission on Economic and Technological Cooperation's report. The two countries agreed to establish a working relationship within respective emergency centers, and the DoE agreed to provide MinAtom with emergency management training and exercise assistance. The Department of Energy also agreed to provide future assistance for technology and engineering at the Situation Crisis Center. In preparation for the date rollover, the DOE helped to monitor all 29 nuclear power reactors in Russia, and provided hardware, software, training, and assistance with contingency plans.

PAROLE: WEB CAMERAS REDUCE TIME

A web camera has eliminated forty-hours-per-month of travel for hearing judge, Carol Bohannan, who used to make the drive from the Arkansas state's Post Prison Transfer Board office in Little Rock. Before the *webcam* was in place, Bohannan and two other hearing judges traveled about 2,000–3,000 miles each per month to revocation proceedings throughout the state. Hearing officers and board members, who hold parole hearings at the state's prisons, drove 132,872 in one year in lieu of the webcams.

Jim Williams, who has been doing revocation hearings for nine years, began to think of videoconferencing as an alternative for the Post Prison Transfer Board after seeing such a system five years earlier in the Bentonville court system. He and his colleagues got a camera from the Arkansas Department of Correction to use for videoconferencing in Benton County. Later, the three hearing officers began using web cameras attached to personal computers to conduct hearings elsewhere.

Ernest Sanders, a hearing officer since 1998, said the hearing judges each spent about $30 of their own money to buy cameras because they did not want to wait for availability of funds for more expensive conferencing equipment. Sanders proclaimed, "We just dial in. They have a camera; we have a camera. We focus, and we're ready to go." About nine county parole offices now have access to videoconferencing systems. Some hearings are in remote areas of the state.

With more than 14,000 parolees in the state, the Post Prison Transfer Board, which has a yearly budget of about $1 million in general-revenue funds, justified $11,000 more to buy video communication equipment.

Within the next year, the state Department of Community Correction plans to use existing funds to buy more video communication equipment for hearing officers to use, said Rhonda Sharp, a spokesperson for the department and the Post Prison Transfer Board. Eventually, she said, Post Prison Transfer Board members hope to adopt the technology as well. Sanders said he doesn't think that conducting hearings over the Internet takes anything away from the process, and the hearing judges have heard no complaints from parolees.

"The technology is so clear you can see facial features and body language," Sanders declares. Williams agrees as he evokes a police officer in a hearing showing him crime-scene photos of a stabbing victim. Recalling how clear they were, even over the computer, still causes him to wince.

Municipal courts in Pulaski County began using videoconferencing equipment in 1999 for arraignments. The system precludes the need to move inmates from the Pulaski County jail for a hearing that may last only minutes. James McMillan, principal court technology consultant with the National Center for State Courts, said he first saw videoconferencing in San Bernardino County, Calif., in the early 1980s. Since then declining prices have allowed more counties to use video communication for short hearings for which a trip to a local courthouse would equate to an extra expense and a risk.

The last time the center conducted a survey on the subject, in 1994, it found that about 600 courts across the nation used video communication. Even with a limited number of counties using the cameras, Bohannan, Williams and Sanders say the averted travel allows more time for completing mandatory reports and reviewing paperwork from parole officers. The hearing judges say the flexibility video communication gives them enables them to respond to emergency hearing requests that can arise when a parolee in a county jail has a disciplinary or medical problem.

NORTH SLOPE BOROUGH SCHOOL DISTRICT: A BIG COLD CLASSROOM

Distance education has benefited rural communities that often face challenges in attracting teachers, especially in specific areas of curricula. Low demand for particular courses, and difficulty in meeting state-mandated requirements compound the problem of delivering rural education. Barrow, Alaska's North Slope Borough School District, provides K–12 service to one such rural community.

The North Slope Borough School District is the largest in the United States. Equal in size to the state of Minnesota, it measures 650 miles from east to west and covers 88,000 square miles. All points are above the Arctic Circle. Scattered among eight villages, 86% of the 1,700 students within the district are Alaskan natives. More than 1,000 of these attend school in Barrow. Others attend village schools where, with only few high school students and very harsh climatic conditions, it is hard to find and retain trained professional staff with proficiencies in all subjects; most teachers need to be generalists. Some high schools include only two or three seniors.

The North Slope Borough School District, a pioneer in distance learning, placed VTel PC-based video communication systems in village schools. The systems offer both face-to-face video communication and the ability to create and store graphics and text files. A multipoint bridge allows simultaneous conferencing so that students and teachers may converse between three or more sites. This replaced a satellite system that provided receive-only video and two-way audio.

Instructors find that students enjoy education over video communication. The medium allows increased social interaction between students that might otherwise meet only once or twice a year. Video communication allows students at the smaller schools to leverage resources and develop new relationships. Eleven interactive video-conferencing sites, 650 computer workstations, and more than 2,500 users use the system. With the advent of the distance learning curricula, national achievement test scores improved steadily throughout the borough.

DEUTSCHES FORSCHUNGSNETZ: RESEARCH & EDUCATION

The Deutsches Forschungsnetz (DFN), in Germany, is a national communications network for institutions of public and private research groups, higher learning, and government facilities. The DFN broadband research network fosters joint research between institutions, distance learning and virtual university programs, communications between staff and associates, and data transmission.

In the late 1990s, the DFN launched the Gigabit-Wissenschaftsnetz (G-WiN), a gigabit pure IP network, to support R&D and the internal and external communications of its members. Early pilots resulted in the following findings:

1. The network data annual growth rate was 2.2 percent and therefore, traffic would grow by a factor of 50 within five years.

2. Multimedia, videoconferencing, and supercomputing were spiking network traffic volume.

3. International traffic was growing at an exponential rate.

4. Demand for quality of service and services integration was growing.

These findings prompted organizers to bridge the G-Win research network with the US Internet2 initiative, which, in turn, added another 100 universities and research facilities to its reach. DFN needed the ability to manage these complex, distant connections and provide bridging for its own growing number of collaborative research projects. With more than 550 member facilities and more than 10,000 network devices, DFN needed an intelligent gatekeeper that could track and supervise which entities received access to specific services, and when.

DFN chose the RADVISION Enhanced Communication Server (ECS) because it could provide videoconferencing services for up to 400 simultaneous participants, manage more than 12,000 devices on the network and hierarchically manage multiple communication servers. Another issue was allowing DFN members to communicate amongst themselves and with outside organizations. Organizations connected by ISDN rather than dedicated IP required a gateway, which DFN addressed with RADVISION's viaIP multi-service platform.

The additions enable DFN to manage IP telephony and multimedia communications intelligently on one network. The ECS allows network managers to set policies and control network resources such as bandwidth allocation. With viaIP, DFN members can conference with up to 400 participants on a single videoconference. It also provided a growth path to ensure the viability of its network into the foreseeable future while enabling interoperation with existing infrastructure.

MEDICAL COLLEGE OF GEORGIA: HEALTHCARE AND EDUCATION OVER DISTANCE

Even the best of physicians occasionally needs a second opinion. When appropriate help is in another town, the solution used to be to fax X-rays and attempt to describe the circumstances of the case over the telephone. Videoconferencing provides a better alternative for the Medical College of Georgia (MCG).

Dr. Jay Sanders took a leading role in one of the country's first Telemedicine applications, at Massachusetts General Hospital in Boston. Subsequently, as Director of Telemedicine for MCG, Dr. Sanders provided the vision to build, "a system that electronically transports a consulting physician from a medical center complex to a patient at a distant health care facility." The MCG's system integrates interactive video communications with remote controlled biomedical telemetry to allow a consulting physician to examine a patient at a satellite location as if the patient were in the physician's office.

The lack of access to specialty care is often a disadvantage of living in an isolated community. Rural hospitals require a minimum number of full beds to support doctors, nurses, technicians, and administrators. Therefore, every patient counts in the hospital's struggle to survive; each time it must move a patient to an urban hospital for specialty treatment, the bed census statistics of the small rural hospital are adversely affected. In addition, transportation of fragile patients presents health risks, and discontinuity of care can result. Rural doctors often suffer from professional isolation, and find continuation of their medical education difficult. A highly specialized form of videoconferencing, Telemedicine, allows rural physicians to access, almost instantaneously, the specialized medical expertise of university doctors. It decreases the recurrence of referrals, provides immediate attention to life-threatening problems, establishes an integrated care network, and provides enhanced quality of care in rural situations. Even so, the patient's cost is significantly reduced.

Videoconferencing enables the MCG to improve care and delivery of medical services, and to reduce travel time and expense. It enhances patient referrals, consultations, and post-operative examinations. MCG's vision included emergency-room-to-emergency-room support, emergency psychiatric consultation, supervision at community hospitals, benefits consultations, recruitment, administrative meetings, and clinical research collaborations. Specialties such as cardiology, dermatology, neurology, and ophthalmology all employ telemedicine.

Through videoconferencing interns, residents, fellows and students exchange published papers, discuss patient cases, and analyze prognoses throughout patients' illnesses. In this way, videoconferencing improves the quality of patient care by fostering new relationships. Moreover, by using the system to provide staff with information on such diverse subjects as women's health issues and financial planning, the institution believes videoconferencing has improved morale and saved more than $10,000 for every twelve hours that it is used. In addition, referring physicians earn credit for participating in tutorials with consulting physicians who are also members of the institution's faculty. The system records consultations on videotape for subsequent observation by review panels.

Through distance learning and individualized laboratory work, MCG allows remote medical professionals to participate in classes without traveling. This is important to Georgia hospitals because state law requires a medical technologist to be on duty on the hospital premises at all times. Cameras in classrooms at both remote and primary sites capture and transmit images across special cabling. Instructors provide personal attention by scheduling time after lectures or answering toll-free telephone calls. Video communications foster personal interactivity between students and promote competition between groups of students at remote sites; as little as half a given class may participate at the primary site.

State and federal funding has furnished each facility with recorders, motion cameras, auxiliary cameras, and graphics cameras. The graphics cameras provide whiteboard and transparency projection capabilities, and record live or still graphics for broadcast. Users control the video system, audio system, and the codec (including camera pan, tilt, zoom, and focus functions, as well as transmission of motion and still video) with a touch-screen control panel. Using wireless microphone systems with lavaliere lapel microphone and separate clip-on transmitters, a receiving unit, and an antenna, presenters move freely within classrooms. MCG leverages a sophisticated network of terrestrial lines, satellite dishes, and microwave and cable systems to examine real-time images through viewers that interoperate with microcameras and to broadcast live, real-time, interactive health programming from the MCG campus to distance learning facilities throughout the state.

The State of Georgia worked with five LECs, AT&T, and Sprint to bridge service boundaries, and required vendors to provide an automated scheduling system to manage the system. Instructors provide schedules to DOAS, which programs the system to connect sites at set times. The state also required vendors to provide a flat billing structure for every location in the state, a point-to-point or multi-point environment, and the ability of any site to talk to any other site.

THE ABORIGINAL PEOPLES VIDEOCONFERENCE

In the fall of 1991, the remote Aboriginal communities of the Tanami, consisting of Yuendumu, Lajamanu, Willowra and Kintore, made a decision to install a six-point videoconferencing network for links between villages, to Darwin and Alice Springs, and to other points. The Tanami Network consists of PictureTel codecs linked via Hughes earth stations to the AUSSAT 3 satellite that serves Australia.

Dating back more than 50,000 years, the Aboriginal is the oldest surviving culture. In the early 1940s, the Australian government established Aboriginal communities in the desolate outback region known as the Tanami desert. Unfortunately, outside intervention began to disrupt the complex network of information and personal contacts that, for thousands of years, had existed between the area's people. Aboriginal law and traditions started to erode. Young people turned from their elders and family members to television. Alarmed, the Tanami community employed videoconferencing to resurrect tribal tradition in which culture and knowledge passed from the elders to the young. The Tanami Network now plays a role in preserving the age-old culture and heritage by remotely conducting ceremonial activities and other Aboriginal business.

The development of the videoconferencing network began in 1990 with a successful trial of systems that connected Lajamanu, Yuendumu, and Sydney. Over the links, students received education and community spokespersons gained a voice in their government. Aboriginal artists discovered that they could use the system to negotiate art and craft sales directly with buyers all over the world. This eliminated a chain of go-betweens that had, for years, exploited the Aborigines' isolation.

After a successful trial, videoconferencing was widely deployed. The philosophy behind the $200,000 project consists of three aspects. First, the Aboriginal people will have local control of the systems and network. Second, the project will be but one part of an attempt to enhance ceremonial and family links. A program of community visits, workshops that elders conduct, and proactive government involvement will further social communication that is vital to the Aboriginal community. Other technologies such as telephone, facsimile, radio, local video production, and regional broadcasting are being integrated. This includes the display and promotion of arts and crafts directly to overseas markets, the delivery of educational, medical and governmental services from remote locations, support of community detention and bush and urban court proceedings, and the strengthening of community, family and ceremonial links.

HOWE PUBLIC SCHOOLS LEARN WITH VIDEO

Howe Public School District, in LeFlore County, Oklahoma, is comprised of an elementary, middle, and high school. Collectively, the schools serve a population of only 389 students and fall into the state's highest category for poverty and child abuse reporting. Howe is a 90% E-Rate district, which means that more than 75% of the student body qualifies for the free and reduced lunch program. Moreover, the district's budget barely met teacher salaries and operating expenses and thereby left nothing for technology and non-core classes. Consequently, the district had no network, no modern computers, and no plans for introducing technology into the curriculum. Howe is isolated from educational opportunities and cultural institutions such as museums, theaters, or zoos. Many of Howe's students have never left Oklahoma, and a large percentage does not graduate. The district's challenge was to provide students in an impoverished, rural locale with the educational experiences and resources that students in urban areas enjoy.

With $571,000 in grants, Howe first built a robust IP network. With that network in place, Howe expanded course offerings by installing Tandberg videoconferencing technology. Howe's low student population had been a formidable obstacle for schools to providing elective classes such as foreign language. Howe leveraged videoconferencing to provide district-wide classes. Two or three students in each rural school, together, comprised a large enough group to make such classes viable. As a result, Howe was able to add classes such as Spanish I, Spanish II, AP Psychology, Physics, and Art. The district now uses its video lab continuously seven hours a day, five days a week to provide such classes.

Howe is also now connected with cultural and educational institutions to provide experiences that otherwise would not have been available. Through Tandberg's Connections program, Howe high school literature students have collaborated with the staff at Rutgers University to stage a production of Shakespeare's *A Midsummer Night's Dream*. Howe third graders have chatted with staff at the Indianapolis Zoo's Penguin Exhibit. The system enables the district to introduce other Howe students to new places, ideas, and experiences.

Howe also found its videoconferencing system invaluable in providing professional development opportunities for teachers and other staff, who leverage it to take graduate courses from state universities without traveling. Lance Ford, IT Director for Howe, proclaimed, "The distance learning program at Howe has allowed this district to create a stronger curriculum, expand the educational opportunities for its students, and provide an exceptional educational program for the benefit of not only the students and teachers, but also the community in general."

KAPI' OLANI MEDICAL CENTER

Kapi' olani Medical Center faced the challenge of managing high-risk pregnancies when doctors only visited the island once each week. Flying to Honolulu took a high financial, physical, and psychological toll on families.

Based in Honolulu, Hawaii, Kapi' olani Fetal Diagnostic Center (FDC) is the only tertiary-level hospital for women and children in the Pacific Basin. It specializes in care of high-risk pregnancies. Unfortunately, the large coverage area, the distance between the islands, and the limited number of specialist medical personnel have been obstacles to providing fast access to the specialized care these patients require. Consequently, the medical staff has tolerated expensive and time-consuming travel between islands and locations to examine and meet with patients and colleagues.

To transcend this circumstance, Kapi' olani Medical Center deployed the first tele-ultrasound network in Hawaii, which included the Tandberg HCS III product line. Eventually, the network will comprise ten sites on four islands with a hub site in the FDC in urban Honolulu. The application required retrofitting existing ultrasound machines to enable perinatologists to see a fetus remotely with video monitors. Perinatologists can subsequently provide advice and care to patients, those patients' regular obstetricians, radiologists, and sonographers 24x7. With the new system, they provide ready access to the specialized services that maternity patients require. A grant from the US Department of Commerce Technology Opportunities Program, support from Kapi' olani Health, and the Harry and Jeannette Weinberg Foundation funded the $1.8 million project.

More than 6,000 health problems can be detected prenatally; ultrasound is a primary detection tool. In a high-risk pregnancy, monitoring a fetus by ultrasound can be particularly important. The goal of the Fetal Tele-Ultrasound Project is to extend patients' access to this critical pre-natal care.

XEROX PARC EXPERIMENT: MEDIA SPACES

According to Sara Bly, Steve Harrison and Susan Irwin, the authors of *Media Spaces: Bringing People Together in a Video, Audio and Computing Environment*, a media space is a "technologically created environment" that emerges from a "concern for both the social and technical practices of collaborative work and from an effort to support those practices." In the mid-1980s, XEROX PARC's System Concepts Laboratory (SCL) was geographically split between Palo Alto, California and Portland, Oregon. The intent was to "maintain a single group and explore technologies to support collaborative work." The media space had to support not only cross-site work but also provide the necessary social connections between workers to allow them to "be together."

At the time of the media space experiment, the SCL had a hierarchical management structure, but lab members "regularly attended staff meetings and most decisions were made by consensus." SCL's way of working was acknowledged as both professional and social. The group accepted the split as a challenge to integrate access across distance to information, computing, and social interaction including "casual interrupting, gossiping and brain-storming."

Media space put cameras and microphones in offices and common areas on both ends (Portland and Palo Alto). A fixed two-way 56 Kbps video link connected the sites. The link included video compression equipment, an audio teleconferencing system, and consumer-quality video cameras and monitors in the commons area of each site. Microphones were interspersed throughout each building. The link was dedicated and was, therefore, always active. The portable video equipment was sometimes moved from the commons into individual or private offices.

Eventually the Design Methodology researchers expanded on the initial prototype by adding the ability to reconfigure the cameras, monitors, and microphones in each office using a crossbar switch. Participants could walk to a panel and push a button to change the audio and video arrangement to match their current need and activity. Just as often, participants would change the arrangement at random so that anyone might be seeing and talking to anyone else on the other end. The goal was to approximate the casual encounter and move through a range of socially significant contacts in any given day.

Users controlled their availability across the medium by controlling cameras (on, on but focused on something other than the person in the office, and off). Individuals took responsibility for controlling their own personal and private space boundaries.

What did the media space project learn about work over distance? First, that technology can enable the social relationships of workers separated by space. The frequent and regular use of the media space for awareness, informal encounter, and culture sharing indicate that technology can support more than task-specific communication. The media space project showed that a group could and did maintain itself as a single community that only distance separated. Participants routinely referred to all members across sites as, "We."

The group learned that different settings require different media spaces. During design, thought should be given to "not only the objects of the system but also to their placement," how they are accessed, and how well they are integrated into the ongoing organization of work life. Open, continuously available video and audio is not appropriate for every environment or culture. The challenge is to build systems that reflect the changing needs of user communities. One size does not fit all in electronically-mediated connections.

The unique needs of the SCL led the researches to uncover a need for a shared drawing surface. Users not only wanted to see what their coworkers had written, but also wanted to amend what had been written or drawn, to collaborate on problems and solutions. This aspect of the project resulted in Xerox's LiveBoard, which it introduced in 1993. It is a 3- by 5-foot backlit active-matrix liquid crystal display (LCD) powered by an Intel processor. LiveBoards allow workers in different locations to simultaneously view and edit the same document.

The group also noted that typical "marketplace technological offerings" including videoconferencing and desktop video devices did not offer the flexibility of media spaces because they were not integrated into a computing environment. "Most videoconferencing requires that one 'go' someplace; it is not an integrated part of the office itself." The study went on to note that it would be "interesting to explore the possibilities for desktop video in supporting activities like peripheral awareness, chance encounters, distributed meetings, and discussions and envisionment exercises."

The researchers of the project noted the importance of media literacy and how participants progressed from computing vs. computing augmented with video and audio. This progression illustrates an evolution "toward a unified field of audio, video, and computing." Finally, the project led the research team to consider whether "awareness of people and activities in a working group had value independent of other mechanisms that might be available for collaboration." In other words, just thinking about co-workers in another area or department might be the first step toward working together; and seeing and hearing them from time to time might stimulate thinking.

CHAPTER 3 REFERENCES

"Tennessee Telehealth Connection." December 2004.
http://www.polycom.com/company_info/1,1412,pw-4087-4088,00.html - 129.3KB

Deutsches Forschungsnetz, Deploys Videoconferencing Services to over 550 Facilities with RADVISION's Videoconferencing Infrastruture Solutions. www.radvision.com January 2004.

"Howe Public Schools." *Tanderg.* 2004. www.tandberg.net

"Kapi' olani Medical Center For Women and Children: Extending Care." *Tandberg.* 2004. www.tandberg.net

Johnston, Margret. "U.S., Russia Set Up Y2K Video Link." *Computerworld*, December 3, 1999 04:24 PM.

"InfoValue Video-on-Demand, Multicasting Solutions Chosen for New York State School System." *Business Wire*, November 22, 1999 p0039.

Baard, Mark. "CNN News Hounds Use Digital Technology To Report, Edit, Produce," *Macweek* June 3, 1996 Volume 10 Number 22.

Bly, Sara A., Harrison, Steve R., and Irwin, Susan. "Media spaces: bringing people together in a video, audio and computing environment," *Communications of the ACM*, January 1993 pp. 28-50.

Chute, Alan G., Ph.D. and Elfrank, James D. "Teletraining: Needs, Solutions..." *International Teleconferencing Association 1990 Yearbook*, June 1990.

EDGE on & about AT&T. "Videoconferencing: SunSolutions unveils industry's first complete desktop product for workstations," November 1, 1993 p. 7.

Fogelgren, Stephen W. "Videoconferencing at Management Recruiters." A document provided courtesy of MRI, Cleveland, Ohio, 1993.

Gold, Elliot M. "Industry-Wide Video Networks Thrive," *Networking Management* November 1991 pp. 60-64.

Halhed, Basil R. and Scott, D. Lynn. "There Really Are Practical Uses for Videoconferencing," *Business Communications Review* June 1992, pp. 41-44.

Hayes, Mary. "Face To Face Communication–Desktop videoconferencing helps utility respond to potential nuclear emergencies," *Information Week* December 9, 1996, Issue 609.

Lach, Eric. "Flagstar's Snappy Solution—Videoconferencing Helps Mortgage Banker Speed Loans," *Communications Week* September 18, 1996, Issue 629.

Lane, Carla, Ed.D. *Technical Guide to Teleconferencing and Distance Learning* Applied Business teleCommunications San Ramon, California, 1992 pp. 125-193.

McMullen, Barbara E. and McMullen, John F. "Global Schoolhouse," *Newsbytes* April

29, 1993, p. NEW 04300011.

Newcombe, Tod. "Georgia Spans Education, Medicine Gap with Two-Way Video," *Government Technology* 1995.

Persenson, Melissa J. "Electronic field trips for the '90s," *PC Magazine* February 8, 1994 p. 30.

PictureTel Application Notes and VTEL and Panasonic User Profiles were used to develop several of these applications.

PictureTel Corporation. "This Way to an Eye-opening Exhibit of Creative Uses for Videoconferencing," 2268/APPS/496/15M.

Rash, Wayne. "Virtual meetings at the desktop. (Fujitsu Networks Industry...)" *Windows Sources* January 1994 p. 235.

Sanders, Dr. Jay. "Three Winning Videoconferencing Applications," Advanstar's VIEW '93 Conference, Videoconferenced between Houston, Texas and Atlanta, Georgia, November 1, 1993.

"Soldiers Take Their Offices To The Battlefield," *Government Computer News* April 15, 1996.

Toyne, Peter. "The Tanami Network," As presented to the Service Delivery and Communications in the 1990s Conference, Sydney, Australia, March 17-19, 1992.

"United States Largest School District Uses Videoconferencing to Bridge Gaps," *Business Wire* (an Information Service of INDIVIDUAL, Incorporated) October 26, 1992.

4

THE BUSINESS CASE

Talk of nothing but business and dispatch that
business quickly.
Placard on the door of the Aldine press,
Venice (1490)

In an IBM Systems Journal article, Peter G. W. Keen set forth a process for senior managers who must deal with the challenge of ensuring that business processes, people, and technology are meshed within an organization. He noted that an executive's most important contribution to the organization is to, "clarify the firm's business imperatives that are based on knowledge anchors and linked to its vision and strategic intent." In a discussion of information technology (IT), Keen continues, "The key step in the business and technology dialog is to link business imperatives to IT imperatives." It is on this note that we begin our discussion on making the business case for video communication.

To sell a technical concept to senior management, one must link that concept to business imperatives. One must explicitly interpret how the technology can provide a competitive edge, reduce costs, (and) or provide the business with some other significant, tangible benefit. Although this point is not a new one, it is often an overlooked one. To one who is familiar with a technology, that technology's benefits are conspicuous. Such a perspective complicates objectivity and makes it too easy to assume that others will see that technology's value and will, consequently, embrace it and pursue its benefits. To those who address the realities of business on a daily basis, there are too many valuable projects and too few dollars to accommodate all, or even most of them. Consequently, for video communication to become a vital part of the communications infrastructure, its champions must describe it in terms of how it can enable or facilitate business imperatives.

The best business case for video communication is not a *technology-oriented* one because technology has no inherent value. Information technology is valuable only as in collecting, storing, accessing, or processing data. In business, IT provides value only as an extension of the organization's infrastructure and serves as a foundation upon which the business can reduce costs, increase revenue, or otherwise serve its stakeholders. In the vernacular of the sales profession, the features and advantages foster credibility, but the benefits make the deal. An example of a feature might be the ability to integrate video communication with collaborative tools. An example of a benefit might be greater availability of critical personnel. However, the benefit is either business opportunities gained or (expressed in dollars) dollars saved.

Just as a company's mission statement and culture provide a basis for its objectives, infrastructure is the foundation on which a company builds its business applications. Infrastructure links all parts of the business and provides a mechanism for sharing policies, customer information, product information, market intelligence, or regulatory requirements. Infrastructure enables the business to operate as a coherent whole. Therefore, in making the case for a technology such as video communication, project vision must be rooted in an active business model, and one should consider technologies' infrastructure components.

The priorities of senior managers include increasing market share, improving efficiency, reducing operating costs, and fostering agility in a rapidly changing economic climate. Consequently, one must frame one's video communication proposal in those terms. Most, if not all, top-level managers have endorsed IT investments that have not met expectations. Most are, consequently, skeptical. The business case for video communication must clearly inform these executives of cost reductions or markets gained. They will want to know whether competitors, customers, or suppliers are making similar or complementary infrastructure investments.

Government and public sector enterprises, like their private sector counterparts, must increasingly leverage limited resources for maximum return. Improving the quality of service to the public and accomplishing more with less is critical. A number of agencies are using, and many more are considering the use of, video communications as a service delivery tool. Since the early 1990s, federal government agencies have used video for internal communications. The Social Security Administration has used video communications to resolve claims disputes and disability cases. Other agencies, such as Health and Human Services, have leveraged video communication via kiosks and the Internet.

When developing a business case for a technology with which one's organization is inexperienced (and about which it may entertain misconceptions), one must hone skills in areas such as:

- Understanding the enterprise's market, industry, and challenges
- Knowledge of standard business concepts and accounting practices
- Internal partnering, investigation, and needs-analysis
- How video communications can enhance competitive position
- Vendor alliances and support from key business partners.

KNOW YOUR BUSINESS

The first step in implementing video communication as a business solution is becoming familiar with the organization's market, industry, and challenges. IT professionals increase stature by putting business first and technology second. The key to success is thinking strategically, thoroughly understanding the enterprise and its market, and developing strong relationships with managers in functional groups outside IT. Enthusiasm for a technology rarely compensates for failure to appreciate fundamental issues. A prepared IT professional will have studied the company's prospectus, annual report, or SEC filing (in the case of a publicly held organization). Thorough work involves scanning trade magazines and publications for articles about the company, and consulting administrative staff for insight into the executive agenda. One must base the business case on the company's mission, value statements, strategic goals, and business objectives.

Most organizations promote mission and values, strategies, goals and objectives. Selling technology to upper management generally requires drawing synergies with these. Only when one understands the organization's (or business unit's) objectives, can one think in terms of the results that matter to that organization.

An organization's mission describes its purpose and its market. Values, the abstract ideas that guide thinking, provide the basis for the mission statement. Together, mission and values drive the goals and organizational structure. While mission and values remain consistent throughout the organization, goals differ between functional areas. Goals are future states or outcomes that the organization desires. An organizational goal might be to increase market share or to improve customer service. By contrast, an objective is a specific, measurable, target.

Following are examples of goals:

- Goal: Enhance competitive position through improved communications
- Goal: Improve quality standards on a global (national, regional) basis
- Goal: Increase the global market share of product X
- Goal: Improve product time-to-market schedule
- Goal: Improve productivity within the corporation
- Goal: Get closer to the customer/ improve customer service
- Goal: Retool the organization in response to our changing market.

Here are some sample objectives:

- Objective: Include European management in 200x planning meetings
- Objective: Reduce customer complaints by 20% by year-end
- Objective: Establish operations in Singapore and Australia by Q4

- Objective: Get FDA approval on projects X and Y within nine months
- Objective: Improve the output of widgets per worker by 10% by June, 200x
- Objective: Establish programmed account sales plan by Q3. Hire managers
- Objective: Train X% of our personnel on Y by March 200x.

In most companies, a material capital expenditure requires some form of needs-analysis, projected outcome, and project proposal. Surveying customers and functional groups such as departments or teams is crucial to preparing such documents. In the needs-analysis phase, it is important to keep the discussion focused on the customer's situation, goals, and challenges. It is especially important to listen for a manager's dissatisfaction with a current manner of achieving a goal. The focus should be on understanding objectives and gauges of success; it is best not to suggest solutions at this point.

In interviewing each customer group, it is helpful to develop a needs analysis matrix of objectives and problems that video communication might solve, and note issues related to time, distance, teamwork, and collaboration. It is important to listen closely when discussing project management or heightened demands on scarce personnel. Another area to explore is communication difficulties between isolated workgroups (headquarters and the field, marketing and engineering, engineering and service). Any obstacle that relates to getting the job (e.g., training and retooling) done over distance is a potential project driver.

Note that there are two kinds of objectives: mandatory and desirable. Mandatory objectives, as Harry Green points out in the *Irwin Handbook of Telecommunications Management*, "Are the conditions that an alternative must satisfy to be acceptable." On the other hand, Green defines desirable objectives as "The conditions we would like the alternative to satisfy, but their lack will not disqualify." For example, it might be mandatory that a video communication system interoperate with an existing system at a customer's location. A desirable objective might be that establishing a connection would be as easy as dialing a telephone, when a slightly more complicated connection method might be acceptable. It is helpful for the needs analysis matrix to include critical problems and attributes that an acceptable solution must have, and note characteristics that are valuable but not compulsory.

Needs-analysis will result in a catalog of problems to which one can explore video communication solutions. For a small, centrally-located team that provides sales, training, or engineering support to a large number of field offices, one can offer a productivity-enhancing tool that can also minimize "burnout." Customers in highly competitive or volatile environments might find that video communication improves response time to fast-breaking events. Channel managers might use it to maintain and improve customer, distributor, partner or value-added reseller (VAR) relations. Sales may use it for product demonstrations or to add executives or specialists. Video communication can involve a broader base of people in strategy setting and decision-making. It has a record of reducing product time-to-market and helping to avert late penalties. Video communication is a *solution* that can only provide value as it solves a business problem.

Because video communication is such a powerful tool, it is a mistake to narrow the focus to reducing travel expense unless that is the top business objective. A senior manager may not be concerned with the cost of a plane ticket (and probably will not mind earning frequent-flyer miles). She is more concerned with business results, productivity, and personnel management. When it comes to spending scarce capital funds, she will not invest funds just to reduce a travel budget unless savings are exorbitant. She will invest in technology that enables her organization to succeed in core business areas.

THE PROJECT PROPOSAL

As with virtually any business initiative, proving-in and selling video communication starts with a business case or project proposal. The proposal outlines the project vision in terms of a project overview, customer description, justification, and goals. It proposes a solution and describes the result of the solution's implementation. It offers a plan of action (tasks for implementing the solution), considers the technical impact, and provides a cost-benefit analysis.

The cost-benefit analysis documents the project's total development cost. It typically includes costs relating to equipment purchase, network installation or enhancement, outside consulting and contracting, user training, and internal IT personnel hiring, training, and deployment. Documentation of personnel costs should include equipment installation and operation in total and broken down by department or cost center. Using a company-specific project cost worksheet is ideal; using some form of cost estimate worksheet is essential.

The proposal should include a statement about net gain in operations. That figure summarizes the difference between current annual operating cost and projected future annual operating costs offset by the implementation of video communication. Operating costs are not limited to the IT organization; they include all direct cost savings to the corporation. Revenue saved by improving productivity, and revenue gained by taking product to market early and eliminating travel could be included in this category.

The proposal should also list additional indirect benefits. Examples should include *soft-dollar* benefits such as improved service and account management for a key customer, more up-to-date product information, distributed training, and enhanced decision making through inclusion of more people in the decision-making process. It is not always possible to attach a dollar value to these intangible advantages.

The proposal should also include specific evaluation criteria that the Chief Information Officer (CIO) or other senior executives will use to assess the impact of video communication on the organization. Those criteria will include outside influences. Several different types of outside influences apply to video communication. For instance, t\he federal Clean Air Act, Americans with Disabilities Act, and Family Leave Act could all prompt implementation of video communication tools within a company. Competitive pressures can also prompt the installation of video communication. Recall how Management Recruiters Incorporated (MRI) used video to reduce costs associated with executive search activities. Certainly, many of MRI's competitors use video communication as a result of MRI's success. In still other cases, a key customer or business partner may request that an organization install video communication systems. Identifying such outside influences, and detailing the demands of each may prove invaluable in winning approval for video communication.

Evaluation criteria should also describe the cost recovery period. Dividing total development cost by the net gain in operations derives this figure. From a financial perspective, a good investment is one with a positive net present value—that is, one in which value exceeds costs. Most companies will not consider a project if they cannot recover costs within eighteen months. Nevertheless, an organization may be able to justify video communication as a defensive investment (e.g., to *not lose* market share), or to execute a critical transformation—although it may be hard to prove this beforehand.

A project proposal or business case should end with a statement of recommended action. That statement should reflect IT assessment of facts (as well as a statement of desirable information that is not known) and corporate activities. If video communication is viable, this section should present a recommendation that upper management approve funds and schedule the project (along with a proposed schedule). Some authorities suggest that an IT manager prepare a purchase order request in advance to anticipate and avoid delay.

When presenting a recommendation, the very specific business benefit offers the most punch. For instance, rather than saying:

> *We recommend the purchase of video communication because it can increase the number of contacts between marketing and engineering and reduce travel expense by $76,000. Travel savings will pay for all costs related to video communication within fourteen months.*

It is better to say:

> *We recommend the purchase of video communication. It can increase the number of contacts between the marketing group in Georgia and the engineering group in Minnesota by 30% (80 additional group contacts per year) while improving productivity by 22% (38 trips eliminated with each representing 10 hours of lost productive time). This will allow the firm to accelerate the introduction of product X by nine months, and represent anticipated revenues of $78,000 per month. In*

addition, the tool will reduce travel expense by 12% ($76,000). Travel savings will pay for all costs related to video communication (capital and operational) within fourteen months, and the firm can expect ongoing savings of approximately $50,000 annually.

SELLING A RECOMMENDATION

Some video communication sales require little effort on the part of the IT professional, as follows:

A senior-level project sponsor understands, and is ready to defend, the need for video communication. It is easy to articulate how the organization will apply the technology to solve one or more business problems because the company has evolved to the point that managers perceive the value of such strategic investments.

However, most cases are more challenging. Often mid-level managers perceive an application and a need for video communication, but key decision-makers remain unconvinced. In such an instance, familiarization through trials and demonstrations can prove a critical step in the justification process. In all cases, it is better to use video communication to support actual business meetings rather than to use canned demonstrations. One should test everything and conduct at least one *dry run* before *going live*—resolve lighting and audio issues, and become familiar with the management software and adept at moving cameras. The more mission-critical the meeting, the greater the risk, and any risk should be a measured risk. Presenting to every potential project champion at once is dangerous; the consequence of a calamity is too high. When possible, address each meeting attendee individually to understand his or her priorities and reservations.

As key individuals become comfortable with the technology and understand the potential benefits, they can become champions. However, they are only likely to undergo this transformation if they perceive the technology to address their business priorities. The likelihood of such potential champions (especially non-technical ones) to accept technology increases with the degree to which the technology achieves the goal transparently. Video communication demonstrations should emphasize product maturity and simplicity, and the degree to which the market is adopting it. With that said, it is important to note that honesty is crucial. The demonstration should reflect the actual experience as closely as possible. Managing expectations is a part of maintaining integrity and ensuring project success.

Some video communication sales require an extended adoption period. Many companies are slow to change, and many place a high value on *pressing flesh* and communicating face-to-face. Some corporate cultures simply do not embrace technology. Compared to accounting and marketing, technology is a very new discipline in the corporate organization. Moreover, organizations that do not allow for the difficulty in measuring strategic value are likely to spend an inordinate amount of time trying to catch-up with competitors. Highly qualitative measures such as teamwork, innovation, and customer satisfaction can be every bit as important to the long-term health of an organization as quantitative financial measures, which rely on conventional, entrenched measurement methodologies.

By presenting video communication as a business enabler—such as the telephone or email—one may traverse the initial reluctance to *add more technology*. Most senior managers would agree that telephone systems, computers, and fax machines are necessary. Video communication is becoming another *must-have* communications tool. Seventy-two percent of companies with 1000 or more employees already use room-based systems; sixty percent of all companies use personal video systems. Video communication has moved to the mainstream, and it is important to note this when proposing a video communication system.

TIPS FOR COLLECTING DATA

Bell Labs, the former research arm of AT&T, conducted early studies to determine how video communication can improve communication and, by extension, work. Although the findings were not conclusive, the Association of Computing Machinery published them in 1994, and they serve as a pioneering resource.

Although web-based travel planning is prevalent, corporations still deploy travel agents or departments. Travel professionals are people-oriented service providers who know not only who travels most frequently (individuals and departments), but also why they travel and can, therefore, serve as valuable resources in creating a video communication proposal. They can prepare data on frequently traveled city-pairs and impart the average per-trip cost of air travel, lodging, car rental, and incidentals. In-house travel departments usually know the size of the entire travel budget, and can usually attribute it by department. Travel professionals, especially those who work on a commission-basis, may view video communication as competition, so soliciting help from travel professionals may require diplomacy. In-house travel departments may be more cooperative.

One can calculate direct savings if one knows the total amount the organization spends on travel. The savings that will typically result can be crucial to the business case. Often, such savings will cover system depreciation and operating costs. Travel savings can offset video communication costs or make various secondary benefits essentially cost-free. It is risky, however, to promise more than 15% travel reduction without flawless historical data and a senior-level manager who is committed to using it whenever possible.

To build upon travel savings, one can engage with functional managers to formulate assumptions about the value of productive time recovered, and how the organization can reinvest salvaged time. If employees could meet more often via electronically mediated meetings, one should quantify that benefit in dollars and include it in the proposal. If the organization could complete projects more quickly, one should quantify that benefit in dollars and include it in the proposal. The proposal should also create assumptions to quantify the cost of *not* meeting.

If there is a high concentration of employees with key skills that are in demand in remote locations (e.g., trainers, sales support, engineers, project managers), the monetary value of increasing their reach can be invaluable in proposing video communication. Other questions to consider might include:

- How could the organization invest reclaimed travel time to increase revenue?
- Could employees spend more time with customers?
- What is the return on responding more quickly to problems and opportunities?

In constructing a business case, one can start planning for the procurement of video communication. Contacting prospective suppliers early is useful as most actively gather and disseminate genuine applications information. Nearly all track the success of their customers' projects, and some sponsor *Best Practices* contests in which customers submit detailed information on projects and results.

Some vendors offer innovative tools and white papers that address the value of video communication. They can provide information for proving-in video communication. Vendors want such efforts to succeed and are usually great allies. An applications statement is a crucial component of any structured procurement document (e.g., request for proposal or request for information). The degree to which one clarifies and emphasizes objectives (to both customers and suppliers) is the degree to which one is likely to succeed in proposing video communication. It is worth the effort.

CHAPTER 4 REFERENCES

Shurley, Traci. "Web Cameras Cut Travel Time in Parole Cases." *Arkansas Democrat-Gazette*. Little Rock. November 26, 2004.

Camp, Robert C. *Benchmarking: The Search for Industry Best Practices that Lead to Superior Performance* Quality Press, Milwaukee, Wisconsin, 1989 pp. 17-21.

Caruso, Jeff. "Mission: impossible?" *CommunicationsWeek* July 1, 1996 n617 p. 29.

Dickinson, Sarah. "Videoconferencing: hard sell, soft dollars," *Data Communications* May 1996 v25 n6 p. 35.

Dudman, Jane. "Have you been framed?" *Computer Weekly* February 8, 1996 p. 38.

Eccles, Robert G., "The Performance Manifesto," *Harvard Business Review* January-February, 1991, pp. 131-137.

Gilder, George. "Into the Telecosm," *Harvard Business Review* March-April, 1991 pp. 150-161.

Green, James Harry. *The Irwin Handbook of Tele-Communications Management*. Dow Jones-Irwin, Homewood, Illinois, 1989, 68-78.

House, Charles H. and Price, Raymond L., "The Return Map: Tracking Product Teams," *Harvard Business Review* January-February, 1991, pp. 92-100.

Porter, Michael E. "Competitive Strategy; Techniques for Analyzing Industries and Competitors," *The Free Press*, A Division of Macmillan Publishing Co., Incorporated pp. 7-83.

Robinson, Teri. "Where it all begins: desktops: end-user applications drive the technology," *CommunicationsWeek* March 18, 1996 n601 p. S7.

Simon, Alan R. *How to be a Successful Computer Consultant*, 2nd Ed. McGraw Hill, San Francisco, California, 1990, pp. 101-114.

Smith, Laura B. "By the numbers," *PC Week* June 17, 1996 v13 n24 p. E1.

Synnott, William R. *The Information Weapon* John Wiley and Sons, Incorporated 1987, 93-122.

PART TWO

TECHNOLOGIES AND STANDARDS
FOR
VIDEO COMMUNICATIONS

5

THE HISTORY AND TECHNOLOGY
OF ANALOG VIDEO

When we stop talking about the technology,
that's when it will be here.

Dr. Norman Gaut

While the advent and adoption of video communication seems destined, its story is an improbable one filled with twists and turns. Although knowing video communication's history is not likely to affect the success of its implementation, it will foster an appreciation of the work that went into getting us where we are now and the challenges that previous implementers needed to overcome. The latter part of this chapter outlines the evolution of broadcasting techniques, why there are different television systems in different parts of the world, and what those systems are. We also examine how video communication technologies *trick the eye*, how cameras work, what scanning is, and how, together, the technologies capture and display a video communications signal.

To people born in the western world after World War II, television is not a marvel of technology but practically a birthright. It is an object of fascination only when we see ourselves on it and, when we do, our experience with the technology causes us to want the best possible quality. This provides a challenge since we grant video communication systems only a fraction of the bandwidth we provide for television. First demonstrated in the 1920s, television is the predecessor of video. As is true with most technology, it is nearly impossible to establish the precise time at which TV was *invented.* Centuries of experiments combined successively to produce the telegraph, the radio, the television, room-based interactive video communication and, now, virtual anywhere-to-anywhere interactive video communication.

We are still living in the age of discovery, and gifted engineers and scientists in all parts of the world are pushing the limits of video compression, transmission, and applications. This golden age of technology began when early scientists applied the principles of electromagnetic waves to communications over distance.

RADIO—SOMETHING IN THE AIR

The Europe of the nineteenth century was the center of the developed world. Communication between the great cities of Europe (e.g., London, Paris, Berlin) controlled commerce. Moving information faster between these centers meant achieving competitive advantage. In 1842, Samuel Morse devised the telegraph code, and introduced an advanced tool for accelerating trade. Almost overnight, people were transmitting messages between Berlin and Paris on the world's first long-haul telegraph system. By the twentieth century, telegraph networks linked all the continents. Huge companies, the first of the global corporations created to operate for a profit across national boundaries, owned and operated those networks.

Wires were expensive to place and cable was costly to maintain. Therefore, a wireless transmission method appeared, in the late 1860s, when James Clark Maxwell, a Scot, mathematically proved that electromagnetic waves exist and that they travel at about the speed of light. Twenty years later, a German, Heinrich Hertz, devised an apparatus for detecting and producing electromagnetic waves that demonstrated the validity of Maxwell's theory. We characterize an electromagnetic wave by its frequency, which we commonly diagram as a sine wave (Figure 5-1). We generally portray sine waves as a continuous series of S-shapes placed on their sides. A conceptual base line, (zero crossing point), bisects the 'S.' One full 'S' represents a wave cycle. Each cycle has a positive polarity when the wave rises above the zero crossing point and a negative polarity when it drops below.

Hertz's breakthrough was fundamental to electronic communications. For a period after he introduced his scientific work, people commonly referred to electromagnetic waves as Hertzian waves. The frequency of their oscillations was measured in cycles. Later, in the 1960s, the term *hertz* replaced the word *cycles* in describing the number of complete cycles that occur in a second. Throughout this book we will use terms such as kHz (in which "K" symbolizes *kilo,* or thousand) and MHz (in which "M" symbolizes *mega,* or million).

In 1891, an Irish-Italian, Guglielmo Marconi, began experimenting with the newly discovered potential of Hertzian waves. He proceeded to explore *aerials* (antennas), and proved the importance of a grounded connection to a wireless transmission system. The same year, Frenchman Edouard Branly perfected an apparatus for intercepting wireless impulses. The device intercepted Hertzian waves in the air and caused them to ring an electric bell. A Brit, Sir William Crookes, proved the theoretical feasibility of "telegraphy through space" in an article published in the *Fortnightly Review*. In 1895, Thomas Edison patented a system of induction telegraphy that included elements that later advanced wireless communications. In 1831, Michael Faraday (another Englishman) discovered *induction* by showing that a current in one wire could produce a current in another wire.

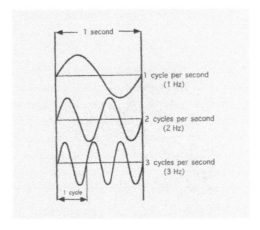

Figure 5-1. Sine Wave (Michel Bayard)

Finally, in 1895, Marconi amalgamated all the principles just mentioned, added some of his own science and applied for the first patent for a wireless telegraph. It became the basis for his Marconi Wireless Telegraph Company, the world's first commercial radio service. In 1901, Marconi was able to receive the three dots of the letter "S" in a transoceanic broadcast between England's Cornwall coast and Canada's Newfoundland. In 1907, he established the first regular commercial transatlantic wireless telegraph service.

In 1901, Dr. Reginald Aubrey Fessenden, a professor at Pittsburgh's Weston University, took another huge stride forward when he transmitted and received the human voice without wires. Fessenden, instead of interrupting a wireless wave with signal bursts, altered the inherent characteristics of a sine wave (in this case, its amplitude), by superimposing another signal over it. The carrier or sideband, over which the second signal traveled, was a wave of extremely accurate frequency. Fessenden's approach, today called amplitude modulation (AM, as in *AM radio*), became the foundation for broadcast technologies.

Fessenden filed a series of patents that described continuous transmission and formed the National Electric Signaling Company. In 1906, working with a Swedish-born scientist, Ernst F. W. Alexanderson, he created an alternating current generator and used it to power a radio transmitter located at Brant Rock, Massachusetts. From that location, they transmitted the first broadcast of sounds (as opposed to letters) on Christmas Eve, 1906. It started when Fessenden tapped out a series of CQ signals used to gain the attention of radio operators. Once done, he broadcast a live Christmas concert complete with music and readings from the Bible. At the end of the historical broadcast, Fessenden requested that anyone receiving the signal contact him. Many did, including representatives of the United Fruit Company. Its wireless operators on banana boats in the West Indies heard the broadcast. United Fruit, who envisioned the possibilities of ship-to-shore communications, immediately placed an order for a system.

A month later an American, Lee De Forrest applied for a patent on a three-element amplifier that he called the Audion tube. An improvement on an earlier two-element tube, it generated the needed power to push a radio or television signal over great distance. De Forrest and his assistant were testing the Audion in a Manhattan laboratory when they impulsively decided to conduct an experimental broadcast of the William Tell overture. An astonished wireless operator in Brooklyn Naval Yard, four miles away, who could not understand why he was getting music through his headset intercepted the transmission, and the radio tube was born.

By the end of World War I, radio tubes were mass-produced. The term *radio* derived from the fact that transmitters radiate signals in all directions and, by 1912, the term replaced the word *wireless*. It was around that time that David Sarnoff, the legendary promoter of American television, began his rise through the ranks of the Marconi Wireless Company.

Sarnoff emigrated from Russia to the US in 1900. In 1919, Sarnoff left Marconi to join the newly formed Radio Corporation of America (RCA). By 1920, RCA had launched two radio stations, KDKA of Pittsburgh and WWJ of Detroit; they were soon making regular broadcasts. Within the next two years, radio experienced explosive growth that only that of its cousin, television (and much later, the Internet), surpassed.

SPINNING DISKS AND SHOOTING ELECTRONS

During this same era, the foundation for today's visual communications systems emerged. No one individual deserves the credit for inventing the television—as was true with radio, it was a collective, although not always cooperative, effort. The process may have started in the 1870s when a couple of British telegraphy engineers discovered some intriguing properties of the element *selenium*. In their experiments, selenium conducted electricity better when it was exposed to a light source than when it was left in darkness. They developed this discovery into a system in which they place a source of light at one end of a circuit and selenium at the other. If an object passed between the light and the selenium, it would interrupt the flow of electricity. By combining a group of these circuits, they used selenium to create a mosaic of light.

Many inventors, not the least of whom was Alexander Graham Bell, began experimenting with selenium. In 1880, Bell invented something he called the photophone. It used selenium to transmit a visual display of the frequencies that comprised the human voice. It is astonishing that Bell overlooked the potential of image transmission. However, Bell's priority was to leverage another medium to simulate sound so that his deaf wife could better communicate.

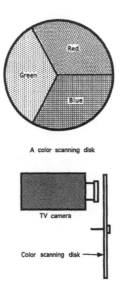

Figure 5-2. Nipkow Disk (Michel Bayard).

Nipkow, a German university student in Berlin, became entranced with the idea of image scanning. He developed a spherical disc that he perforated with holes. He positioned the apertures of the disk such that, as the disc revolved at a specific speed, each hole swept over a segment of a picture (Figure 5-2). When rotated between an object and a light source, it reproduced a crude electrical representation of the original object by responding to variations in light. The concept provided a framework for reproducing images with electrical signals, but it was too slow to offer passable picture quality. Nevertheless, the invention earned Nipkow, in 1883, German patent number 30,105, "The master patent in the television field."

In 1893, Archibald Campbell-Swinton, a Scottish engineer, made an address to the Roentgen Society in which he outlined plans for an electronic television system that was remarkably similar to that which the British Broadcasting Company (BBC) adopted, in 1937. Campbell-Swinton never made a television or similar apparatus; his work was purely theoretical.

In 1907, a German professor, Boris Rosing of the Technical Institute of St. Petersburg, became interested in understanding how images could be produced using electrical means. His research led him to develop a prototype device—the first *cathode ray tube* or CRT. Using a rotation-mirror image to transmit a signal that it received with a cold-cathode picture tube, it produced crude outlines of shapes. Magnetic coils generated the scanning signals that the tube required. It was the world's first glimpse of television.

Although Rosing's CRT produced only faint patterns with no light or shade, it attracted the attention of one of his students, Vladimir Kosma Zworykin. Fascinated with the CRT, Zworykin continued to perfect it after immigrating to America, in 1917. While employed at Westinghouse, in 1923, he filed a patent for a television receiver. By 1924, he was able to demonstrate the technology—a seven-inch CRT with electrostatic and electromagnetic deflection to steer an electron beam. This first picture tube was so crude that it could barely transmit a single character. Nevertheless, it proved a fundamental component of the television system that Zworykin proceeded to invent.

Next, Zworykin developed an improved image-scanning device. It replaced Nipkow's *mechanical* disc with an electronic one. Combining the Greek words *icon* (image) and *scope* (to watch), Zworykin named the invention an *iconoscope* (Figure 5-3). It was a CRT with a lens on an angled shaft. At the back of the device, an electron gun focused a beam of electrons on a target that consisted of a mica plate that was covered with tiny photosensitive cesium-silver droplets. The beam of electrons varied in accordance to the light captured from the original image. The beam energized the cesium-silver droplets as it passed over them. The change in their charge induced a current in the signal plate, and varied in relation to the lights and shadows along each line of the picture. The result was a coded signal that traversed a transmission path to a television's picture tube, which decoded it. The original image was recreated from behind, and appeared on the screen as a series of glowing dots that formed a picture.

In 1925, Zworykin filed a patent application for his electronic television system. That same year, an American inventor named Charles Jenkins used mechanical means (Nipkow's disc) to produce a system that transmitted a picture of a waving hand. The signal was carried via radio between Washington D.C. and Philadelphia. RCA acquired the rights to most of Jenkin's inventions when the Great Depression, in 1930, claimed Jenkins Television Corporation as one of its many casualties.

Across the ocean, a Scot, John Logie Baird, was also experimenting with television. In April 1925, he exhibited a working prototype that produced a decidedly low-definition picture with eight lines of resolution. The demonstration featured two images: a wooden ventriloquist's dummy named Bill (who did little but sit quietly for the event), and the live moving head of His Royal Highness, the Prince of Wales, the future King Edward VIII. As with Jenkins' television, Baird's system relied on Nipkow's disc to produce electro-mechanical images. What it lacked in electronic sophistication, it compensated for with results. Unlike Zworykin's transmissions that were crude outlines of a scanned object, Baird's system transmitted images complete with gradations of light and shade.

Thus began the race for television. Baird was often at the head of the pack. In 1926, he offered the first formal public broadcast of his television system. Viewers could clearly see the head of Bill the dummy, back on the set for another go.

It was a higher-definition Bill; Baird's screen now boasted 30 resolvable lines. Moreover, having just secured a radio license from the Post Office (which was also in charge of broadcasts), Baird was transmitting from his own experimental television station, 2TV, in Harrow, England.

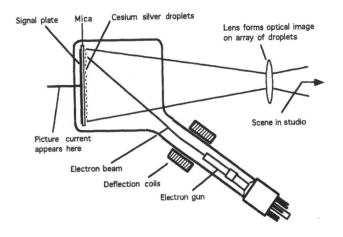

Figure 5-3. Iconoscope (Michel Bayard)

With the help of an investment firm (that apparently promoted the misconception that Baird had a monopoly on television), Baird formally launched the Baird Television Development Company, in 1927. Meanwhile, in America, researchers at Bell Labs developed an electromechanical television with motor-controlled scanners to capture images and an array of neon tubes to, collectively, form a display. On April 9, 1927, they presented the technology to the public across in a transmission between Washington D.C. and New York, a distance of more than 200 miles. Word of the demonstration reached England on April 28, the day after Baird's public offering had closed.

Shareholders furiously claimed that Baird's had misled them, but he redeemed himself when he sent pictures by telephone line over a distance of 435 miles between London and Glasgow. Although the images flickered unsteadily, it was a major achievement. In 1928, Baird dispatched 30-line pictures between England and New York through an amateur short-wave radio station with a 35-foot antenna. The New York Times proclaimed, "Baird was the first to achieve television at all, over any distance. Now he must be credited with having been the first to disembody the human form optically and electrically, flash it piecemeal across the ocean and then reassemble it for American eyes. His success deserves to rank with Marconi's sending of the letter S."

In 1928, Baird demonstrated color television by transmitting images of blue flowers and red berries in a white basket. He used a transmitter with three interlaced spirals, one with a red filter, another with a green one, and a third with a blue filter. He thereby demonstrated the first transmission of a RGB (red, green, blue) component television signal. Next, Baird recorded video frequencies, noting the distinctive *sound* of a subject. He eventually filed a British patent for his "Phonovision." The device incorporated discs and played like a record to produce a recognizable picture—it foreshadowed the DVD.

In 1928, Baird astonished the public by demonstrating a three-dimensional (3D) color television. On the receiving screen, two images were displayed that were separated by one-half an inch. One corresponded to objects as seen by the right eye, the other as seen by the left. When viewed through a stereoscopic device, the two images merged.

The BBC, still focused on radio, reacted negatively to a new technology from Baird that complicated their existence. However, in 1929, they allowed Baird to use their medium-wave transmitters to broadcast experimental television programs. The morning broadcasts each lasted thirty minutes. Sound and vision alternated every two minutes. In 1930, Baird and the BBC achieved simultaneous transmission of picture and sound. In 1931, Baird helped the BBC make its first-ever-scheduled outdoor transmission. Viewers in a London theater saw the Derby broadcast direct from Long Acre, filmed from a van situated next to the finish line. As the winner, Cameronian, galloped past the post, viewers heard the roaring crowd, almost as if they were there themselves. The event was such a huge success, that the BBC repeated it in 1932.

On August 22, 1932, the BBC inaugurated a 30-line television system. It produced a pinkish picture that was received horizontally. A lift-up lid, with a mirror on the underside, displayed the image. One was installed at 10 Downing Street and, by 1933, almost 10,000 TV receivers had been sold in the UK. Nevertheless, Baird was frustrated with TV's general progress. He was restricted by British regulations to the use of medium-wave transmission. This limited him to thirty lines of resolution with only 12.5 frames sent each second. Later, as ultra short waves were introduced, he increased his resolution to 600 lines with a higher frame rate.

In the U.S., television was still experimental; no sets were installed in homes. In 1930, RCA demonstrated large-screen television at a theater in New York. The screen consisted of more than 2,000 small lamps that each contributed to a particular element in the picture. Elsewhere, the big news was radio. Its popularity had exploded, and stations proliferated, and transmissions began to encroach upon one another. All the activity over the airwaves prompted Congress to create the Federal Radio Commission through passage of the Radio Act of 1927. In 1928, the FRC established classes of stations by geographic zone. It was too little too late. General Electric was operating what one might describe as an experimental TV station, WGY, based in Schenectady, New York. It broadcast the first TV drama that year. However, the Act neither addressed television broadcast, nor asserted jurisdiction over telegraph and telephone carriers.

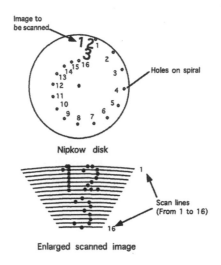

Figure 5-4. Field Sequential System (Michel Bayard).

In 1929, Vladimir Zworykin and David Sarnoff, both born in Russia, met for the first time. Zworykin was interested in joining RCA to pursue television technology. However, as a condition of employment, he required Sarnoff to renounce electro-mechanical television and commit to electronic methods, even though they still faced many unsolved problems. Sarnoff asked Zworykin to estimate RCA's cost to prepare electronic TV for commercial use. Zworykin placed it at $100,000, and Sarnoff consented. Over a twelve-year period, RCA's development of television exceeded Zworykin's estimate and cost nearly $50 million. Nevertheless, Zworykin prospered at RCA and basked in the luxury of a large research budget and supportive executives. In that fertile environment, he perfected his science. There was only one real problem. In Provo, Utah, a youth named Philo T. Farnsworth was far ahead of everybody. He was busy filing important patents.

Farnsworth, unlike Zworykin, did not have a sponsor with deep pockets. Self-taught but brilliant, he received sporadic funding from small investors. In 1928, he demonstrated the world's first all-electronic television system. He based it on a camera that Farnsworth called an "image dissector." By 1939, he had coupled the camera with a sophisticated picture tube to intensify the dissected image.

The system was far more sensitive than was anything RCA could conceive and develop, but Farnsworth would not sell his patents. RCA took to the courts in a strategy that eventually cost more than $2 million. Another $7 million went for research in an effort to get around Farnsworth's technology. Eventually, aware that they would fall behind in the global race for television, RCA gave Farnsworth his price. At age 34, he retired to a farm in Maine. Zworykin and Sarnoff, both Russians, went down in history as the fathers of American television. At the 1939 New York World's Fair, the National Broadcasting Company (NBC) inaugurated the first RCA "high-definition" television set, which boasted 441 lines of resolution.

In 1933, Franklin D. Roosevelt requested that his Secretary of Commerce convene an inter-agency committee to study how the U.S. should regulate electronic communications. The airwaves of the day were awash with signals. The committee recommended that Congress establish a single agency to regulate all interstate and foreign wire and radio communications including television, telegraph, and telephone. Congress, in turn, passed the Communications Act of 1934, and the Federal Communications Commission (FCC) began operation on June 11, 1934. One of its important tasks was to manage the airwaves, which it did by allocating spectrum between competitors. The FCC also inherited the task of approving television standards.

TELEVISION STANDARDS AND TECHNOLOGY

The year 1934 marked the beginning of the standards wars. England was in the process of investigating a new television system—something with more than Baird's 30 lines of resolution. There were two contenders: Baird's system, now producing a mechanically-scanned non-interlaced 240-line television picture, and Marconi-EMI's all-electronic system that interlaced two fields to produce 405 image lines. The BBC placed the two systems in direct competition in a series of trials. In May 1937, it selected Marconi-EMI's system as its standard, and broadcast George VI's coronation to 60,000 people.

By 1937, RCA had rectified their television system but faced new problems. The television industry could not agree on a common standard. In 1936, the Radio Manufacturers Association (RMA), which RCA backed, appointed a committee to approve a system. In 1937, they rubber-stamped RCA's 441-line 30 frames per second (fps) proposal. However, RCA's iron grip on the industry stirred controversy and, in 1940, the FCC chairperson intervened to order additional study. It downgraded NBC, which had been broadcasting since the 1939 World's Fair, from a *licensed station* to an *experimental station* until the difficulties could be resolved.

A flurry of hearings ensued. On April 30, 1941, the FCC approved an amplitude-modulated monochrome television system. It defined a video frame to contain 525 scanning lines. Each second, thirty such frames were to be transmitted and synchronized with frequency-modulated audio. The system was analog; the electron flow that it sent as a signal was analogous to the original waveform.

As part of the standard, the FCC determined the range of frequencies to allocate to each television channel. They specified that monochrome television signals be modulated onto radio frequency (RF) carrier waves with bandwidths of 4.2 MHz. Each channel would occupy 6 MHz of bandwidth. Audio information would be transmitted above video starting at 4.5 MHz. Unused bandwidth above the audio and below the video would protect the signal from interfering with other channels.

On July 1, 1941, the FCC licensed NBC and the newly formed Columbia Broadcasting System (CBS) to operate commercial television stations in the U.S. They promptly began broadcasting.

In 1945, the FCC allocated spectrum space for 13 television channels, all in the very-high frequency (VHF) band that occupies the spectrum between 54 and 216 MHz. Also interspersed in this band were FM (frequency modulation), mobile and emergency radio, and air navigation signals. Eventually, it eliminated channel one and left only twelve channels. Soon afterward, U.S. entry into World War II interrupted all television progress. The federal government banned manufacture because to conserve resources for the war effort. It was only after the war ended, in 1946, that television broadcasting resumed in earnest.

In 1952, the FCC established ultra-high frequency (UHF) television channels, which it numbered 14–83, and extended from 470 to 890 MHz. The FCC also reserved TV channels for non-commercial educational stations, specified mileage separation distances to reduce the potential of interference between TV stations, and made city-by-city television assignments.

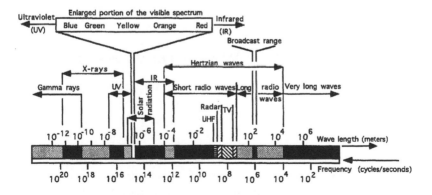

Figure 5-5. Frequency Chart: X-rays to Radio Waves (Michel Bayard)

The monochrome image of 1950s TV sets needed frequent adjustment. Viewers had to fiddle with vertical and horizontal hold, twist knobs to juggle light-dark contrast and fuss with the antennae to eliminate ghost-like images and snow. Nevertheless, television achieved overnight acceptance. Market penetration accelerated from 6% of US households in 1949, to 49% in 1953. This was an average growth rate of more than 300% for the first five years of commercial operation.Although post-war US viewers tolerated black and white (monochrome) TV, they wanted color. It had been available in Europe for years. Americans were not to be outdone. The FCC started exploring ways of adding color to the existing monochrome system and, in 1950, tentatively endorsed what CBS had developed and referred to as a *field-sequential* system. In this system, a monochrome camera scanned an image, as with black-and-white TV. Unlike monochrome filming however, it placed in front of the lens a color wheel that it divided into transparent RGB sectors (see Figure 5-4). The timing of the rotation was such that a colored sector appeared in front of the lens for exactly one field. The resulting sequence of fields corresponded to the color components in the image. At the television receiver, synchronized with the color field, another color-filter disk rotated in front of the picture tube.

The system was bulky, noisy, and prone to wear. It required exact synchronization between the rotating filter and the fields presented on the picture tube; mis-timing would cause a color shift. However, the biggest problem with field-sequential color was incompatibility with the monochrome system of the day. The color system generated 24 fps rather than the monochrome system's 30, and (like European systems) it generated 405 scanning lines rather than the monochrome system's 525. To identify a solution, they established, in 1950, the National Television Systems Committee (NTSC). Three years later, in December 1953, the NTSC defined a color-telecasting standard that was compatible with the monochrome system. The standard, over time, has changed how Americans view the world.

FROM LIGHT WAVES TO BRAIN WAVES

To understand the NTSC system of color telecasting, it is helpful to understand how eyes process light. Simply stated, eyes capture patterns of light and send them to the brain for interpretation. Television exploits the eyes' properties and is patterned after them.

The human eye's structure is similar to that of a camera. At the front of the eye is the transparent cornea. It captures a picture of the outside world and forms an image. Behind the cornea is the iris, a circular muscle that expands and contracts to control the amount of light that enters the eye. It is similar to a camera's automatic exposure system. The lens focuses an image by changing shape to accommodate differences between close and distant subjects. Behind the lens is a fluid-filled compartment on the surface of which is the retina. The retina consists of more than one hundred million tiny photoreceptors called rods and cones. Rods, which are exceedingly sensitive, respond to brightness only, and thereby describe an image to the brain in shades of gray. Cones, which are concentrated toward the center of the retina, respond to color, specifically, light of three different wavelengths that we know as red, green, and blue. Blending these three colors in the brain derives all other colors.

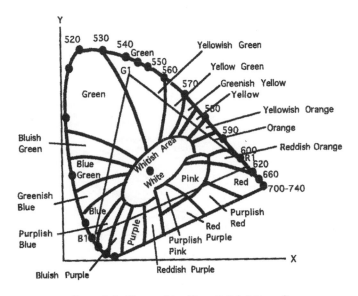

Figure 5-6. Chromaticity Chart (Michel Bayard).

Each rod and cone is connected to an individual fiber in the optic nerve. When we scan an image, the rods and cones send messages to create in the brain a high-resolution *light mosaic* that the brain interprets as an image.

Visible light—color—is but a small interval in the spectrum of electromagnetic radiation. This radiation, which fills the universe, possesses dual characteristics of waves and particles. The particles, called photons, contain energy. For the purposes of video communications, energy is measured in *electron volts* (eV). The energy of visible light varies from two to four eV.

The particles that comprise electromagnetic radiation are assembled in waves. A signal's frequency can be determined by measuring the number of wave-crest cycles that pass a fixed point in one second. Although frequency is measured in hertz, it is not practical to use the traditional system to measure light because the frequencies are too high. Therefore, we discuss these waves in terms of their lengths. Light waves span 380–770 nanometers (10^{-9} meters), in other words, between 16 and 28 millionths of an inch. We describe these various wavelengths by giving them color names—violet, blue, green, yellow, orange, and red. The exact length of a wave determines its color, but most visible lights we see are blends of three primary colors—red, green, and blue (RGB). These colors are additive primaries (in contrast to the subtractive primaries cyan, yellow, and magenta that are used in printing).

In 1931, the Commission Internationale de l'Eclairage (CIE) set forth international standards for color measurement. It defined the wavelengths of colors: Red 700 nanometers (nm), green, 546 nm, and blue, 435.8 nm. In 1950, the FCC redefined the standard to better adapt it to the phosphors used in TV picture tubes. The FCC's red is 610 nm, green is 534 nm, and blue is 472 nm. The NTSC recognized the FCC's color definitions when it developed the 1953 standard for color telecasting.

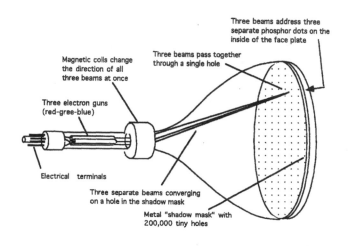

Fiigure 5-7. Scanning in the Picture Tube (Michel Bayard).

A chromaticity chart, shown in Figure 5-6, demonstrates that, in television, blending the primary colors can produce any color. However, it is not possible to create primaries themselves from other primaries. They are unique.

SCANNING AND FRAMING

Of course, the process of capturing an image for a television or video application involves the use of a camera. As do eyes, cameras scan a scene to detect, capture, and transform light waves into electricity.

The first all-electronic camera was the image-orthicon tube that RCA introduced in 1941. The more compact and simple vidicon tube was developed in 1951. It trained an electron beam, a scanning point, on an image. The beam swept the image left-to-right and top-to-bottom to measure the scene's photosensitive surface.

Conventional optical techniques are used to focus the light onto a target area at the rear of the tube's faceplate. The back of the target is coated with a transparent, electrically-conductive substance, and a photoresistive coating. The light of an image passes through the transparent conductor, strikes the photoresistive coating, and thereby produces a charge that varies in proportion to the light that strikes it. An electron beam behind the target measures the fluctuating charges and translates them into a series of voltages that are proportional to the changing light intensity. Heating of a cathode creates the electron beam that does the measuring. It is focused using principles of static electricity. Electromagnetic deflection coils that are placed around the neck of the vidicon tube move the beam vertically and horizontally.

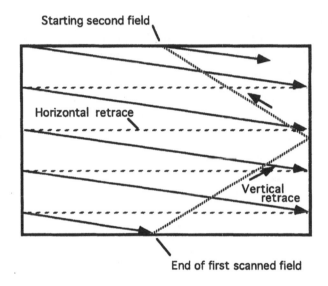

Figure 5-8. TV Raster Scan (Michel Bayard).

The optic nerve is a huge number of separate fibers that each carries a minuscule part of the image to the brain. All operate in parallel, and the brain receives a single image that consists of many parts. Video transmission networks do not operate like the optic nerve; they could never cost-justify parallel communications. Processing images with human vision techniques would require hundreds of thousands of channels. Each would have synchronized with all others. Parallel communication is not feasible over distance; it is too complex and costly to have separate paths for each information bit.

Color cameras separate light waves into RGB components using prisms, color filters, or dichroic mirrors. After separating RGB signals, older systems used multiple camera tubes to apply these principles. Today's cameras utilize solid state chips known as charge-coupled devices (CCDs) to store the electrical charges that make up the light's intensity. The charges are transferred to another layer called a frame buffer where they are formatted. Next, a formatted electrical signal, with amplitude directly proportional to the amplitude of the light, is sent across a transmission medium or network for a viewer's screen or monitor to displays.

Video communications technologies rely on dividing an image into lines containing picture bits and transmitting them serially—one after the other. The video image is, therefore, a mosaic of light, and the *tiles,* or points of light, are called picture elements. The term is abbreviated to *pel* in broadcast TV parlance. Those in the fields of telecommunications and computing know them as *pixels.*

A video monitor image is constructed before our eyes one pixel at a time. What looks like a continuously moving picture is actually a series of still images called frames. When a series of still image frames arrives fast enough, the images fuse in our minds and appear to be moving. The faster the frame rate, the better the fusion, and the smoother the motion will appear.

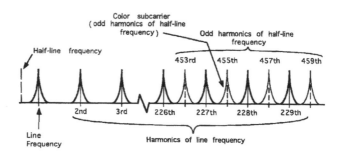

Figure 5-9. Color Television Spectral Detail (Michel Bayard).

In the early 1820s, Peter Mark Roget, a doctor who specialized in vision (and the father of the Thesaurus) began to experiment with the properties of human vision. In 1824, he delivered a paper to the Royal Society in which he introduced the concept of "persistence of vision." One can define it as the brain's ability to retain the impression of light after the original stimulus has been removed. The brain holds the light for about one tenth of a second, although it starts to fade almost immediately upon receipt. Therefore, if images are sent at a rate of at least 10 fps, they appear to be continuous, though somewhat jerky. Somewhere between 10-20 fps, the jerkiness disappears. A new image is presented long before the old one appreciably fades. The brain fails to perceive that the video monitor is actually just *blinking* images.

A frame of video is displayed by reversing the camera's scanning process. A process called raster-scan paints images on a screen. When the cathode gets hot, it causes electrons to stream from a device called an electron gun. The beam of electrons, sometimes referred to as a reproducing spot, continuously bombards the inside surface of a monitor. The beam traces its course in a series of thin lines, and strikes phosphor dots that coat the inside of the monitor. When hit with an electron, an individual phosphor emits light.

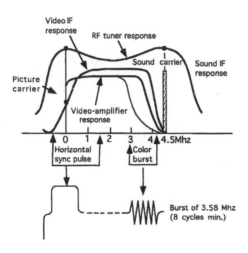

Figure 5-10. Color Burst (Michel Bayard).

Frame rates differ between video applications. Television standards specify that video frames be sent at a rate of between 25–30 per second, depending on the standard. Scanning systems reflect the electrical systems that exist in a country. In Europe and other parts of the world that Europeans colonized, the national electrical system operates at 50 Hz. In the U.S. and most of the rest of the world, the electrical system delivers alternating current at a rate of 60 Hz. These differences caused television systems to evolve dissimilarly. The U.S., Mexico, Canada, and Japan (and 19 other countries) use the NTSC system. Europe, Australia, Africa and parts of Asia and South America use a television system that specifies 25 fps. The number of lines of resolution varies from 625 in England and much of Western Europe, to 819 in France, Russia, and Eastern Europe.

Although we have just asserted that North American television incorporates a rate of 30 frames per second, adding color to the television signal requires that the rate drop to 29.97 fps (most people round it to 30). Each 525-line video frame is divided into two halves, each of which is called a field. One of the 262.5-line fields contains the odd scan lines, and the other contains the even ones. During display, first the odd-line field then the even field is painted in a process called field-interlacing. Interlaced fields of video are sent at a rate of 59.94 fps. Why divide an image in half? Why not send complete frames with the odd and even lines combined? The answer is in the eye of the beholder. Although the brain retains the perception of light for one-tenth of a second, the image starts to fade instantly. If complete video frames are sent at a rate of 30 per second, the eye senses the change in light intensity and images appear to flicker. For most people, images fuse at frame rates faster than 40 per second. To solve flicker problems, each image is displayed twice using a shutter speed of 60 images per second.

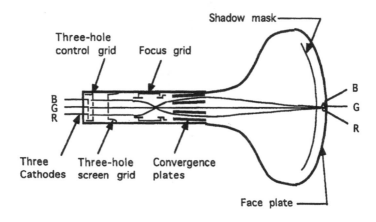

Figure 5-11. Color Picture Tube (Michel Bayard).

Starting at the top-left, the spot sweeps horizontally to the right, and completes the task in about 50 microseconds (µs). At that point, for 10 µs, the beam shuts off and the electron gun executes a carriage return or a *horizontal blanking interval.* After the blanking interval, the beam starts up again on a new line. The process continues until the spot reaches the lower right-hand corner of the screen where it shuts off again during the vertical blanking interval. Next, *retrace* occurs as the reproducing spot returns to the top of the screen. This process continues left to right, top to bottom, synchronized by oscillators that keep the electron operating strictly on schedule. As the horizontal oscillator drives the electron stream laterally, the vertical oscillator moves slowly downward. In that manner, still-image frames create the illusion of motion.

Video frame rates vary enormously and, as they increase, the bandwidth necessary to transmit them also increases. The more the video frames per second, the smoother and more natural the appearance of the motion sequences. Video communications systems designers aim to make motion sequences appear as smooth and natural as the ones viewers have come to expect from television, but they rarely get to work with comparable bandwidth or service quality.

IN LIVING COLOR—NTSC

Acquaintance with the color television system that the NTSC approved in 1954 facilitates comprehension of video communication technology. Before it devised the NTSC system of color telecasting, the committee considered various solutions.

Members soon realized they could not expand on the existing monochrome approach by using separate RGB subcarriers. This would have been a synchronization nightmare and would have required that the bandwidth allocated to each TV channel be greatly expanded to convey not only luminance but also red, green, and blue signals. This was not practical. Nor was it acceptable to simultaneously broadcast a program in both monochrome and color. They required a hybrid system that could superimpose color information (chroma) over the brightness (luma) signal.

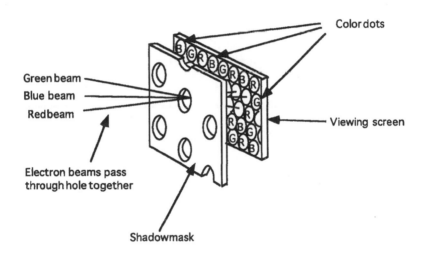

Figure 5-12. Shadow Mask (Michel Bayard).

The system that the NTSC selected used this approach and added hue (the identifiable color) and color saturation (color intensity, e.g., pastel pink, bright red, rose) to the luma signal. It did this by exploiting luminance's occupation of the 4.2 MHz subcarrier in bursts of energy at harmonic multiples of the horizontal scanning rate. This creates frequency clusters with empty spaces in which hue and saturation signals can be inserted. Hue and color saturation are thus inserted in these empty spaces using a technical concept described as frequency interleaving. The NTSC selected the frequency of the color subcarrier 3.579545, rounded 3.58 MHz, to achieve compatibility with the installed base of monochrome television sets.

The NTSC color television system is detailed in a document prepared by the Electronics Industries Association (EIA). This highly technical document, known as RS-170A, describes how color is added to the monochrome signal.

The NTSC system, and any other system that combines luminance and chrominance to represent images, is called *composite video*. In the case of NTSC, the composite signal is achieved not only by frequency interleaving but also through three different modulation techniques (phase, amplitude, and frequency). Each of the three conveys different elements of the signal. Audio is frequency-modulated, as is the luminance information contained in the signal.

Color television is encoded through a process of *color differencing*. Subtracting luminance information, referred to as Y, from each of the primary color signals, R, G and B, forms signals. Only two of the three color-difference signals are needed; the third is redundant. NTSC uses (R-Y) and (B-Y) that are combined to create two new signals, *I* and *Q.* The formula for deriving I and Q is:

$$I = 0.74(R\text{-}Y) - 0.27(B\text{—}Y)$$

$$Q = 0.48(R\text{-}Y) + 0.41(B\text{—}Y)$$

The I and Q signals are used to modulate two color subcarriers at the same frequency but 90 degrees out of phase—an approach that is also referred to as *quadrature modulation.* This technique exploits the fact that the polarity of an alternating current is constantly reversing. Phase is measured in degrees. The full cycle of a sine wave describes a 360-degree arc. The sine wave crosses the base line at zero volts, achieves its highest positive voltage at 90 degrees, dips down to touch the base line again at 180 degrees, achieves its lowest negative voltage at 270 degrees and finally returns to the zero volt base line at 360 degrees. Phase-modulation electrically alters this pattern of positive and negative undulation to convey information—in this case, the color of a pixel. Phase is hard to determine without a phase reference, so a short eight-cycle burst of the 3.58 MHz color subcarrier frequency serves as a phase reference. This color burst, which is transmitted during horizontal blanking as a reference for establishing the picture color, allows synchronization. The luminance and the modulated color signals are added to produce a final signal. This YIQ signal modulates the RF carrier for broadcast over the air.

When received at the set, the YIQ signal has no color content. The receiver's electronics translate it back into the three different wavelengths that comprise the RGB signal. Each color signal charges a separate electron gun. The application of magnetic deflection coils varies the aim of the three guns. The coils control the vertical and horizontal stream of electrons. Another method is used to control which gun energizes which phosphor dots. A metal plate, pierced with more than 200,000 tiny holes, the shadow-mask, sits between the electron guns and the viewing screen. It conceals the two unwanted dots in each triangular group of three so that they converge on only one hole. This confines the beam to the phosphor dots of the proper color—for instance, the red signal strikes only the red dots and the blue signal strikes only the blue dots. *Blending* the output of multiple electron guns produces variations of red, green, and blue. Red and green combine to produce yellow. Blue and green combine to produce cyan. Red and blue combine to produce magenta. Red, green, and blue combine to produce white and gray. Black is the absence of light, and white is the presence of all colors in equal volumes.

The NTSC system allowed coexistence of monochrome and color television. Both receive luminance information, but the chrominance information is lost to monochrome sets. Today, the NTSC system seems convoluted; the common gag is that NTSC is stands for *Never The Same Color*. The NTSC system is an awkward method for adding color to a video signal, but the approach was born of necessity. The committee had to live within the constraints of the existing television system and did so by fitting color into the monochrome signal's frequency band. With television a new technology, they could not pay for progress with obsolescence. Instead, they found a way to preserve the existing installed base of monochrome sets.

In Germany, the Telefunken found that the NTSC standard caused phase errors introduced during signal propagation that, in turn, resulted in uneven color reproduction. To cancel out the phase errors, the Germans inverted the color signal by 180 degrees on alternate lines. More than 40 countries including the U.K., Germany, and Western Europe (except France) adopted phase alternate line (PAL). The French adopted an FM system to transmit color television information that eliminated the phase error problem. The French system SECAM (Sequential Couleur Avec Memoire), separates color information into red and blue and sends it by two alternating carriers. A memory circuit within the TV set remembers the signals and uses them for faithful color reproduction.

The NTSC system has seen few changes since 1953. Modulating the already compromised signal onto an RF subcarrier amplifies the system's inherent problems. Once propagated in free space, it is fair game for interference from a wide variety of random signals—a phenomenon known as radio frequency interference or RFI. Cable television eliminates most RFI problems, which may explain why the system has lasted so long.

In spite of its glitches, color TV still relies upon the NTSC system because the U.S. was slow to approve a digital television standard, which is gradually replacing the NTSC system.

HDTV

By the year 2010, the FCC may ban analog TV transmitters from the airwaves if the 98% of Americans (who own most of the world's 600+ million television sets) either upgrade to HTDV or buy digital adapters for existing sets. HDTV sets display engaging wide-screen, high-definition images with about twice as many scanning lines as NTSC. HDTV also offers simultaneous broadcast of multiple channels of sound. HDTV offers unprecedented image, sound, and interactivity.

HDTV is a logical successor to NTSC in part because it offers the sense of involvement that audiences appreciate in the wider screens that theaters offer. HDTV offers greater visual detail, but it achieves much of its effect from its wider horizontal dimension. By offering almost twice as much detail, HDTV allows viewers to sit closer and thereby sense even more of a peripheral visual field. The experience seems more real when the screen extends to one's periphery.

Japan was the first country to transmit high definition television. It did so in the early 1980s, after which it reported its success at an international meeting of telecommunications engineers in Algiers. The Japanese system, MUSE (Multiple Sub-Nyquist Encoding), was a variant of its NHK HDTV standard for direct broadcast satellite service. Although MUSE is analog, the U.S. worried that Japan might overtake it in the global market for broadcasting equipment. Nevertheless, the FCC was slow to endorse an HDTV format. It was not until the Japanese actually demonstrated MUSE HDTV to American broadcasters, that the broadcasters themselves asked the FCC to develop an HDTV system for the U.S.

The FCC responded, in 1987, by creating its Advanced Television Systems Committee or ATSC, which oversaw the competition to create a domestic HDTV system. A variety of companies each worked on an analog version of the specification. Then, abruptly, General Instruments demonstrated a digital system. Shortly thereafter, the ATSC asked HDTV competitors to cooperate. Should the FCC approve one competitor's strategy, all others could lose their investments. That risk, along with the prohibitive costs of going it alone, offset general reluctance. What is more, none of the four contenders could claim technology leadership; each led in one area or another. Therefore, in May 1993, they formed a cooperative effort known as the *Grand Alliance.* From the time of its formation until April 1995, the group (which included AT&T, General Instrument, MIT, Philips, David Sarnoff Research Center, Thomson Consumer Electronics, and Zenith Electronics) worked cooperatively on the specification.

The Grand Alliance based conventions for HDTV upon the International Standards Organization (ISO) Motion Picture Experts Group (MPEG) MPEG-2 specification. They derived scanning formats from computer-friendly progressive scanning and television-friendly interlaced scanning modes. The specification included MPEG-2 transport and compression technologies and quadrature amplitude modulation

(QAM) or a variant, vestigial-sideband (VSB) modulation. The system specified Reed-Solomon Forward Error Correction (RSFC), which is common in CDs, hard drives, telephone modems, and digital transmission systems. Audio encoding was specified via six-channel, CD-quality digital surround sound.

With the completion of a specification, the Grand Alliance (GA) began laboratory tests. They found VSB superior to QAM, and modified the specification accordingly. At the end of all their proof-of-concept testing, the GA documented their HDTV specification and submitted it to the ATSC for approval. On April 12, 1995, the ATSC approved the GA's Digital Television Standard for HDTV Transmission (formally known as ATSC A/53.) ATSC A/53 uses AT&T's and General Instrument's video encoder, Philips' decoder, Sarnoff/Thomson's transport subsystem and system integration expertise, and Zenith's modulation sub-system.

In 1995, the computer industry belatedly noticed that ATSC A/53 was not computer-friendly. It was biased toward interlace scanning (as opposed to PC friendly progressive scanning), used an aspect ratio that was not compatible with computer monitors (16:9), and formatted pixels in an awkward fashion. In a last-minute effort to stop the FCC's final approval of ATSC A/53, Microsoft, Apple, Compaq, and others formed the Computer Industry Coalition for Advanced Television Service (CICATS.) CICATS led the charge against interlace, and submitted a standard based on the work of Gary Demos, a winner of an Oscar for technical excellence in filmmaking. Progressive scanning offers an observed resolution of about 70% of maximum vertical resolution because pixels fall between the (three) scanning lines. Interlaced scanning offers only 50% of maximum vertical resolution unless the image is still. Moreover, interlacing produces serrated edges, and flicker along horizontal edges and improperly aligned frames.

From July 1995 until November 1996, the Grand Alliance and CICATS remained at a standoff. Finally, when it appeared that the FCC was prepared to do nothing rather than approve a hotly contested standard, the broadcasters compromised. The Grand Alliance agreed to include progressive scanning in with the interlace specification. The FCC approved the modified HDTV standard in December 1996. In April 1997, it adopted an expedited timetable, ground rules, and a preliminary channel allocation; in February 1998, it approved a final channel plan. At that point, the FCC mandated that all HDTV outlets be broadcasting by 2003, and announced that it would repossess all analog channels by 2006. By 2000, the number of HDTV stations grew to nearly 100.

Despite HDTV's startling improvement, it is enhanced definition that consumers were buying as of 2005. According to Displaysearch, in the third quarter of 2004, 54% of all plasma sets shipped worldwide were 42-inch EDTVs. In the US, according to NPD Group, more than 60% of plasma sets sold were EDTVs. After so much investment in HDTV, manufacturers are stuck anticipating whether consumers will remain satisfied with enhanced-definition television rather than moving up to HDTV as more programming becomes available.

Broadcast television, especially sports events, night-time dramas, and comedies are commonly available in high-definition. It is also available on a small proportion of cable and satellite channels. Today, more than 50% of TV shows are broadcast in HDTV, and more than 80% of USA households get five or more DTV channels. The industry is hoping, and some industry analysts believe that, as HD programming becomes more common, EDTV popularity will yield to HDTV. If that is the case, TV upgrading will be here for a while. Where at least 85% of households have digital sets, federal mandate requires television stations to convert to digital broadcasting by 2007. However, it is not that simple. By 2008, 37 million US households are expected to receive high-definition programming. By 2006, more than 14 million US households are likely to own HDTV capable hardware, up from only 8.7 million by 2004. Those adoption numbers do not appear to justify the FCC's goal of a full transition to the new standard by 2010.

Much of HDTV technology is or will be interchangeable with video communication technology. We will want to use cameras, monitors, chips and transmission systems of one for the other. Products such as HDTV Wonder™ (www.ati.com), a PCI card for analog TV, free-to-air digital TV, and full quality free-to-air HDTV reception, are preparing the PC to become the first HDTV device in many homes. They offer personal video recorder capabilities, with the controls to watch, pause and record all forms of TV to a hard disk, to CD, or to DVD. The interest that HDTV will generate will do wonders for the acceptance of video communication, and the enhanced resolution could foster acceptance. To explore what digital television may do for video communications, in the following chapter, we will address the basics of digital coding and compression.

CHAPTER 5 REFERENCES

Kerner, Sean Michael. "TV 2.0: HDTV, Program Guide and DVR Outlook." October 13, 2004. http://www.clickz.com/stats/sectors/hardware/article.php/3421201

Colker, David. "Blurred Reception for HDTV." *LA Times*, January 8, 2005.

Abramson, Albert. "Pioneers of Television—Vladimir Kosma-Zworykin," *SMPTE* Vol. 90, July 1981 pp. 580-590.

Bussey, Gordon and Geddes, Keith. "Television, the First Fifty Years," *Philips Electronics and the National Museum of Photography, Film and Television* (Prince's View, Bradford, West Yorkshire, England), 1986, pp. 2-5.

Frezza, Bill. "Digital TV limps to the starting line," *Network Computing* Oct 15, 1996 v7 n16 p. 35.

Fink, Donald G. and Lytyens, David M. *The Physics of Television* Anchor Books Doubleday & Company, Incorporated Garden City, New York, 1960 pp. 17-82.

Herrick, Clyde. "Principles of Colorimetry," Reston, Virginia. *Reston Publishing, Incorporated A Prentice-Hall Company* 1977, pp. 12-65.

Leopold, George. "Fate of HDTV is now in lawmakers' hands," *Electronic Engineering Times* March 25, 1996 n894 p. 1.

Leopold, George. "First HDTV broadcast studio set," *Electronic Engineering Times* April 1, 1996 n895 p. 1.

Leopold, George. "New HDTV format offered," *Electronic Engineering Times* March 11, 1996 n892 p. 22.

Noll, A. Michael. *Television Technology: Fundamentals and Future Prospects* Artech House, Incorporated Norwood, Massachusetts, 1988 pp. 9-151.

Yoshida, Junko. "HDTV's story: fits, starts and setbacks," *Electronic Engineering Times* May 27, 1996 n903 p. 20.

Yoshida, Junko. "High definition still elusive," *Electronic Engineering Times* June 10, 1996 n905 p. 35.

6

CODING AND COMPRESSION

*"God made integers, all else is the work of
man."*

Leopold Kronecker, 1823 - 1891

We experience the natural world through our senses. Our eyes and ears capture light and sound and then pass the sensations to our brains for processing. In much the same way, cameras and microphones detect and respond to the electromagnetic waves associated with light and sound and pass them, in analog form, for manipulation by audiovisual processing systems. Some audiovisual processing systems are analog (conventional broadcast television), and others are digital. The primary concern of this book is *digital* video communications; hence, this chapter begins with a discussion of digital encoding techniques.

We refer to the process of converting a signal from analog to digital as *encoding*. For years, telephone companies have been encoding analog signals to transmit them digitally. Digital transmission systems have replaced most analog telephone networks in the industrialized world. The reasons include improved transmission quality, cost, increased bandwidth, and efficiency. Once a Telco converts a signal to a stream of bits, it can easily shuffle that bit stream with other bit streams, regardless of source or content, for better use of telephone lines (often figuratively referred to as *pipes*).

Video and audio signals benefit greatly from analog-to-digital conversion. The universe contains innumerable electromagnetic waves—not all of them relate to any particular signal. That so many frequencies exist in free space does not matter to human eyes and ears that can only detect a limited range of sights or sounds. However, it does become a problem when these sights and sounds are modulated electronically onto a carrier wave.

A carrier wave is an electrically constant energy path that one uses to transport a signal from one place to another. Modulation is the process of altering this stable carrier wave in response to fluctuations in the signal being carried. Analog modulation techniques perform well in controlled environments or over short distances. The signal, when it first is superimposed onto the carrier, is strong and unadulterated. However, because travel steadily weakens a signal, a signal that travels any distance requires amplification. If amplification were to restore the signal to its original pristine state, it would affect no harm. Unfortunately, frequencies abound and, unless a transmission system is shielded, interference contaminates the signal. The envelope, that originally contained only the signal, collects random waveforms that are not associated with the original signal, but which occupy the same frequency band. Amplification boosts both the signal and these random waveforms, which we refer to as *noise.*

Noise affects video signals. They may arrive on a screen full of *artifacts* (undesirable elements or defects in a video picture). Image aliasing (subjectively intolerable distortions in the picture) is another noise-related problem. Unlike analog signals, digital ones are resistant to noise. The signal is coded and sent as a string of zeros and ones. Sufficiently distinct phenomena (the absence or presence of light—positive or negative voltage levels) distinguish these binary values. In digital systems, a repeater (a corollary to an amplifier in analog systems), can accurately distinguish between a zero and a one. The integrity of the signal is extremely high because, with today's transmission systems, bit errors are quite rare.

Digital transmission preserves the quality of an original signal. However, that is not the only reason digital technology is so widely accepted today. A bigger reason— the real reason— is that computers are inherently binary, or *digital.* They do not process analog data. To leverage the enormous power that computers can offer practically every application, one must convert sounds and pictures from their natural analog state to a digital (or binary) one. Once they have been digitized, these streams of audio and visual information can be compressed, and moved economically across telecommunications networks.

AUDIO ENCODING

Like conventional television, digital video consists of a series of image frames that are synchronized with an audio track. The sounds are interleaved with the video data during the process of capture and transmission.

Audio and video must each be digitized. Digitizing a signal is a multi-step process that starts with sampling. Sampling is the process of measuring slices of an analog signal over time, and at regular intervals. Typically, "X" represents the number of samples taken in a second; X represents a number that is twice the highest frequency of the analog signal.

In 1928, a Bell Labs engineer named Harry Nyquist presented a paper that laid out a technique for sampling an analog signal in such a way that its digital representation would be *lossless*—that is, functionally identical to the original waveform. Nyquist's Theorem, the gospel of digital encoding, states that the sampling rate must be twice that of the highest frequency present in the analog waveform. Not following Nyquist's Theorem during the analog to digital (A/D) video conversion process, often leads to *aliasing*. Aliases show up as moiré (wavy patterns that look like plaid or watered silk) and color shifts (where rainbow hues appear in an object of supposedly uniform color).

In the analog world, the bandwidth of a signal is the difference between its highest and its lowest frequencies. It is the highest frequency with which sampling is concerned. The metric for the frequency of sound waves is hertz (Hz); an expression of the number of complete cycles that occur in one second. Although those with the most acute hearing can perceive frequencies as high as 20 kHz (20,000 hertz), the public telephone network does not convey them because they do not occur in speech. The human voice typically produces frequencies between 50 Hz and 4 kHz. Since this range is where most of the information in human speech is concentrated, telephone networks were specifically designed to carry those frequencies. Thus, the analog local loop—the connection between an end-user and the telephone company—has a bandwidth of 4 kHz. Nyquist's Theorem instructs that this signal must be sampled at a rate of 8,000 samples per second. The device that does the sampling is a *codec* (coder-decoder) and it forms the basis for much of today's digital telephone network.

The basis of the North American public switched telephone network is an extremely precise 8 kHz (Stratum) clock. This clock produces reference pulses that disseminate across all networks in North America. This clock orchestrates analog signal *sampling*, in accordance with a digital modulation technique known as *Pulse Code Modulation* (PCM).

In 1938, Alec Reeves, an English employee of ITT, developed the concept of PCM. Unfortunately, in the 1930s, PCM was not deployable in the public network. The pulse generation equipment of the day relied on vacuum tubes that were too large and power-hungry to make wide-scale implementation feasible. In 1947, a group of Bell Labs physicists were experimenting with a germanium crystal they had placed between two wires. Unexpectedly, a signal that was traveling across the wires began to amplify to 40 times its original strength. This discovery of the transistor—from which all integrated circuit technology evolved—paved the way for the digital world in which we now live.

PCM is the most common method used to convert an analog audio signal to a digital one. An audio channel is sampled at uniform intervals 8,000 times per second. As a sample is taken, its voltage is compared to a set of numeric values. Since the sample is going to be encoded using eight bits, one of 256 integers can be used to represent it. The integer most closely resembling the sample is used to represent it. This step is quantizing. The step that follows quantizing is *encoding*. In this step, the selected integer is expressed as an eight-bit word, a BYTE Streams of bytes are transmitted across networks at a rate of one every 125 microseconds. As each byte reaches its destination, a codec reconverts it to its analog form.

PCM-encoded voice requires a transmission line with a speed of 64 kilo-bits-per-second (Kbps). The signal is sampled 8,000 times per second and each sample is encoded using eight bits (8,000 x 8 = 64,000). This 64 Kbps channel is the fundamental building block of digital networks, although many telephone carriers rob 8 Kbps from a circuit and use that 8 Kbps to carry timing information.

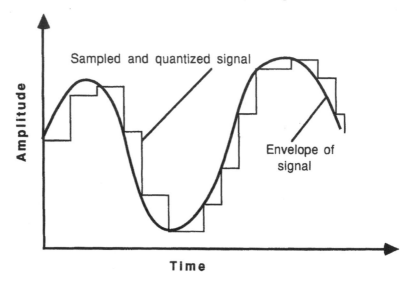

Figure 6-1. Pulse Code Modulation (Michel Bayard).

The 64 Kbps data rate forms the basis of the Integrated Services Digital Network (ISDN) bearer (B) channel. It is *multiplexed* with 23 other PCM-encoded channels into a wide pipe known as a DS-1 (T-1). DS-1 is a digital transmission link with a capacity of 1.544 Mbps (in which M represents million). The bandwidth of a DS-1 is divided into channels using a technique called time division multiplexing (TDM).

The North American digital hierarchy multiplexes or *muxes* four T-1 signals together to derive a T-2 or DS-2 (6.312 Mbps channel) and seven DS-2 pipes together to create an even higher-speed multiplexed signal called a DS-3. DS-3 is the term most commonly used to refer to the 44.736 (usually rounded to 45) Mbps pipe that is used for high-capacity transmission. The digital multiplex system we have just described has largely been replaced by *Synchronous Optical NETwork* (SONET) transmission systems.

Europe uses a system similar to T-1 that is known as *E-1/CEPT* (typically referred to as E-1). E-1 is a 32-channel pipe in which two channels (13 and 30) carry only network signaling and synchronization. Phone companies base their networks on digital carrier. That is to say, they connect their switching centers together using channels packaged in accordance with the digital hierarchy specification. The specification for all digital networks is D4 framing; it describes the basic DS-1 signal including the framing bit.

We digress from audio and video coding specifically into networks for good reason. T-1 pipes are often installed in conjunction with videoconferencing *room* systems when T-carrier is not already installed for voice or data services. Multiples of the 24 channels (six is typical) are commonly used for video communication applications. The remainder is available for other services such as outbound and inbound long distance or access for various data communications services.

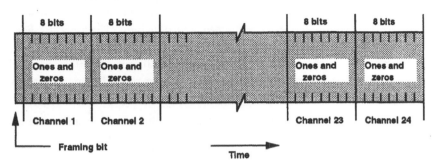

Figure 6-2. T-1 Frame Including Framing Bit (Michel Bayard).

The public switched telephone network was designed for audio signals. Dealing with audio is easy; human speech possesses characteristics that permit efficient coding. The range of frequencies is narrow and sounds occur at the lower end of the electromagnetic spectrum. Coding audio information is easy in comparison to coding video signals.

VIDEO ENCODING

Video consists of patterns of light. This light is a manifestation of energy that occurs at the upper end of the human-defined electromagnetic spectrum. In this sphere, waves are very short, frequencies are very high and encoding techniques must be very sophisticated.

The human eye, as we said in the last chapter, incorporates photoreceptor cells (cones) that capture this light. There are three types of cones; the human visual system exists in a *color space* that recognizes brightness, hue, and saturation.

Because the human visual system is three-dimensional, we must use three different components to describe a video image. This three-dimensional representation of light also expresses images in terms of their brightness, saturation, and hue. A center dividing line or brightness column is the axis. Along that line, no color exists at all. Hues (colors) form circles around this axis. A horizontal axis describes the amount of saturation—the depth of color. Highly saturated colors are closest to the center and less saturated colors are arranged toward the outer edges. Chrominance is the combination of hue and saturation that, taken together with luminance (brightness), define color.

95

Color encoding systems used for video must transform color space; they must change it into a different format so that it can be more readily encoded and compressed. This conversion of color space exploits the characteristics of the human visual system. The human eye accommodates brightness and color differently. We perceive changes in brightness (luminance) much more readily than we do fluctuations in chrominance. Hence, video-encoding systems devote more bandwidth to the luminance component of a signal. The symbol Y represents this *luma* component. White light is the composite of equal quantities of red, green, and blue light. Therefore, to determine how much red, green or blue light is present in an image, one can develop a system that subtracts color (chrominance) information from luminance information. This is exactly what video coding systems do. There are many different variations but, essentially, they all rely on *color differencing.*

Color differencing is the first step in encoding the color television signal. Subtracting the luminance information from each primary color forms the color difference signals: red, green, or blue. Color difference conventions include the SMPTE format, the Betacam format, the EBU-N10 format and the MII format. Color difference signals are not component video signals—these are, strictly speaking, the pure R, G, and B waveforms.

There are two types of video signals, component and composite. In component video, the Y or luminance signal is recorded and transmitted separately from the C (hue and color saturation) signal. Because the luminance signal is recorded at a higher frequency, more resolution lines are available to it. Because the eye is less sensitive to chroma signals, those signals are recorded at lower frequencies and transmitted using less bandwidth. Component video is not a standard, but rather a technique. It results in greater signal control (keeping signals separate makes them easier to manipulate) and higher image quality. Component video is often called *Y/C* video.

NTSC and PAL television systems use composite video. In composite video, brightness, hue and saturation are combined into a single signal. In its analog form, the chrominance signal is a sine wave that is modulated onto the luminance signal, which acts as a subcarrier. In NTSC systems, the YIQ trichromatic color system is used. The two color differences are first combined, using a technique known as quadrature modulation. Next, this color information is modulated onto the 4.5 MHz Y subcarrier with the phase of the sine wave describing the color itself and the amplitude describing the level of color saturation. *Frequency interleaving* is the name of this technique.

A digital image is a grid that contains a two-dimensional array of values representing intensity or color. To achieve this format, an analog image must first be converted to the 2-D space, using pixels to convey its resolution. If the image is black and white, the values used might be 0 and 1. Eight bits might be used for a gray-scale image. True color requires at least eight bits each for R, G, and B; it also requires three frames of information, one for each color component.

Multiple video-encoding schemes correlate to multiple divergent applications. Still video applications demand different requirements than do moving video applications. Digital video can be stored on a typical CD-ROM or on a powerful video server, and delivered in broadcast (one-way transmission) format. However, digital video is not always stored; video communication is an interactive real-time digital video application.

COMPRESSION: KEEPING ONLY THE CRITICAL BITS

Compression, the art and the science of compacting a signal for transmission, is far more sophisticated today than it was ten-years ago. Nevertheless, despite today's fast networks, a mismatch persists between the bandwidth needed to transport digitized video, and the bandwidth available to carry it. Different approaches to encoding a video signal produce different results, but virtually all techniques result in uncompressed bandwidth requirements of 25 Mbps or more. Networks that offer this capacity are rare and expensive.

Compression would be superfluous were bandwidth available in limitless quantities and at extremely low cost. Economy mandates compression. If the additional cost to compress is lower than the price of necessary additional bandwidth to carry an uncompressed signal, customers will compress. In the early 1980s, both costs were very high and, consequently, video communication was not very popular. As time went by, video codecs became more powerful. By the mid-1980s, they reduced the bandwidth required for a conference to below T-1 speeds.

Visual compression techniques usually involve manipulating pixels—the medium of exchange in video communications. As a codec compresses, it replaces the data of the original signal with complex mathematical models. The goal is to eliminate as much of the signal as possible without destroying its information content.

Compression involves two-steps. Anything that is compressed must be decompressed. A trade-off exists between the time it takes to compress an image and how much compression one achieves. Time-consuming compression techniques do not work for video communication applications, which must move quickly to produce the illusion of fluid motion.

In video communication, there are two primary methods for categorizing compression. One technique, redundancy elimination, is based on eliminating signal duplication. The other is based on degrading the picture and sound slightly, preferably in areas where the human eye and ear are not particularly sensitive. It is *quality reduction*. Before we get into the details of these techniques, we must introduce some new terms.

One fundamental way to categorize compression is to look at the result. After decompression, are the data identical to those that constitute the original signal? If so, the compression is *lossless*. If compression is achieved by permanently discarding parts of the signal (ideally, components not critical to its interpretation), the technique is *lossy*. When reversed, lossy compression does not produce a signal that is identical to the original but, rather, approximates it. Lossy techniques can produce very high ratios of signal compression, but do so at a considerable price. That price is reduced image quality or sound fidelity (or a combination of both). Generally, signals degrade as compression increases. Interactive video compression uses both lossy and lossless techniques. However, lossless methods usually occur at the end of the process and involve expressing strings of zeros and ones using *shorthand* codes.

Another method for categorizing compression is based on the methods used for decompression. Is decompression an exact reverse of the compression process? If so, the compression is *symmetric*. If the techniques used to compress are more compute-intensive than those used to decompress are, compression is *asymmetric*. Real-time, interactive video communication applications use symmetric techniques. Video playback applications commonly use asymmetric compression. Most of the resource goes into compression so that receiving devices of limited sophistication can decompress cheaply and easily.

Even symmetric compression does not divide the work equally. Real-time encoding requires complex math, so the sending codec does more work. However, in interactive videoconferences, both codecs are simultaneously sending and receiving.

INTRAFRAME AND INTERFRAME COMPRESSION

Frames of motion video are full of duplication, as one can demonstrate with a strip of movie film. One will discern little difference between the successive frames. The background is static, and action takes place incrementally. Moreover, within a single frame many long sequences of constant pixel value exist (e.g., walls, tables, clothing, and hair tend to comprise masses of uniform color). Efficient codecs do not waste bandwidth on repetitive information. They code the unique parts of the signal and describe, using an algorithm, how patterns repeat.

Video codecs that compress data within a single frame of video are *intraframe* codecs. They commonly divide a video frame into blocks and then look for redundant data. When they find it, they eliminate it; they attempt to reduce the content of that frame to *entropy*. Other types of codecs compare multiple frames of data. These are *interframe* codecs. They, too, try to achieve entropy through compression.

What is entropy? It is a scientific term that refers to the measure of disorder, or uncertainty, in a system. Good compression techniques eliminate signal duplication by using shorthand methods that the coding and decoding devices on each end readily understand. The only part of the signal that must be fully described is the part that is random, and is thereby impossible to predict.

Redundancy-elimination compression first worked on achieving entropy in text messages. Text files consist of only about 50 different characters, some of which appear often. The letters E, T, O, and I, for instance, collectively account for 37% of all character appearances in an average text file. David Huffman, a researcher at MIT, noted and subsequently exploited their repetition. In 1960, Huffman devised the first text compression algorithm. The idea behind Huffman encoding is similar to that of Morse code. It assigns short, simple codes to common characters or sequences and longer more complex ones to those that appear infrequently. Huffman encoding can reduce a text file by approximately 40%. A lossless compression technique, it is used to reduce long stings of zeros in many standards-based video compression systems including H.261, H.263, and MPEG.

Other lossless compression techniques followed. Developed in the 1970's, Lempel-Ziv-Welch (LZW) compression focused not on the characters themselves, but on repetitive bit combinations. LZW coding builds a dictionary of these commonly used data sequences and represents them with abbreviated codes. Originally developed for text, another technique called run-length coding, compresses at the bit level by reducing the number of zeros and ones in a file. Sending only one example, followed by a shorthand description of the number of times it repeats, indicates a string of identical bits. When LZW and run-length coding are used in video compression, it is generally after lossy techniques have done their job.

Compressing moving images is much more demanding than compressing text. A frame of video contains vastly more information. Fortunately, there is usually a correlation between neighboring pixels. Collocated pixels can be neighbors in terms of space (they sit adjacent to each other) or time (they occupy the same space but in different frames). Intraframe compression techniques eliminate spatial redundancy. Interframe techniques eliminate temporal redundancy. Intraframe compression relies on transform coding (DCT, for instance). Interframe compression is largely a subtractive process: pixels contained in a frame of video are subtracted from a previous frame's pixels. The difference is information—the entropy that results from the interframe coding process.

Early forms of video compression relied solely on interframe coding. In the 1960s, Bell Labs began experimenting with a variation of PCM called differential pulse-code modulation (DPCM). DPCM uses "differencing" to compare successive video frames. Codecs (one transmitting and one receiving), use identical methods to predict, using a past frame, what the pixel composition of a present frame will be. The transmitting codec, after making its guess, checks its accuracy by comparing its prediction, (stored in memory array), with reality. It subtracts reality from its prediction, encodes the *errors* and transmits them to the receiving codec. Because the receiving codec is making identical predictions, it knows what to do with errors, which it interprets as corrections. It makes the changes, displays the signal, and then stores the amended frame as the basis for the next prediction. The prediction and correction process is continuous; each new frame updates the previous one.

DPCM is still used in standards-based videoconferencing compression, although it is generally optional (by optional, we mean that a decoder must decode a DPCM compressed image, but a coder is not required to encode it). DPCM works because, on average, only about 9% of the pixels in a moving sequence change between frames. In 1969, a process known as conditional frame replenishment (CFR or CR) was developed. CR recognizes that information (what is not already known) is present in a scene only when motion occurs. Moreover, information is present only in the part of the frame that contains motion. CR compression techniques transmit only the changes in the part of the frame that comprise motion. It works well when the motion in a scene does not exceed 15% between frame intervals.

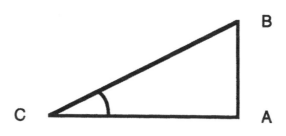

Sine of angle ACB = AB/BC

Figure 6-3. Sine (Michel Bayard).

In 1972, two engineers recognized that, in a moving sequence, there is less correlation between pixels that occupy the same place in subsequent frames than there is between pixels that are close to each other but shifted to a different position. After reflection, it became clear that motion vectors play a significant role in pixel prediction. As objects move, pixels shift left, right, up, or down. If a codec were prepared to recognize the direction and rate of pixel motion, it would do a much better job of predicting where each pixel might end up in successive frames. This discovery led to another form of interframe compression called motion compensation. It improved the image compression factor by 50–100% when compared to CR.

The basis of *motion compensation* is the observation that moving objects are best described by re-mapping their original position to a new position. For instance, a pointer used to call attention to a figure on a whiteboard moves in a predictable fashion. Only the motion vector requires retransmission because the receiving codec already possesses a mathematical representation of original object.

In other words, in motion vector prediction, the codec must send information about the position of a subject, but not about the subject itself.

Intraframe encoding came along in the mid-1970s. It is compute-intensive; and uses more sophisticated algorithms than all but a few interframe codecs. Two different intraframe coding techniques are typical used today. One, which is common to almost every standards-based video encoding technique today, is the discrete cosine transform, better known as DCT. The other, which is more prevalent in proprietary codecs, is *vector quantization* (VQ).

DISCRETE COSINE TRANSFORM

Any student of image compression will encounter DCT; it forms the basis for three very important video communications compression standards. One is the Joint Photographic Experts Group (JPEG) standard, which addresses still-image compression. Another is the Moving Picture Experts Group (MJPEG) standard that compresses TV-quality or near-TV-quality moving pictures (typically for playback applications). Furthermore, the ITU-T's H.26x Recommendations, the primary standards-based videoconferencing codecs, rely upon DCT.

The DCT is a technique that converts pixel intensities into their frequency-based equivalents. To do this, it applies numeric transformation to the data contained in a pixel block. After the DCT is complete, a block of video data is described in terms of frequencies and amplitudes rather than a series of colored pixels. Frequencies describe how rapidly colors shift. For instance, high frequencies denote very rapidly shifting colors such as edges where one color stops and another starts. Amplitudes describe the magnitude of a color change associated with each shift.

To understand DCT, one must possess some knowledge of sines and cosines. In trigonometry, the sine is the ratio between the side opposite an acute angle in a right triangle and the hypotenuse (the side of a right-angled triangle opposite the right angle). Examine Figure 6.3 for a straightforward picture. A sine wave is a graphical representation of the sine ratio. Remember that the polarity of an AC power source is constantly changing from positive to negative. The waveform produced from alternating current can be plotted. The result is a sine wave that undulates with the shape of the wave as it continues from zero polarity to its peak positive value, then back to zero and beyond, to its peak negative value and then back to zero.

Webster describes a cosine as "the sine of the complement," the ratio between the side adjacent to a given acute angle in a right triangle and the hypotenuse, as depicted in Figure 6.4. It is the reciprocal of a sine. How does discrete cosine transform work? When the information in a picture is processed, using a technique called *spatial-filtering*, it is transformed into a series of frequencies that mimic the eye's perception of light and color. The high-frequency portions of the picture are the areas of greatest change; for instance, complex patterns in clothing and black letters on a white page. The low frequency elements are the parts with relatively little change across an area such as blank walls and other objects of uniform color and shape. The DCT uses the reciprocal of those frequencies to describe a pixel block.

The DCT process begins by dividing a video frame into eight-by-eight pixel blocks that each contains a set of waveforms of different horizontal and vertical frequencies. These waveforms, when scaled by numbers called coefficients then added together, represent any of the 64 sample values (a coefficient is nothing more than a multiplier). For instance, in the formula $a(y + z)$, "a" is the coefficient. Y could be the sine and z the cosine. A coefficient is the least common denominator; that is how DCT uses it.

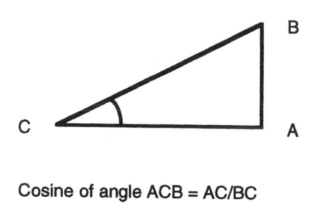

Cosine of angle ACB = AC/BC

Figure 6-4. Cosine (Michel Bayard).

After dividing a picture into blocks of pixels, the DCT scanning procedure starts in the upper left-hand corner of a block and finds the average luminance for the entire block. This value is the *DC coefficient,* and it is the first number produced by the transform. The DCT scanning process continues and as it does, 63 more values are produced. The second and third are expressions of the lowest frequency in the horizontal and vertical directions. As scanning proceeds, it concentrates the low-frequency components, which describe large areas of uniform color, in the upper left corner of the array. The high-frequency components, describing detail shifts in the image (edges and color changes), are concentrated in the lower right.

Since there are usually more flat surfaces than sharp edges in a picture, the first few values will probably be non-zero. However, as the zigzag scanning continues the frequency increases. The coefficients produced become smaller and smaller. By the time the sixty-fourth value is measured, it is an expression of the magnitude of the highest frequency in both the horizontal and vertical planes—a value that tends to zero after the quantization process is applied.

The resulting 64 DCT values are arranged in a matrix. This all occurs before the actual compression process; what started as 64 eight-bit pixels resulted in 64 eight-bit coefficient values, most of which are zero or almost zero. In the process of quantization, the almost-zero values become zero; they are not important because the human eye would not perceive them anyway. The most important value, the DC coefficient, is quantized and coded using an 8-bit code.

After quantization, a bit stream is formed from the pixel block. This bit stream is ready for compression. Compression in H.261 and H.263 codecs takes the form of run-length and Huffman encoding. In run length encoding, the codec compresses a string of identical bits by sending a bit pattern followed by a token that indicates the number of times the pattern repeats. Huffman encoding then converts the most frequently repeated tokens into the shortest bit strings.

To summarize, the human eye, like the human ear, is more responsive to low frequencies than to high ones. In audio companding, PCM codes low frequencies with more precision than higher ones. DCT works the same way. The low frequencies in image encoding contain the picture's detail. DCT compresses by giving priority to that which humans need most for image processing.

VECTOR QUANTIZATION

Used in a number of proprietary compression techniques (most notably, Intel's Indeo) vector quantization is a less compute-intensive alternative to the DCT.

Vector quantization is a lossy compression method that uses a small set of data to represent a larger one. Stated differently, codecs that use vector quantization share a *codebook*. This codebook contains an array of possible values, and shorthand codes for conveying them.

As the name implies, vector quantization quantizes vectors. A vector is a mathematical representation of a *force*. In video coding, this force represents a color (in other words, a frequency). Not only must we consider frequency when dealing with images, we must also consider the intensity of that frequency (its amplitude). Luminance determines amplitude, the relative saturation of a pixel's color.

In vector quantization, we analyze a block of pixels to discover its vector. This vector is then used to select a predefined equation that describes it in a more efficient way. Codecs can be tuned so that the quantization process considers, not just a pixel block's vector, but also how this vector relates to those of surrounding blocks within the image. In most pictures, colors (and their corresponding saturation) are concentrated in certain areas. These colors are associated with objects (e.g., walls, white boards, clothing, and paper). A codec can select a relatively small subset of the representative vectors in an image and use this subset to approximate the vectors of all other pixel blocks. If a pixel block's vector, once measured, falls into the "red" codebook, it can be described using an abbreviated code. That code, which might be only a few bits long, is interpreted by the decoder as a "red" with a given hue and saturation.

VIDEO COMPRESSION

Video compression techniques do not strive for total quality, just something good enough. Good enough, of course, depends on the application, and quality is, therefore, subjective. However, all types of lossy compression leverage the human tendency to consider context and thereby interpret what one sees or hears when minor bits of information are missing.

Video compression techniques exploit the functioning of the human eye. The eye-brain team is quite sensitive in some ways and quite tolerant in others. For instance, the eye adjusts very well to variations in relative and absolute brightness, but not to lack of focus. It is sensitive to distortions in familiar images (e.g., people's faces), but can tolerate minor flaws in motion handling as long as the overall movement looks realistic. Viewers want the parts of the picture that convey critical information—numbers, letters, and the eyes of other people—to be clearly discernible, but blurring in a picture's background is not distracting.

Techniques that sacrifice video quality for reduced bandwidth fall into two categories. One method is to approximate instead of sending the actual information. DCT approximates luminance. Motion compensation uses motion vectors to approximate how an object will move from one frame to the next. These techniques reduce quality, but the approximations are good enough and, since most humans cannot perceive the loss, trivial.

Another method of quality reduction works on picture resolution. Resolution depends on pixels, which are the conveyers of video information. The more pixels, the more detail. Detail is, for some applications, critical. The more detail in an X-ray, the better. Other applications require less detail. Seeing the pattern on the tie worn by a person in a videoconference is not important. Capturing the messages conveyed through body language and gesture is important. As long as quality-reduction does not impair the conveyance of important information, we overlook it. Most participants in a videoconference become accustomed to the reduced quality image and cease to notice it—particularly if the audio quality is good.

Different video formats define different picture resolutions. For instance, color VGA video specifies a graphics mode that is 640 pixels wide by 480 pixels tall. Super-VGA is 1,024-by-768 pixels. In both formats, a colored pixel is defined using 24-bits. The CCIR (an international standards body that is part of the United Nations) defines a digitized version of NTSC in its Recommendation CCIR-601 in which 525 vertical lines are each painted with 720 horizontal pixels (For more on CCIR-601, please refer to Appendix C). HDTV in North America will have yet a different resolution, as will that little window on the PC that contains video clips downloaded from the World Wide Web.

To match image-sharpness with application requirements, one can use various pixel reduction techniques. One approach is to shrink the viewing window using a technique known as *scaling*. Scaling works because smaller viewing areas require fewer pixels. The image displayed in a quarter-screen PC monitor is painted with far fewer pixels than the 27-inch monitor on a videoconferencing room-system requires.

Reducing the frame rate is another way to compress a motion video sequence. Although the NTSC standard calls for 30 frames per second, one rarely gets 30 in a videoconference, especially when using a desktop system. When there is little or no motion taking place on the distant end, one may receive 15 to 20 frames per second, possibly even more. However, the moment someone moves the camera, or gets up and walks across the room, a flood of new bits must be shipped to the receiving end and the frame rate drops substantially.

Dropping the frame rate allows transmission bandwidth to be significantly compressed with little corresponding degradation. Unlike a car race, most business related video applications include limited motion. Nevertheless, even business meetings can generate more motion than a low-bandwidth channel can accommodate. A person who is rising from a chair and walking to a whiteboard will translate to a large amount non-zero information that must be sent. It may begin to exceed the channel's bandwidth. At that point, some of the signal must be discarded. As more and more people get up and walk around, the codec gets frantic. It eventually falls behind, discarding excessive numbers of frames. The picture degrades to a point at which it is annoying. This is a much greater problem at low bandwidths (POTS or ISDN BRI) than at speeds of 384 Kbps, or especially 224 Kbps, because the codec has less room to maneuver.

Other quality reduction techniques are sometimes used in video compression. One we have not yet mentioned has to do with color. If a color coding system were to represent a video image using only one bit per pixel, the picture would be monochrome. Bits could take only two values. A zero would represent an absence of light while a one would portray light of a single frequency, white or red, for instance. A coding system of this type has a *single bit plane*.

Additional colors are added to the system with multiple bit planes. The total number of planes used to represent an image is a bit-map. The more bits in the map, the better the resolution. With a two-bit-plane, four color values can be represented: 00, 01, 11 and 10. Bit planes containing eight bits can represent 256 discrete values. When R, G, and B are each assigned eight bits, millions of color combinations can be created (256^3). Each time an additional bit is added to a bit plane, the picture's resolution doubles. Unfortunately, so does the bandwidth required to transmit the uncompressed image.

AUDIO COMPRESSION

As we asserted earlier, human speech has attributes that lend themselves to efficient compression. Compression can eliminate pauses between words. Amplitudes rise and fall predictably. We also instructed that PCM coding results in eight-bit samples. While this is true, we have omitted a step. Samples are quantized at 2^{12} (4,096 values) and then compressed to eight bits (256 values) using logarithmic encoding. This process is called *companding*; it is a contraction of the words *compress* and *expand*. Companding is performed in accordance with either μ-Law (pronounced "mu-Law") or A-law; μ-Law is used in North America and Japan, and A-Law is used elsewhere. When compressing voice, logarithmic modeling works better than linear schemes. This is because humans are more sensitive to small changes at low volume than changes of the same magnitude at higher volumes.

ITU-T Number	Range of Compression	Description of Encoding Technique (Comments)	Used with ITU-T Codec Recommendations
G.711	56–64 Kbps	Pulse Code Modulation (PCM) of voice frequencies	H.320
G.722	64 Kbps	7 kHz audio-coding within 64 Kbps (PCM)	H.320, H.322, H.323
G.722.1	24 or 32 Kbps	Modulated Lapped Transform (MLT)	H.320, H.323
G.723	5.3 and 6.3 Kbps	Multipulse-Maximum Likelihood Quantization (MP-MLQ)	H.322, H.323, H.324
G.728	16 Kbps	Low-Delay Code Excited Linear Prediction (CELP)	H.320, H.322, H.323
G.729	8 Kbps	Conjugate-Structure Algebraic-Code-Excited Linear-Prediction (CS-ACELP)	H.323
G.729A	8 Kbps	CS-ACELP	H.323

Figure 6-5. CCITT/ITU-T Audio Encoding Recommendations.

The CCITT (now known as the ITU-T) has standardized PCM as G.711 audio. Using G.711 coding, speech and sounds with bandwidths of approximately 3.4 kHz are converted to 56 or 64 Kbps bit streams.

A slightly more recent ITU-T standard, G.728, encodes 3.4 kHz audio frequencies into 16 Kbps bit streams. It encodes speech to fit a simple, analytical model of the vocal tract. It generates *synthetic* speech that is very similar to the original waveform. The *code-excited* version LPC goes a step further to compute the errors between the original speech and the LPC model. It transmits both the model parameters and a highly-compressed representation of the errors. CELP results in compression that is remarkably faithful to the original signal. Indeed, G.728 audio virtually replicates G.711 quality at one-quarter the bandwidth.

There are many different audio-encoding standards. The ones that relate to video communication are listed in the chart below, along with an indication of with which ITU-T video compression standard they are associated (for a thorough discussion of video compression standards, refer to Chapter 7).

DIGITAL SIGNAL PROCESSORS

Compression methods can be categorized in one additional way. Some do their work in software only, while others use a combination of software and dedicated processors. Hardware-assisted codecs have one significant advantage over their software-only cousins: they are much faster.

Codecs are usually implemented on digital signal processors (DSP). In the past, DSPs were software-based algorithms that ran on mainframes. Engineers soon found ways to move these processors off the CPU and onto less expensive silicon. In the early 1980s, DSPs became available as specialized high-performance coprocessors. Today, DSPs are fine-tuned to the point that they can execute even complex multiply-accumulate (MAC) instructions in a single clock cycle. Because of DSPs, the cost of multimedia on the desktop has declined to the point that it is an attractive option. One multimedia application that makes heavy use of DSP chips is video communication.

DSPs come in two flavors: *function and applications specific integrated circuits* (FASIC) and *programmable*. Programmable DSPs are multimedia-processing engines that perform multiple tasks. For instance, they can be programmed to compress either audio or video, using a wide variety of proprietary or standards-based algorithms. Some programmable DSPs process millions of operations per second (MOPS) while others process billions (BOPS). The ability to process different media (audio, video, data) at high speeds is important in video communication applications because, increasingly, a videoconference has video, audio, and graphics content. Moreover, POTS-based videoconferences also require modulation to convert from a digital to an analog format for transmission.

FASICs are also used in video communication applications. They are referred to by the function/application for which they are designed. There are MPEG-1 FASICs, H.320 FASICs, and audio FASICs (e.g., G.723 or G.726 chips).

Many companies manufacture programmable DSPs and FASICs. These include (but are not limited to) C-Cube, IBM, Intel Corporation, and Lucent Technologies. There are many important software-only codecs on the market, too. These include White Pine's Enhanced CU-SeeMe, Intel's Indeo, H.324, and others.

At the time of this writing, the leading video communication codec choices conform to the ITU-T's various codec recommendations. These recommendations are referred to as the H.32X family of videoconferencing codecs. They include compression algorithms that are designed for transmission over ISDN-B channels, over LANs, and over POTS. The next chapter presents a thorough discussion of these codecs and the standards that define them.

CHAPTER 6 REFERENCES

Brooks, John. *Telephone—The First Hundred Years*, Harper & Row, Publishers, New York, New York, 1975, pp. 202-223.

DeBaldo, Paul. "Compression Technologies and Techniques," *Teleconference Magazine* (San Ramon, California) March/April 1983, pp. 10-11.

Green, James Harry. *The Business One Irwin Handbook of Telecommunications,* 2nd Ed., Business One Irwin, Homewood, Illinois, 1992, pp. 109-133.

Iinuma, K. and Ishiguro, T. "Television Bandwidth Compression Transmission by Motion Compensated Interframe Coding," *IEEE Communications Magazine* July 1982, pp. 24-30.

Lucky, Robert W. *Silicon Dreams, Information, Man and Machine,* St. Martin's Press, New York, New York, 1989, pp. 37-348.

Mokhoff, N. "The Global Video Conference," *The Institute of Electrical and Electronics Engineers, Inc., Spectrum* November 1980 vol. 17, pp. 45-47.

Newton, Harry. *Newton's Telecom Dictionary*, Flatiron Publications, New York, New York, 1991, p. 630.

Rocca, F. and Zanoletti, S. "Bandwidth Reduction Via Movement Compensation on a Model of the Random Video Process," *IEEE Communications Magazine* Comm. 20, 1972, pp. 960-965.

7

STANDARDS FOR VIDEOCONFERENCING

"The future of proprietary algorithms for
videoconferencing is as bright as the future of
EBCDIC for data exchange."
        Dave Brown, Network Computing

THE ITU

To accomplish successful interchange, information must be represented in a mutually agreed-upon format. The process of defining the specifics of that format is *standards-setting*. Perhaps the most important international standards-setting body in the world is the United Nation's International Telecommunications Union (ITU). An intergovernmental treaty organization, the ITU formed in 1865. Its objective was to publish telecommunications-oriented recommendations after studying technical, operational, and tariff-based issues.

The part of the ITU that promotes interoperability in wireline telecommunications networks is the Telecommunications Standardization Sector, or ITU-T (formerly the Comité Consultatif International Téléphonique et Télégraphique, or CCITT). A sister organization to the ITU-T is the ITU-R (in which R represents radio). The ITU-R (formerly the Comite Consultatif International des Radiocommunications—CCIR) coordinates wireless (mainly broadcasting) standards and is currently working to achieve global harmony in the area of HDTV, among other things.

Good things are happening within the ITU-T. The new name indicates an expanded interest—its orientation now plainly extends beyond Europe. Whereas the CCITT originally worked in four-year cycles, the ITU-T has adopted an accelerated standards-setting schedule. The approval process, formerly a face-to-face arrangement, is now largely conducted via electronic communications. Provided a study group is in consensus, it can achieve final approval in months. Also speeding things along are the efforts of non-accredited standards groups such as the International Multimedia Teleconferencing Consortium (IMTC). These ad-hoc bodies develop specifications and introduce them to the ITU-T for approval. This increase in global participation and speed has helped expedite many technologies to market. Video communication is no exception.

THE HISTORY OF THE PX64 STANDARD

Europe was the first part of the world to embrace and implement video communication as a tool for stimulating commerce. In 1979, British Telecom initiated the idea of a European visual service trial. The BT trial started in 1982 and ran concurrently with the European Visual-teleconference Experiment (EVE) that was promoted by the Conference of European Postal & Telecommunications Administrations (CEPT). Both projects required standards-based video codecs, and thereby prompted the CCITT, in 1980, to form the COST 211 Specialists Group. COST was an acronym for Co-Operating for Scientific and Technological research. In 1983, the Specialists Group defined a codec—a single chassis equipped with roughly 40 circuit boards. The specification was formally adopted by the CCITT in 1984 as Recommendation H.120 (Codecs for Videoconferencing Using Primary Digital Group Transmission). It incorporated the standard PAL television signal that most Western European countries embrace.

H.120 multiplexed audio, video, and signaling using the G.732 Recommendation (Characteristics of Primary PCM Multiplex Equipment Operating at 2048 Kbps). It offered interchangeable compression techniques: one optimized for normal face-to-face business meetings, and the other—that incorporated an approximately 1.5 second transmission delay—optimized for sending colored graphic images. H.120 provided for both monochrome and color operation, for multipoint switching between video codecs, and for signal encryption. Eventually, H.120 was expanded to include the SECAM broadcast standard. The new recommendation was named H.130 (Frame Structures for Use in the International Interconnection of Digital Codecs for Videoconferencing or Visual Telephony). Upon H.130's adoption, networks of standards-based publicly-available video communication studios were constructed throughout Europe.

Unfortunately, the *international* H.130 standard fragmented the global video communication market. European codec manufacturers embraced it, but U.S. manufacturers could not deploy it as it was developed for E-1 (as opposed to T-1) networks. To address the problem, the CCITT established, in 1984, a new specialists group. Dubbed the Specialists Group on Coding for Visual Telephony, it was a genuinely international effort that included North American members, enjoyed strong European support, and was chaired by a representative from Japan.

Working on two standards simultaneously, the group met more than ten times between late 1984 and 1988. One of the standards was referred to as Mx384; M stood for *multiples,* the x was pronounced *by* (i.e., M-by-384), and the 384 represented transmission speed in Kbps. The group expected Mx384 to leverage the internationally-available ISDN H0 standard for switched 384 Kbps digital dialing. The second standard was known as Nx64 in which "N" referred to a variable per-second frame rate from 1–30, and "64" referred to the 64 Kbps ISDN B-channel over which a compressed video signal would travel.

By September 1988, the Specialists Group had completed the Mx384 standard. Unfortunately, it was already obsolete. One year earlier, an American manufacturer had introduced a product that delivered adequate picture and sound quality at 224 Kbps. Although the Specialists Group recommended that the CCITT General Assembly formally ratify Mx384, none of the world's codec manufacturers ever built products that conformed to it. Instead, they waited for the Nx64 standard (which was later renamed Px64).

During the process of developing Px64 (pronounced *P times 64*), Study Group XV (the term Study Group replaced the term Specialists Group) had to consider many alternative compression algorithms. No single method offered the efficiency and flexibility to meet the group's goals, so they combined techniques. Unable to afford the time or expense required to model every coding iteration, the Group used computer-based simulation systems that consisted of digitally sampled video sequences. To simulate picture-coding algorithms, they applied various computer programs to the sequence. After compression, they viewed and evaluated the results, made changes, and re-applied the simulated coding process. Eventually, Study Group XV selected a hybrid-coding scheme that became the foundation for H.261, the "Video Codec for Audiovisual Services at Px64 Kbps." H.261, and five other recommendations, were submitted to the CCITT during its July 1990, meeting in Geneva. In December 1990, the CCITT approved all six recommendations, and formally declared them the H.320 family of standards.

H.261

Study Group XV considered many issues as they developed the H.261 specification. First, they recognized that the market for video communication services would benefit from a global standard. Second, since ISDN B channels were internationally deployed, they found these digital circuits, used individually or aggregated, to be the ideal transport mechanism. Third, they knew that a compression specification would not achieve broad acceptance if it were not adaptable. By adaptable we mean two things: flexible enough to permit various elaborate and inexpensive implementations and versatile enough to incorporate new technologies as they become available.

The H.261 standard meets these criteria. It accommodates globally disparate digital network and television standards. Many, if not most, H.261-based implementations use ISDN B channels, just as the CCITT predicted. The products built on H.261 are diverse. Some implement the Recommendation's optional procedures for delivering superior resolution and motion handling while others omit them to reduce cost. H.261 has proven to be flexible enough to accommodate ongoing advances in technology. As chips get faster, frame rates can increase and motion handling can improve without imposing fundamental change on the standard.

H.261 can be a video system's sole compression method. It can also be ancillary and thereby used as a substitute for a proprietary algorithm when two dissimilar codecs must interoperate.

H.261 and CCIR Recommendation 601

The Consultative Committee for International Radio developed Recommendation 601 as an international specification for the analog-to-digital (A/D) conversion of color video signals. An H.261 source coder manipulates non-interlaced pictures in accordance with a portion of that Recommendation. To understand H.261, one must possess some knowledge of CCIR-601's characteristics.

CCIR-601 Characteristics	NTSC 525/60	PAL/SECAM 625/60	CIF	QCIF
Scanning Technique	Interlaced	Interlaced	Non-Interlaced	Non-Interlaced
Fields sampled per second	60	50	30	30
Sub-sampling direction	Horizontal	Horizontal	Vertical and Horizontal	Vertical and Horizontal
Luminance resolution	720 H x 485 V	720 H x 576 V	352 H x 288 V	176 H x 144 V
Chrominance resolution	360 H x 485 V	360 H x 576 V	176 H x 144 V	88 H x 72 V
Chroma Sub-sampling	YCbCr 4:2:2	YCbCr 4:2:2	YCbCr 4:1:1	YCbCr 4:1:1

Figure 7-1. CCIR-601 Picture Formats.

CCIR-601 addresses the incompatibility of the world's three television formats (NTSC, PAL and SECAM) by defining two picture structures that will work with any of them. These are the Common Intermediate Format (CIF) and the Quarter Common Intermediate Format (QCIF). In Figure 7-1, the four CCIR-601 picture structures are presented, along with the techniques used to derive them.

Since H.261 is concerned only with CIF and QCIF resolutions, in our discussion of CCIR-601, we will address only CIF and QCIF image formats.

Both CIF and QCIF are compromises that use the frame rate of North American TV (30 fps), and a resolution that easily adapts to that of European television. Compliance with CIF (also called full-CIF or FCIF) is optional under H.261, while compliance with QCIF is mandatory. At one-quarter the spatial resolution of CIF, QCIF is best suited to small screen (20" and below) and "talking head" (PC window) video communication applications. CIF is much more important to large screen (group system) applications.

The CCIR-601 A/D conversion process begins with frequency filtering. A loop filter separates the frequencies contained in non-interlaced CIF or QCIF images into horizontal and vertical functions. The output of the filter is a two-dimensional array of values.

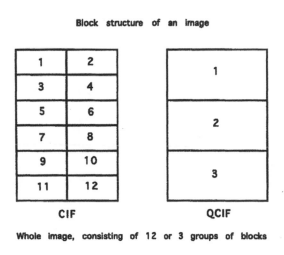

Figure 7-2. CIF and QCIF (Michel Bayard).

The values that result from frequency filtering are of two basic types: luminance and chrominance. The human visual system is more perceptive to subtle changes in light than to subtle changes in color. For that reason, the CCIR-601 conversion process allots higher spatial resolution for luminance than for chrominance, specifically for every four luma (Y) samples, two color-differenced samples (Cb and Cr) are taken. This method, known as chroma subsampling, results in a format known as YCbCr 4:1:1.

As we discussed in Chapter 6, quantization follows sampling. Once scanned, the uncoded YCbCr data is quantized according to CCIR-601. Luminance (grayscale) information is quantized using one of 220 different levels. Color-difference information is quantized as one of 255 different levels. These quantization levels are the same as those used in the DCT (which, in H.261 encoding, is the next step after quantizing).

One of two different types of compression is applied to the 8-by-8 pixel blocks: intraframe and interframe. It is possible to perform both intraframe and interframe coding on a pixel block but only intraframe coding is mandatory in H.261. Intraframe encoding relies on the DCT (see Chapter 6). Interframe encoding uses DPCM, and motion compensation techniques (also described in Chapter 6). H.261 codecs that perform both DCT and motion compensation are *hybrid coders.* Of course, H.261 codecs operate in pairs or multiples. Each has a coder and a decoder component. The coder on one end talks to the decoder on the other. Between them flows a bit stream that is arranged into 512-bit frames. Each frame contains two synchronization bits, 492 data bits that describe the video frame, and 18 error-correction code bits. These 512 bits are then interlaced with the audio stream and transmitted over Px64 channels in which 'P' represents the number of 64 Kbps channels that are available for transmission.

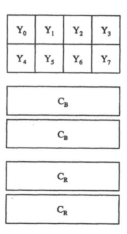

Figure 7-3. YCbCr 4:1:1 Format (Michel Bayard).

H.261 Options

As we have stated, H.261 is a permissive standard. It allows for a great deal of flexibility. This flexibility can confuse video communication system buyers (and, for that matter, video communication system vendors).

As we mentioned above, QCIF, a lower-resolution picture format, is mandatory under H.261 and CIF is optional. CIF and QCIF can code at varying frame rates: 7.5, 10, 15, or 30 per second. As we said in the previous chapter, dropping frames is common as a means of compressing data for transmission. A codec that produces only 7.5 fps is, technically speaking, just as "H.261 compliant" as a codec that produces 30 fps. This vagueness in the standard allows a manufacturer to slow the frame rate to deliver CIF resolutions (which come at the expense of motion handling).

On the other hand, no matter how many frames are sent in a given second, a receiving codec *must* decode up to and including 30 frames per second. Therefore, an evaluator can only assess the result of a coder's capacity by viewing the decoder. Stated another way, when one compares interoperating codecs, what one sees on the near end is the result of the work of the codec at the far end.

Motion Compensation

As we just said, H.261 specifies, as an option, motion compensation. Motion compensation is performed *in addition to* intraframe transform (DCT) encoding, which is mandatory. While intraframe coding removes an image's spatial redundancy, interframe encoding works on temporal (frame-to-frame) redundancy. In all cases, H.261 encoding will start out with an intraframe-coded image that has been compressed using a combination of DCT, Huffman and run-length encoding. It must do this in order to have an original frame of video that can become the basis for the motion compensation process.

Block Structure of an Image

1	2	3	4	5	6	7	8	9	10	11
12	13	14	15	16	17	18	19	20	21	22
23	24	25	26	27	28	29	30	31	32	33

Group of blocks, consisting of 33 macro-blocks

Figure 7-4. Subdivision Of A Frame Of Video Into GOBS (Michel Bayard).

An encoder that compresses using motion compensation presumes that successive frames of video will contain similarities. Thus, it models current frames after previous ones. The codec, working at the macroblock layer, compares a previously encoded frame (a reference frame) to an actual one. As it does so, it looks for motion vectors (a directional movement of groups of pixels). Upon seeing movement, the coder tells the decoder about which pixels are moving and their line of direction. Both codecs then assume this motion will continue. The coder transmits only the prediction error—motion that did not correspond to what was predicted. When the prediction error is minimal, it is not transmitted at all.

When prediction error *is* transmitted, the decoder adds it to the previously coded image, and thereby creates a mathematical representation of the original block. The encoder stores this and uses it to predict the next image. Were this done indefinitely, it could result in cumulative error. For that reason, one out of every 32 blocks is encoded using intraframe techniques, and thereby provides a fresh image.

The requirement for an H.261 codec to decode motion vector information is mandatory, but an H.261-compliant codec does not have to send such information. Sending codecs that *do* use motion vectors often produce superior pictures at the receiving end, in comparison to sending codecs that do not. That is because they are generally three to four times more efficient (allowing better use of bandwidth).

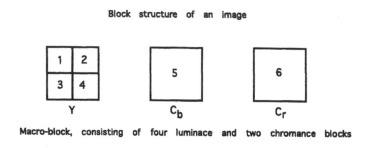

Block structure of an image

Macro-block, consisting of four luminace and two chromance blocks

Figure 7-5. Macroblocks (Michel Bayard).

Forward Error Encoding

The CCITT included forward error correction (FEC) as an option in H.261. Forward error correction can reduce the number of transmission errors in a video stream but it does so at the expense of an increase bit rate. The H.261 standard allows a decoder to ignore an encoder's FEC codes, which provide little value in instances in which H.261 applications traverse networks with very low bit error rates. FEC is rarely implemented in H.261 coders or decoders.

Video Multiplexing

Video multiplexing is integral to H.261. It allows a H.261 decoder to accurately scan and decode the flood of bits that represent video data. Multiplexing is performed as a step in the H.261 process by the video multiplex coder portion of the H.261 codec.

An H.261-compliant multiplex encoder must present its output bit stream in accordance with the requirements of a hypothetical reference decoder, described in Annex B of H.261. Various aspects of this hypothetical reference decoder (HRD) are set out in the standard: how it *clocks* to the encoder; what its buffer size is; how often the buffer is examined, what is stored in the buffer and precisely when stored data must be removed. The HRD's buffer size, by definition, depends on the bit rate of the connection. If the bit rate is relatively high, the buffer is relatively large.

Slower transmission speeds translates to fewer frames delivered in a given second, thus the buffer can be smaller. A HRD is designed to accept the maximum flood of bits permitted in the H.261 Recommendation. The "high water mark" occurs when a high-resolution (CIF-capable) codec is sending images over a very fast transmission link (30 B channels) at a rate of 29.97 per second. The HRD must buffer and decode this fast-moving bit stream without dropping frames. It must also be able to buffer reference frames used during motion compensation. All H.261-compliant decoders must conform to Annex B.

H.221 FRAME FORMAT
AT 384 Kbps

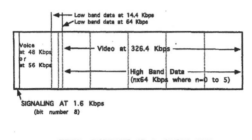

Figure 7-6. H.221 Frame Format at 384 Kbps (Michel Bayard).

Layered Block Encoding

An H.261 encoder multiplexes the bits it passes to a HRD by breaking a picture into four hierarchical layers. A picture (a frame of video) represents the highest layer. This picture has been subdivided into groups of blocks (GOB). Each GOB is sub-divided into macroblocks and each macroblock contains 8-by-eight pixel blocks. Sub-dividing an image into successively smaller blocks permits orderly coding and orderly transmission. Start codes define the two highest layers: a picture is indicated by a 20-bit start code; a GOB is declared by a 16-bit code. At the macroblock layer and below, codewords are used.

A macroblock uses a variable length codeword to indicate where it sits within a GOB. A pixel block uses a codeword to indicate which coefficient is to be used during reverse-transform. At the end of each pixel block there is also an end-of-block marker.

A frame of video is broken into GOBs, each of which contains a specific number of luminance and color-differenced pixels. In a GOB, luminance is conveyed using 176 vertical pixels and 48 horizontal pixels. The pixels that convey color difference (Cb and Cr) are spatially arranged with reference to luminance. There are more GOBs in a CIF-resolved image than in one that is QCIF-resolved. The higher-resolution CIF picture requires 12 GOBs; QCIF needs only three.

A GOB is subdivided in to 33 macroblocks. A macroblock represents luminance using 16-by-16 pixel blocks (four 8x8 blocks). The four luminance blocks are overlaid with two color-differenced (CR and CB) chrominance blocks, which cover the same area as the luminance blocks but at half their resolution.

At the bottom of the hierarchy is the block layer. A block is comprised of eight-by-eight pixels. These pixel blocks are encoded using the DCT and, possibly, using motion compensation (predictive encoding).

We have taken the reader step-by-step through the coding specification laid out in the ITU-T's H.261 specification. We conclude our discussion with this final thought: Prior to the December 1990 approval of H.261, customers displayed unwavering brand loyalty to specific codec manufacturers. In 1991, the first H.261 compliant codecs made their debut. By June 1992, all major North American, European and Japanese codec manufacturers supported the standard. The H.261 codec is, today, the fulcrum of video communication technology. Even at 112 Kbps or 128 Kbps speeds, H.261-based systems reliably interoperate. They yield acceptable resolution, audio, and motion handling with a degree of quality that is more than adequate for most business purposes.

THE REST OF THE FAMILY

The formal title of the H.320 Recommendation is "Narrow-band Visual Telephone Systems and Terminal Equipment." The H.261 Recommendation is only part of the original H.320 family, which was ratified by the CCITT in December 1990. Also included in the original specification are H.221, H.230 and H.242—and two audio encoding specifications, G.711 and G.722.

In November of 1992, the ITU-T added several new recommendations to the H.320 standard. One is H.231, which defines a multipoint control unit for linking three or more H.320-compliant codecs. A second is H.243, which describes control procedures between H.231 multipoint control units and H.320-compliant codecs.

In March 1993, the ITU-T approved G.728, which defines how H.320-compliant codecs will handle 16 Kbps toll-quality audio. G.711, G.722 and G.728 are discussed in Chapter 6.

H.221

"Frame Structure for a 64 to 1920 Kbps Channel in Audiovisual Teleservices" is the formal name for H.221. Its framing recommendation, perhaps the second best known of the ITU-T's H.320 family, defines the frame structure for a video bit stream across one or more 64 Kbps B channels. It acknowledges the ITU-T's HO (switched

384 Kbps) dialing and other recommendations such as H11 (switched 1.536 Mbps) and H12 (switched 1.920 Mbps is available only in Europe). H.221 also defines a protocol that allows a single B channel to be divided into sub-channels to carry video, voice, data, and control signals.

H.221 specifies synchronous operation; the H.221 coder and decoder handshake and agree upon timing. Synchronization is arranged for individual B channels or bonded connections when contiguous B channels are combined. After synchronization has been established, H.221 sends fixed-size frames of 80 bytes across the network.

Information about synchronization is carried over a service channel. H.221 acquires framing at the start of a session by sending some number of bytes in a pattern. The "service channel" is a bit pattern that is spread-out bit-by-bit over dozens of bytes. Within the service channel are two signals, FAS and BAS. FAS, *Frame Alignment Signal*, delineates H.221 frames.

BAS stands for Bit-rate Allocation Signal. BAS codes are used to indicate the capabilities of the audiovisual devices using the B channel's capacity. There are hundreds of BAS codes that define many different functions that are used to combine B channels into high-bandwidth services at H.0 and other rates. They allow a receiving device to inverse multiplex an incoming signal, and to understand the control and indication information coming across the channel. They define how a channel is to be shared between audio, video, and signaling, and indicate how audio is being encoded.

H.230

The formal name for this specification is "Frame-Synchronous Control and Indication Signals for Audiovisual Systems." H.230 Recommendation defines how frames of audio, video, user data, and signal information are multiplexed onto a digital channel. It defines control and indication information for those various data types and supplies a table of BAS escape codes that define the particular instances when control and indication information must be exchanged between sending and receiving devices. The H.230 standard primarily addresses control and indication signaling. H.230 distinguishes four categories of control and indication information: video, audio, codec and MCU interoperation, and maintenance.

H.231

The formal name for this specification is "Multipoint Control Unit for Audiovisual Systems using Digital Channels up to two Mbps." H.231 defines the network configuration for an H.320-compliant multipoint control unit (MCU). The MCU is used to bridge three or more H.320-compliant codecs together in a simultaneous conference. The procedures and functionality of a H.231 MCU are defined in a companion standard, H.243.

H.242

The formal name for this specification is "System for Establishing Communication Between Audiovisual Terminals Using Digital Channels up to two Mbps." Recommendation H.242 defines the protocol for establishing an audiovisual session, and subsequently disconnecting that session. During an H.242 call set-up, a terminal indicates its capabilities to other terminals.

H.242 defines signaling procedures for identifying a compatible mode, for initiating a conference using that mode and for switching between modes throughout a call. Besides that very important exchange, H.242 covers basic sequences for in-band information exchange. It details procedures for recovering from faults; it specifies procedures for adding channels to, removing channels from, and transferring calls into the session. This Recommendation also works with H.221 to permit a digital channel to accommodate audio, video, and data across a range of data rates. Signaling procedures allow codecs to identify how different portions of the channel are allocated.

H.243

The formal name for this specification is "System for Establishing Communication Between Three or More Audiovisual Terminals Using Digital Channels up to two Mbps." Recommendation H.243 is a companion standard to H.231. It describes procedures and functionality for multipoint video communication. Included in the features are chair control/directorship, determining which site will broadcast at any given time, muting video and audio from specific terminals as desired, etc.

THE H.32X FAMILY

The ITU-T has developed, and is continuing to develop, extensions to H.320. H.320 now belongs to a family of standards known as H.32X. In this family, along with H.320, are H.310, H.321, H.322, H.323 and H.324.

The H.32X family is truly a family. Each group of specifications recognizes the other members and each includes similar functionality, albeit with differences, which are largely due to the unique characteristics of the networks, they were designed to operate over.

H.310

H.310 targets HDTV-class video quality (e.g., CCIR-601 and above) using the International Standards Organization (ISO) Motion Picture Experts Group's MPEG-2 video compression algorithm (we address MPEG 2 in more detail in Appendix B).

The H.310 specification employs the MPEG-2 coding system to convert an analog video signal for transmission over a cell-relay-based digital network, specifically, a network based *Asynchronous Transfer Mode* (ATM). IBM did much of the preliminary work on the H.310 standard, and worked closely with the ITU-T to refine it to the point of approval.

H.310 is made up of three sub-parts. One is H.262 ("Information Technology—Generic Coding of Moving Pictures and Associated Audio Information") better known as MPEG-2. Another is H.222.0 ("Information Technology—Generic Coding of Moving Pictures and Associated Audio Information: Systems"). The MPEG-2 standard includes a system layer. The system layer allows MPEG-2 to keep data types synchronized and multiplexed in a common serial bit stream. H.220 adapts H.310 to work within the system layer confines. The third sub-part is H.222.1, "Multimedia Multiplex and Synchronization for Audiovisual Communication in ATM Environments." The MPEG-2 system layer specifies a format for transmitting audiovisual information over a data communications network. The format, known as a transport stream, relies on clock reference information to maintain synchronization between related audio and video streams. H.222.1 describes how MPEG-2-encoded transport streams flow over ATM.

H.310 allows various G.7XX audio-quality standards—there are no constraints because ATM has plenty of bandwidth to accommodate any audio coding algorithm. The H.310 standard also references Rec. H.245. H.245 is a multimedia control protocol that permits H.310-aware audiovisual equipment to establish and maintain broadband audiovisual sessions.

H.321

The H.321 Study Group began developing standards for private ATM networks in 1996. H.321 is a companion standard to H.310. It preserves fundamental components of the original H.320 Recommendation including H.261 and, therefore, maintains a high degree of backward compatibility with it. H.321 functions extend the type of transmission networks that carry H.32X videoconferencing signals to include ATM and broadband ISDN.

As is true of H.310, this specification was still in ITU-T Draft status when this book went to print. For more information, contact the sources listed for H.310.

H.323

The formal name of H.323 is "Visual Telephone Terminals over Non-Guaranteed Quality of Service LANs." It was approved by the ITU-T in October 1996. H.323 defines four components: terminal, a gatekeeper, a gateway and a multipoint controller.

H.323 Components

Terminals include devices such as PCs, workstations and videophones. All H.323-compliant terminals must support voice-data conferencing. Voice-data-video conferencing is optional. Multiple channels of each type are permitted.

A gatekeeper aids in call management. The gatekeeper is a utility that controls videoconference access on a packet-switched LAN. It requires that multimedia terminals register "at the gate," which is accomplished when the terminal provides its IP address. The gatekeeper translates network addresses and aliases to make connections. It can also deny access or limit the number of simultaneous connections to prevent congestion.

A H.323 gateway is different from a gatekeeper. The gateway allows LAN-based H.323 systems to interoperate with other H.32X products. For instance, the gateway could link the H.323 session with an H.320 (ISDN-based) system, an H.321 (ATM-based) system, or a H.324 (POTS-based) system. At the present, most H.323 gateway implementations are concerned with linking H.323 and H.320 systems across a LAN/WAN connection.

A multipoint controller is built into H.323 implementations (typically on the client-side) in which it works to establish multipoint conferences between disparate endpoints. It is the functional equivalent of the H.320's H.231 and H.243 specifications.

H.323 and Internet Protocols

In contrast to H.322, which deals with LANs that have been modified to provide circuit-switched service, H.323 faces the difficult challenge of providing adequate service in packet-switched environments in which latency characteristics are difficult to manage. It copes by borrowing several different protocols from the Internet Engineering Task Force (IETF). These include the Real-Time Transport Protocol (RTP), User Datagram Protocol (UDP), the Transmission Control Protocol (TCP) and the Internet Protocol (IP).

Although H.323 is not aimed exclusively at them, implementations generally traverse IP networks. On IP networks, H.323 audio and video use UDP. When compared to the Transmission Control Protocol (TCP), UDP lacks sophistication. It is a unidirectional protocol that simply divides data into IP packets that it sends with checksums. UDP is better suited to real-time data types than is TCP. TCP strives for reliable delivery and thereby retransmits if it detects corrupted packets or dos not receive acknowledgement of packets. UDP does not retransmit. It discards packets that are corrupted; otherwise, it simply forwards them.

To bolster UDP's lack of reliability, H.323 uses RTP (RTP is also used on the Internet's Multicast Backbone—the MBONE). RTP appends a 10-byte header (that contains a time stamp and a sequence number) to each UDP packet. At the receiving end, H.323 implementations buffer and process UDP packets to capture timing and sequence information and to guard against duplicate and out-of-order packets. Thus, audio, video, and data are generally resynchronized and presented to the user in a coherent stream.

While audio and video use the UDP, H.323 does specify TCP for data conferencing, which is not time sensitive. H.323 also uses TCP for conference control and signaling.

Another IETF protocol, the Resource Reservation Protocol (RSVP), is not an official part of H.323 but products without it may suffer from network congestion problems. The RSVP allows a conference participant to reserve bandwidth for a particular real-time data stream. The reservation is made in the network (in which RSVP-compliant routers receive and grant requests). Although users can ask for bandwidth, there is no guarantee (at the time of this writing) they will get it.

H.323-Compatible Products

Since only the end-points of multimedia applications are impacted by H.323, there is no need to make changes to network interface cards, hubs, routers, or switches—except, perhaps, to upgrade routers and/or switches to make them RSVP compliant. Newer Cisco Systems and Bay Networks products support it, but a bottleneck may occur at any point along the data path at which a device does not.

H.323-compatible products began appearing on the market late in 1996. By the first quarter of 1997, there were many more implementations. In October 1998, the ITU-T ratified H.323 version two (H.323v2). The new version of the standard advanced integration with T.120, resolved limitations in the original version, and specified enhancements to the H.225, H.245, and Q.931 protocols. New features included:

GSM Compression Arbitration—Additional audio capabilities via H.245.

V.Chat Arbitration—V.Chat, an instant messaging protocol, exchange.

Fast Connect or Fast Start—Decreases call setup time by reducing the number of round trip times necessary to establish a session.

Overlapped Sending—Allows connection to be established more quickly by enabling process routing to occur while the calling party is entering the address.

Security—H.235 Specifies enables Authentication, Integrity, Confidentiality, and Non-Repudiation.

Supplementary Services—H.450.1-3 defines the signaling protocol between endpoints for supplementary services, Call Transfer, and Call Diversion.

H.323v2 enhancements include:

Additional Alias Address Types—URLs, Email addresses, and public and private party numbers.

Call Identifier—Creates a unique global identifier to allow all messages associated with a call to be transferable between the RAS protocol in IP networks, and the Q.931 signaling protocol used in circuit-switched telephony networks.

Conference List—Allows a MCU to specify a calling endpoint by sending a Q.931 Facility Message, which contains a list of conferences, to that endpoint.

Conference Requests—Allows for the transmission of new control-oriented H.245 conference request packets.

Dynamic Channel Replacement—Provides a streamlined method of changing modes between codecs that does not require two media decoders.

Empty Capability Set—Enables gatekeeper to re-route connections to an endpoint that supports Supplementary Services (from one that does not).

Endpoint Redundancy—Allows an endpoint to specify an alternate network interface or H.323-defined interface.

Endpoint Type Prefixes and Data Rules—Enables gateways to specify data rates for each supported protocol, and to define prefixes for protocols.

Gatekeeper Redundancy—Enables gatekeepers to specify backup gatekeepers if the primary contact fails.

H.263—Improves video capabilities and provides new structures to improve the H.263 video stream.

Keep Alive—Enables a gatekeeper to verify the status of a user and to act appropriately by allowing an endpoint to register with a gatekeeper and to inform the gatekeeper of all aliases.

Pre-Granted Admission—Decreases call processing time by enabling endpoint pre-verification.

QoS (Quality of Service)—Allows endpoints to set QoS parameters (e.g., RSVP) for media streams.

RAS Retry Timers and Counters—Enhances RAS transaction processing by ensuring that messages arrive within a specified period, that they are re-sent when needed, and that end-users are notified or delivery failure.

Reliable Information Request Response (IRR)—Introduces two new messages, IACK and INAK, to indicate whether messages were delivered.

Request in Progress (RIP) Message—Allows the receiver of a request to specify, when a message cannot be processed before the initial timeout value expires, that a new timeout value should be applied.

Resource Availability—Improves gatekeeper performance by enabling a gateway to notify it of the gateway's present call capacity.

Time to Live—The duration for which a gatekeeper should keep a registration active.

T.120/H.323 Integration—Requirement that endpoints that support both T.120 and H.323 must begin the call with H.323 to standardize and simplify the process.

Tunneling—Enables Q.931 channel to carry H.245 Protocol Data Units (PDUs).

User Input Indication (PDU)—Enables more inclusive DTMF signaling by providing new structures, such as length of tones.

H.324

The formal name of H.324 is "Terminal for Low Bit Rate Multimedia Communication." It was approved in May 1996 as the first standard to provide point-to-point video and audio compression over analog telephone lines (POTS). A shortcoming of H.320 is its requirement that customers use similar devices at all endpoints. H.324 allows users to interoperate across diverse endpoints (IP, ATM, ISDN, POTS, or wireless telephony). In July 1996, sixteen vendors participated in the first formal H.324 interoperability test. More than a dozen participated in subsequent testing.

Whereas H.320 is designed for use with multipoint control units (MCU), H.324 allows users to choose what party they want to see on their monitor through a feature called continuous presence. Continuous Presence allows images to be stacked; it is useful in videoconferences in which one camera cannot adequately capture an entire group at one end. H.320 suffered from incompatibilities between systems that were only technically H.320 compliant. In contrast, H.324 allows users to share video, voice, and data simultaneously over high-speed (28.8 Kbps or faster) modem connections using the application interface, V.80. Within the V.80 standard, *voice call first* allows users with V.8 bis enabled modems to add video at any time during an audio only call. V.80 also allows the transmission of still photographs when a signal is inadequate to transmit motion video.

H.324 incorporates both DCT and pulse code technology to provide enhanced video performance at low bit rate transmission. The ITU-T updated the H.261 standard to H.263; H.263 is H.324's video algorithm.

H.223

The formal name for H.223 is "Multiplexing Protocol for Low Bit Rate Multimedia Communication." Within H.324, H.223 defines the multiplexing of the individual audio and video streams into a single synchronized data stream that suits the bandwidth constraints of POTS. To transmit digital data on analog telephone lines, synchronous modems transmit a High-level Data Link Control (HDLC) protocol that adds a framing protocol layer. Using a standard asynchronous interface, videoconferencing would add another framing layer. Because the H.263 codec performs framing, any additional framing the modem performs is redundant. The V.80 interface mediates by letting the host define the frame boundaries for the multiplexed video and audio signal, but allowing the modem to complete the bit-oriented framing.

H.263

The formal name for H.263 is "Video Coding for Low Bit Rate Communication." H.263 is one of two alternatives for video compression under H.324. The other is H.261, which we have already covered.

Although early H.263 drafts specified data rates of less than 64 Kbps, the final

Recommendation included no such limitation. The coding algorithm of H.263 adds many performance and error recovery improvements to H.261. For instance, it incorporates half pixel precision for motion compensation; H.261 is limited to full pixel precision. H.263 codecs do not require a loop filter, which is required by H.261 codecs.

Like H.261, H.263 makes QCIF picture formatting mandatory. It provides support (as options) for additional frame sizes. One is the CIF, also used in H.261 codecs. Another is Sub Quarter Common Intermediate Format (SQCIF), which is used for very low-resolution images. H.263 compliant codecs compete with higher bit rate video coding standards (such as MPEG) because of the standard's support for two other formats, 4CIF and 16CIF. 4CIF is four times the resolution of CIF, and 16CIF is 16 times the resolution of CIF.

H.324 implementations will most often use SQCIF because it is helpful for low bandwidth video communication. SQCIF resolves an image at 128 horizontal pixels by 96 vertical pixels. Although this resolution is too low for use in many applications, it suffices for communicating over the PSTN with V.90 modems. SQCIF seems to be most useful for providing a brief sample of a more extensive video segment such as a movie or so-called *infomercial*.

H.324 also allows customers to dynamically choose between two low bit-rate audio codecs, Multi-Pulse Maximum Likelihood Quantization (MP-MLQ) or Algebraic Codebook Excited Linear Prediction (ACELP) and, therefore, to choose audio quality. The two standards feature 6.3 Kbps or 5.3 Kbps audio, respectively. The standard leverages T-84 for still-frame graphics.

Finally, H.263 codecs can be configured for a lower data rate or for better error recovery because parts of the data stream's hierarchical structure are optional. The H.263 standard incorporates four negotiable options to improve performance. One is P-B frames, which, much like MPEG frames, provide forward and backward frame prediction. The other three options are Unrestricted Motion Vectors, Syntax-based Arithmetic Coding, and Advance Prediction.

P-B framing allows two images to be coded as a single unit. The term "P-B" is derived from MPEG (see Appendix B). A P-B frame consists of one P (predictive) frame, which is created by using a past frame as a model and one B (bi-directional) frame that is interpolated (created through comparison) from a past P frame and the P-frame currently being computed. Computationally demanding, P-B framing can really compress a video stream down to a low bit rate. The technique works best for simple sequences; it starts to show stress when there is significant frame-to-frame motion.

When an H.263 codec is in the Unrestricted Motion Vectors mode, it can project motion outside the frame and use those vectors (associated with non-existent pixels) to predict movement along the edge of an image (or sequence of images). This mode also allows an extension of the H.261 motion vector range, and allows larger motion vectors to be considered. This is quite useful in the case of rapid camera movement.

Syntax-based Arithmetic Coding can be substituted in H.263 for Variable Length Coding. This allows the compression process that takes place after Huffman encoding to be based on unique aspects of the motion sequence, (rather than relying on a single "codebook").

Advance Prediction is an optional mode with which a codec can use overlapping block motion compensation for creation of P-frames. Four eight-by-eight vectors can be substituted for a 16-by-16 macroblock. The encoder must determine which type of vectors to use. Four vectors generate more bits but give much better prediction. Advance Prediction coding also results in fewer blocking artifacts.

H.264

The formal name for H.264 is "Advanced video coding for generic audiovisual services," and it is also known as "MPEG-4/AVC", "MPEG-4 Part 10," and "H.264L." It is an ITU standard for compressing a videoconferencing transmission. Based on MPEG-4, H.264 uses Advanced Video Coding (AVC) for as much as a 70% reduction in bandwidth over H.263.

Although H.264 created tension between MPEG4 and H.263 advocates, it is a triumph of the ITU-T Video Coding Experts Group (VCEG) and the ISO/IEC Moving Picture Experts Group (MPEG), which collaborated as the Joint Video Team (JVT). H.264, a high compression digital video codec, is identical to ISO MPEG-4 Part 10, and is also known as AVC, for Advanced Video Coding. The group completed the final draft of the first version of the standard in May 2003.

The H.264 name reflects the standard's ITU-T lineage, and the AVC moniker reflects its ISO/IEC MPEG heritage. Multiple names might lead one to infer that such a collaboration is uncommon, but MPEG and the ITU-T previously worked together to develop the MPEG-2 video standard, which is also known as H.262.

The H.264/AVC project goal was to create an easily-implementable standard that could provide good video quality at half or less the bit rates that MPEG-2, H.263, or MPEG-4 part 2 require. The authors intended the standard to facilitate both low and high bit rates and low and high-resolution video not only in ITU-T multimedia telephony systems, but also in RTP/IP packet networks, DVD storage, and broadcast applications. In September 2004, extensions to the original standard, known as the *Fidelity Range Extensions* (FRExt), were completed to support higher-fidelity video coding via higher-resolution color information (including YUV 4:2:2 and YUV 4:4:4 sampling structures) and increased sample accuracy (including 10-bit and 12-bit coding). Included in the Fidelity Range Extensions project are efficient inter-picture lossless coding, support of additional color spaces, residual color transform, adaptive switching between 4x4 and 8x8 integer transforms, and encoder-specified perceptual-based quantization weighting matrices.

H.264/AVC compresses video more efficiently than older codecs and flexibility for a wide variety of network environment and applications. In scenes with back-and-forth scene cuts, uncovered background areas, or rapid repetitive flashing, H.264 significantly reduces bit rate by providing *multi-picture motion compensation*. While prior standards offered referencing to no more than two previously-encoded pictures, H.264 allows referencing to as many as 32.

For motion compensation, H.264 provides *quarter-pixel precision*, which allows for accurate description of the displacements of moving areas (as accurate as 1/8 pixel for chroma). H.264 also allows for more precise segmentation of moving regions, with block sizes as small as 4x4 or as large as 16x16, through *variable block-size motion compensation* (VBSMC). The ringing and blocking artifacts that dog other DCT-based image compression techniques are mitigated in H.264 with an in-loop deblocking filter.

Weighted prediction allows an encoder to specify the use of a scaling and offset when performing motion compensation and enhances performance in such instances as fade-to-black, fade-in, and cross-fade transitions. A new exact-match integer 4x4 spatial block transform enables the H.264 encoder to select adaptively between a 4x4 and 8x8 transform block size for the integer transform operation.

H.264 is a worthy successor to H.263 with an exhaustive catalog of improvements. Other enhancements include intra-frame coding from the edges of neighboring blocks for spatial prediction, context-adaptive binary arithmetic coding (CABAC) for lossless compression for syntax elements in the video stream through, context-adaptive variable-length coding (CAVLC) for the coding of quantized transform coefficient values, *Exponential-Golomb* (Exp-Golomb) code as the variable length coding (VLC) technique for many syntax elements not coded by CABAC or CAVLC, a secondary Hadamard transform on DC coefficients of the primary spatial transform to increase compression in smooth regions.

H.264 enables an encoder to direct a decoder to enter into an ongoing video stream for video streaming bit rate switching via switching slices (SP or SI slices) with an exact match to the decoded pictures in the video stream at that location, even in the absence of reference pictures. H.264 includes additions such as picture parameter sets (PPS) and sequence parameter sets (SPS) to include a network abstraction layer (NAL) definition that provides for video syntax sharing in various network environments. One more feature of H.264 is picture order count, which isolates picture ordering and decoded picture value samples from timing information so that timing information can be carried and manipulated by a separate system without affecting decoded picture content.

H.264 already enjoys wide adoption in videoconferencing, and is likely to find its way into a broad array of products, including broadcast TV in the US. MPEG has integrated H.264 support MPEG-2 and MPEG-4 systems and the ISO media file format specification. The Republic of Korea and Japan have already announced support, as has the Motion Imagery Standards Board (MISB) of the United States

Department of Defense (DoD). The Internet Engineering Task Force (IETF) is defining a payload packetization format for carrying H.264/AVC video with RTP. Europe's Digital Video Broadcast (DVB) standards body approved H.264/AVC for broadcast television in Europe, and it became a requirement for HDTV and pay TV channels for digital terrestrial broadcast television services receivers in France in 2004. H.264 is also included as a mandatory player feature in the DVD Forum's HD-DVD format, which is slated for product deployment by 2006. It is also being built into QuickTime and PlayStation Portable.

H.235

The formal name of H.235 is "Security and encryption for H-Series (H.323 and other H.245 based) multimedia terminals." One improvement that is defined in Revision Two of Recommendation H.323, H.235 was developed to provide a security framework for H.323 and other multimedia systems. It addresses authentication and confidentiality. H.235 can leverage IETF security protocols such as IPSec (Internet Protocol Security) or TLS (Transport Layer Security).

Formerly H.SECURE, H.235 was developed to address the need for organizations to trust the integrity and confidentiality of H.323 calls as they have trusted circuit-switched voice calls. Securing H.323 involves not only protecting the audio or video stream, but also Q.931 (call setup), H.245 (call management), and Gatekeeper Registration/Admission/Status.

H.235 addresses the three fundamental goals of information security, confidentiality, integrity, and authentication. Authentication is the process of identifying and verifying a potential caller/callee, or a particular machine. Authentication is critical in H.323 security since, without it, anyone can represent himself or herself as anyone else and can thereby enter a conference. Moreover, the execution of all other H.235 security options is contingent upon the authentication of nodes in an H.323 call.

To ensure integrity, H.235 validates a packet's data payload to ensure that the payload has not been corrupted or altered while in transit between endpoints during an H.323 call. The process employed in H.235 is similar to that used in other IP communications protocols. However, rather than using checksums and CRCs (cyclic redundancy checks), H.235 data integrity uses an encryption mechanism (a hash) to protect the packet's integrity value. In this approach, the integrity checksum value must be encrypted, but not the actual data payload. This reduces the necessary amount of encryption processing.

H.235's encryption techniques provide a mechanism to conceal data from eavesdroppers. Without the appropriate decryption algorithm and the key to unlock the encrypted packet, an encrypted packet cannot be viewed even if it is intercepted. However, encryption requires processor-intensive computation and thereby adds delay. Therefore, one must balance the risk that the H.323 call may be compromised against the need to emulate circuit-switching quality in the session. Of course, there is no practical way to make a session completely secure, so one must refer to security policies to determine an organization's risk tolerance before deploying H.323.

Non-repudiation is the prevention of any party from falsely denying that they have made a commitment or, in the case of H.235, denial that they participated in an H.323 session. The most conventional example of non-repudiation (repudiation prevention) is the requirement of a signature upon delivery of a package or acceptance of a credit card. This is important for service providers who must accurately charge for their services and validate those charges, but it is also important for any organization that desires to verify who participated in any given H.323 session.

There are two viable RAS authentication techniques: Symmetric Encryption-Based Authentication and Subscription-Based Authentication. Gatekeepers typically authenticate the client. However, clients can also request the Gatekeeper to authenticate itself. Symmetric Encryption-based Authentication is a one-way form of authentication that is used only by Gatekeepers. It requires no prior contact between endpoints. The Gatekeeper uses this form of authentication to validate subsequent Gatekeeper directed requests and to verify that the requests are from a registered node. Symmetric Encryption-based Authentication does not provide a mechanism for the endpoints to validate the Gatekeeper's identity.

Subscription-based Authentication requires prior contact or information exchange between the endpoints—usually through passwords or signed digital certificates. Each endpoint identifies itself with a well-known text string identifier. Subscription-based Authentication allows both endpoints to be authenticated and is, therefore, bi-directional.

The Q.931 protocol uses a reliable IP transport-based mechanism to establish the first leg in a point-to-point session between two endpoints. The H.245 protocol is responsible for capability exchange communication and mode switching. H.235 integrity protection can enhance security by preventing tampering with the Q.931 and H.245 packets, by enabling complete encryption of the Q.931 streams, and by facilitating user authentication to verify end-points during the transmission of the call setup packets.

H.450

H.450 comprises three recommendations for implementation of *supplementary services* to H.323. The formal names of those standards are, "H.450.1 - Generic Functional Protocol For The Support Of Supplementary Services In H.323," "H.450.2 - Call Transfer Supplementary Service For H.323," and "H.450.3 - Call Diversion Supplementary Service For H.323." All were approved in February 1998.

These supplementary services are designed to enable IP networks to provide the telephony features that users have come to expect in conventional circuit-switched networks and, conversely, to enable conventional circuit-switched networks to provide non-conventional services that are offered on IP networks. H.450 enables hybrid IP/circuit-switched networks by enabling interoperability of IP networks with conventional (legacy) telecommunications equipment. Users can establish native H.323 data, voice, fax, or video sessions on a private network without a H.450 gateway as long as both endpoints support the H.450 protocols and the users subscribe to the services. However, the availability of such services in on calls between networks requires such capabilities across the entire signaling path of the call. Moreover, service providers may adopt business models in which they not only charge per call duration, but also per service. Users can attach legacy telephone and fax devices to an H.323 network with terminal adapters that provide similar functionality to the H.323 gateway. Such a device implements the H.450 supplementary services, and communicates with the legacy device with DTMF tones over the analog interface.

The H.450 protocols are simplifying the migration from proprietary PBX and service provider systems to server-based open standards based systems. Some of the basic services are as follows:

Multiple Call Handling—Enables a multimedia client to handle multiple calls concurrently.

N-way Conferencing—Enables establishment of multi-party conferences.

Call Transfer (H.450.2)—Enables manual forwarding of a call to a third party.

Call Forwarding (H.450.3)—Includes Call Forwarding Unconditional, Call Forwarding Busy, Call Forwarding No Reply, and Call Deflection.

Call Hold (H.450.4)—Enables a user to transfer another user to a pending status and to resume the call later. Allows for music or messages on hold.

Call Waiting (H.450.6)—Informs user, who is already on a call (or on multiple calls), that another call is coming in.

Message Waiting Indication (H.450.7)—Enables a user to choose from a variety of methods (e.g., email, voicemail, or page) to be informed that he has received a message.

Name Identification (H.450.8)—Provides the called party with the name of the calling party.

Call Completion on Busy Subscriber and on No Reply (H.450.9)—When a called party is engaged in another session, enables a caller to monitor that called party's endpoint to place the call when the called party becomes free.

Audiographic Conferencing Standards

The H.32X family relies on a series of audiographic standards for data sharing during a videoconference. Chapter 8 covers the ITU's T.120 family of Recommendations that provide audio graphic interoperability during H.32X videoconferences.

SESSION INITIATION PROTOCOL

H.323 does not efficiently accommodate Internet conferencing issues such as wide-area addressing and loop cancellation because it was primarily developed for use over a LAN. Even H.323 Version 2, that supports *zones*, does not scale adequately to the size of an IP-based public telephone network. Consequently, the Multiparty *MUltimedia* SessIon Control (MMUSIC) Working Group of the IETF has developed protocols to support Internet teleconferencing sessions. MMUSIC developed the Session Initiation Protocol (SIP) specifically as a teleconferencing protocol for the Internet, and Session Delivery Protocol (SDP), which SIP uses to convey which codecs an endpoint will support for a session. Concurrently, it developed the Media Gateway Control Protocol (MGCP) as a gateway-to-gateway protocol for use by carriers. Although MMUSIC designed them to support loosely-controlled conferences over the Internet, these three protocols are general enough for managing tightly-controlled sessions.

Like H.323, SIP provides mechanisms for call establishment and teardown, call control and supplementary services, QoS, and capability exchange. H.323 is less prone to interoperability issues because it more meticulously defines supplementary services, and is mandated to be backward compatible and to interoperate with the PSTN. However, SIP provides more flexibility for adding new features and easier implementation and debugging. Since both SIP and H.323 use RTP to carry media flows, the main difference between the two is their methods of handling call signaling and control.

SIP gained early support from Level 3 Communications, MCI WorldCom, and the cable TV industry for carrying IP services and interacting with the PSTN. Comcast, Cox Communications, MediaOne, Rogers Cablesystems Limited, Tele-Communications Inc. (now AT&T), Time Warner, and Le Groupe Vidéotron began sponsoring SIP/SDP interoperability efforts in 1999 in conjunction with CableLabs. CableLabs is a research and development consortium of cable television system operators that represents nearly 90% of the cable subscribers in the United States, 80% of cable subscribers in Canada, and 15% of cable subscribers in Mexico.

The family is comprised of the following protocols:

Session Initiation Protocol (SIP)—For initiating sessions and inviting users.

Session Description Protocol (SDP) and **Session Announcement Protocol** (SAP)—For distributing session descriptions.

SAP Security—For providing security for session announcements.

Real-Time Stream Protocol (RTSP)—For controlling on-demand delivery of real-time data.

Simple Conference Control Protocol (SCCP)—For managing tightly-controlled sessions.

Session Initiation Protocol (SIP)

The Session Initiation Protocol (SIP) is defined in IETF RFP 2543 and was ratified in February 1999. SIP is an application-layer control protocol for creating, modifying, and terminating sessions with one or more participants. It is targeted at Internet multimedia conferences, Internet telephone calls, and multimedia broadcasts.

H.323 was designed to extend H.320 to packet-switched local area networks. It depends upon H.245 for control, H.225 for connection establishment, and H.332 for large conferences. Furthermore, it leverages H.450.1–3 for supplementary services, H.235 for security, and H.245 for interoperability with circuit-switched services. The interrelations between these protocols can be quite complex. Call Forward, for instance, relies on components of H.450, H.225.0, and H.245. Consequently, H.323v1 call establishment may require as many as twelve packets and about 6–7 round-trip times (RTT). In contrast, SIP reuses many HTTP header fields, encoding rules, error codes, and authentication mechanisms to simplify multimedia conferencing over the Internet. Defined in 736 pages, H.323 defines hundreds of elements; defined in only 128 pages, SIP contains 37 headers. One can implement a SIP call with only the headers *To, From, Call-ID*, and *CSeq*, and the request types *INVITE, ACK*, and *BYE*. While H.323 represents its messages in binary, SIP (like HTTP) encodes messages as text. Moreover, SIP allows for HTTP code reuse so that existing HTTP parsers can be quickly modified for its use.

H.323 was not designed to perform loop detection in multi-domain searches. It requires gatekeepers to maintain information about a session (referred to as *state*) for that session's entire duration. For service providers, this is impractical. SIP leverages a loop detection algorithm that is similar to that in BGP that allows a gatekeeper to forget about an established session by using UDP. SIP also allows for *stateful* sessions over TCP. Therefore, SIP scales much more efficiently. In addition, unlike H.323, SIP does not require a central control point that must remain in the session to provide functionality that the session requires.

Members in a SIP session can communicate via multicast, unicast, or a combination of the two. SIP can be used to initiate sessions, or to invite members (or to add media) to sessions that have been advertised and established by multicast protocols such as SAP, Email, news groups, web pages, or directories (e.g., LDAP). SIP does not require a session initiator to participate in any session that she (or it) initiates. SIP transparently supports name mapping and redirection services, and thereby allows the implementation of ISDN and Intelligent Network telephony subscriber services. SIP is an enabling technology for *personal mobility*, or the ability of end users to originate and receive calls and access subscribed telecommunication services on any terminal in any location, and the ability of the network to identify end users as they move. Called parties can redirect callers to any number of locations that are represented by URLs. H.323 can do this to a limited degree, but it does not allow a gatekeeper to proxy a request to multiple servers.

SIP supports establishment and termination of multimedia communications as follows:

User location—Determination of the end system to be used for communication

User capabilities—Determination of the media and media parameters to be used

User availability—Determination of the willingness of the called party to respond

Call setup—*Ringing* or establishing call parameters for called and calling parties

Call handling—Such as transfer and termination of calls

SIP is part of the IETF multimedia data and control architecture. As such, it incorporates RSVP (RFC 2205) for reserving network resources, RTP (RFC 1889) for transporting real-time data and providing QoS feedback, RTSP (RFC 2326) for controlling delivery of streaming media, SAP for advertising multimedia sessions via multicast, and SDP (RFC 2327) for describing multimedia sessions. However, SIP's functionality and operation do not specifically depend upon these protocols, and can be used in conjunction with other call setup and signaling protocols. An end system can use SIP exchanges as a protocol-independent method to determine, from a given address, the appropriate end system address and protocol. SIP can be used, for example, to determine that the callee is reachable via the PSTN, to indicate the phone number to be called, and even to suggest a viable Internet-to-PSTN gateway. Similarly, SIP could be used to determine that a party can be reached via H.323, to obtain the H.245 gateway and user address, and then to use H.225 to establish the call. Although SIP does not offer conference control services (e.g., floor control or voting) and does not prescribe how to manage a conference, it can be used to introduce conference control protocols. SIP does not allocate multicast addresses. SIP does not reserve resources, but can convey to the invited system what information is necessary to do so. It can invite users to sessions with or without resource reservation.

SIP relies upon SDP for notifying an endpoint of which codecs another endpoint supports. Any one of a myriad codecs can be identified with a string name and registered with Internet Corporation for Assigned Names and Numbers (ICANN is the successor to IANA, the Internet Assigned Names Authority). As such, SIP allows for interoperability with any registered (either standards-based or proprietary) codec.

Although H.323 has a significant head start, it may favor to SIP because the latter is specifically designed for the application that is driving the teleconferencing market, Internet telephony. Anyone who is knowledgeable in HTTP will recognize much of SIP, and will find it intuitive. It seems likely that manufacturers will provide SIP as another option in their multiprotocol telephony gateway. If that is the case, the market will get to decide.

CHAPTER 7 REFERENCES

"Advanced Video Coding For Generic Audiovisual Services (H.264)." *International Telecommunications Union (ITU-T)*. May 2004. www.itu.int/

Yoshida, Junko. "H.264 Codec Jeopardizes MPEG-4's Ascendancy." *EE Times*. September 20, 2002. http://www.eetimes.com/story/OEG20020920S0049

Pechey, Bill. "Codec Frees Video Telephony." *IT Week*. April 4, 2003http://www.vnunet.com/analysis/1139947

Dalgic, Ismail and Fang, Hanlin. "Comparison of H.323 and SIP for IP Telephony Signaling," *Photonics East*, Boston, Massachusetts, September 20-22, 1999.

Schulzrinne, Henning, and Rosenberg, Jonathan. "A Comparison of SIP and H.323 for Internet Telephony," *Columbia University Technical Report CUCS-005-98*, January 1998.

Handley, M., Schulzrinne, H., Schooler, E., Rosenberg, J. "Network Working Group RFC: 2543 ACIRI Category: Standards Track SIP: Session Initiation Protocol," *IETF* March 1999.

Korpi, Markku. "Exploring H.450 Needs and Usability," *Telecon West* November 1999.

Brown, Eric. "ADSL Jumps Into the Race," *New Media* October 7 1996 volume 6, number 13.

Grunin, Lori. "Image Compression For PC Graphics: Something Lossed. Something Gained," *PC Magazine* April 28. 1992.

Johnson, Colin. "Tool Compresses Images; Fractal Aid Web Pages," *Electronic Engineering Times* March 25, 1996 Issue 894.

Morris, Tom. "Video standards compared," *UNIX Review* March 1996 v14 n3 pp. 49-55.

Murray, James D. "SPIFF: Still Picture Interchange File Format: JPEG's official file format," *Dr. Dobb's Journal* July 1996 v21 n7 p. 34.

Nolle, Thomas. "Reservations about RSVP," *Network World* October 28, 1996.

Ohr, Stephan. "ITU Effort Eyes Mobile Video Phone," *Electronic Engineering Times*. October 28, 1996 n925 pp. 1-2.

8

STANDARDS FOR AUDIOGRAPHIC CONFERENCING

"Et loquor et scribo, magis est quod fulmine
iungo (And I speak and I write, but more, it's
with light that I connect)."

Giovanni Pascoli, 1911

In the previous chapter, we discussed the ITU-T standards that are oriented toward face-to-face video communication. In this chapter, we will discuss another aspect of conferencing—audiographics conferencing. Originally, the term audiographics referred to facsimile, slow scan television, and 35mm slides. Today, it refers to applications that blend voice communications (point-to-point or multiparty) with PC-oriented graphics, data, and document sharing. The voice portion of the exchange takes place over ordinary telephone lines; the data portion occurs over LANs, WANs, ISDN, the Internet or POTS. The result is a visually enhanced teleconference in which high-resolution images can be created, exchanged, shared and discussed.

When conference participants can conveniently access and share PC-resident documents, group collaboration is a more rich experience. Audiographics conferencing (better known as *document conferencing* or *data conferencing*) provides the ability for two or more users to *gather around* a virtual conference table to observe and collaborate on documents. Regardless of geographic location or computer operating system, conference participants deploy an array of software- and hardware-based tools and utilities to simulate co-location. The emphasis of audiographic applications is on computer documents and data. For that reason, we will use the term data conferencing (it is also slightly more concise than the term *audiographics conferencing* that the ITU-T uses in the specification for applications, services, and protocols).

Data conferencing eases the burden of preparing for a teleconference. Teleconferences are audio-oriented and are still quite common, notwithstanding the widespread acceptance of video communication. Linking distant teleconference

The International Multimedia Teleconferencing Consortium (IMTC) is a San Ramon, California-based non-profit organization dedicated to the promotion of ongoing development and adoption of international standards for multipoint document conferencing (specifically the ITU-T's T.120) and videoconferencing (the H.32X series). Neil Starkey of DataBeam is the long-standing president of the IMTC, and has been instrumental in its success.

The IMTC maintains an information-packed World Wide Web site, which can be reached at http://www.imtc.org. If you look around, you'll find an abundance of useful information on audiographic and video communication standards and applications.

participants often requires a preliminary exchange of faxes, electronic mail messages, and overnight packages. Even with all the effort and expense that goes into preparation, teleconference participants often struggle to "stay on the same page" during a remote meeting. Document conferencing provides the missing link. People can coordinate their thoughts when they collaborate visually.

Many instances exist in which a comprehensive videoconference is overkill, and a telephone conversation is inadequate. The missing element is focal, not facial. Data conferencing provides a place to focus, collect, and store the ideas and experiences that result from a meeting. In practice, data conferencing is being used more in conjunction with audio conferencing than with video communication.

THE HISTORY OF DOCUMENT/DATA CONFERENCING STANDARDS

Standards for document conferencing are relatively new. In November 1993, the Consortium for Audiographics Teleconferencing Standards (CATS), a coalition of sixteen founding companies, announced their intention to draft a suite of data and graphics conferencing standards that would compliment the ITU-T's H.320 videoconferencing interoperability standard. The CATS group, in early 1994, began to promote the ratification and adoption of the T.120 family of standards that the ITU-T was developing. CATS was primarily an U.S. oriented consortium.

A more internationally oriented group existed as the Multimedia Communications Community of Interest (MCCOI). MCCOI was principally concerned with promoting the widespread adoption of H.320 videoconferencing standards. However, a MCCOI committee was looking into the ITU-T's emerging T.120 document conferencing efforts. MCCOI and CATS had some common members and soon the two groups were engaged in informal collaboration.

In 1994, CATS merged with MCCOI. The combined group took on a new name—the International Multimedia Teleconferencing Consortium (IMTC). The IMTC pledged to continue pushing for standards, but to broaden their scope to include document and data conferencing.

140

Meanwhile, a group of 150 vendors from the computer software and hardware, telecommunications, and teleconferencing communities was pursuing a similar but somewhat less standards-oriented agenda. The group, spearheaded by Intel and formed in January of 1994, called itself the Personal Conferencing Working Group (PCWG). The PCWG quickly outlined a comprehensive set of specifications that addressed the special needs of desktop and document conferencing users. Called the Personal Conferencing Specification (PCS), it offered interoperability across a variety of hardware platforms, operating systems, and networks. Central to PCS 1.0 was support for Intel's Indeo compression algorithm and Microsoft's DVI graphics/video interface. The document conferencing specification mirrored the features offered by Intel's ProShare family of personal conferencing products. The PCWG came under fire by opponents (read Intel) who accused it for trying to derail the ITU-T's document conferencing standards effort. The IMTC launched an aggressive effort to force the PCWG to support the de jure standards-setting activities of the ITU-T. In response, the PCWG announced that PCS 2.0 would provide support for the ITU-T's Rec. H.320 and Rec. T.120.

Intel defended the PCWG's decision to ignore H.320 in favor of Indeo. It noted that H.320 was a room-system oriented compression algorithm that was too complex for many desktop systems to implement economically. Indeo, a compression-only codec, can be inexpensively added to desktops. This argument took a blow in February 1995, when Microsoft announced support for ITU-T standards.

In May 1996, the Personal Conferencing Work Group (PCWG) agreed to construct the PCS 2.0 specification such that desktop conferencing products initiate calls to other systems using H.320 and T.120. If, after call initiation, the PCS 2.0 software discovers that the called system also supports PCS 2.0 then the software would communicate in the PCS (Indeo/ProShare) mode.

The PCWG's acquiescence settled a long-standing battle on the desktop conferencing standards-front and cleared the way for wide acceptance of the ITU-T T.120 standard. A number of corporations, that had been waiting to see whether the PCWG or T.120 would prevail, signed orders.

Most desktop video communication products on the market now support the T.120 family of data conferencing specifications.

T.120—AN INTRODUCTION

T.120 is an ITU-T standard that defines point-to-point and multipoint document conferencing over a variety of transmission media. Like H.320, T.120 is an umbrella standard. It is a model that defines a flexible communications infrastructure that enables participants in any teleconference (including an H.32X videoconference) to concurrently view, share and exchange computer files. T.120-based conferences do not rely on services from the H.32X family of recommendations and can just as easily stand on their own (although the T.120 standard does not address the audio aspect of a conference).

Conference participants can use T.120 to share and manipulate information much as if they using a whiteboard in the same room. Sessions can be point-to-point or multipoint. The data exchanged in a T.120 conference can travel between end-points across a variety of network choices. Connections can be established over dissimilar transport services. This allows a single T.120 conference to include consumers connected via POTS, corporate users supported by LANs/WANs, and small office/home office (SOHO) workers linked over ISDN BRI. The conference extends flexibility even further. T.120 conference participants can use virtually any type of desktop hardware and operating system commercially available. Many computer manufacturers now build T.120 support into their operating systems.

T.120 is useful for shared whiteboarding and multipoint file transfer within telemedicine, online chat sessions, on-line, gaming, and virtual reality. While H.32x provides a means of graphics transfer (T.84 / JPEG), T.120 supports higher resolutions, pointing, and annotation. T.120 enables audio bridge manufacturers to add graphics to their products in support of a wide range of applications such as training, project management, brainstorming, and engineering.

Name	Extended Name	Description
T.121	Generic Application Template	Common structure for T.120 applications
T.122	Multipoint Communication Service	Service definition for T.123 networks (implemented in T.125)
T.123	Audiovisual Protocol Stacks	Protocol stacks for terminals and MCUs
T.124	Generic Conference Control	Conference management (establish, terminate, etc.)
T.125	Multipoint Communication Service	Protocol implementation of T.122
T.126	Still Image & Annotation Protocol	Whiteboards, graphic display and image exchange
T.127	Multipoint Binary Transfer	Protocol for exchanging binary files
T.128	Audio Visual Control	Interactive controls (routing, identification, remote control, source selection, etc.)
T.130	Real-time Architecture	Interaction between T.120 and H.320
T.131	Network Specific Mappings	Transport of real-time data used with T.120 over LANs

T.132	Real-time Link Management	Creation & routing of real-time data streams
T.133	Audio Visual Control Services	Controls for real-time data streams
T.RES	Reservation Services	Interaction between devices and reservation systems
T.Share	Application Sharing Protocol	Remote control protocol
T.TUD	User Reservation	Transporting user-defined data

Figure 8-1. T.120 Standards Family

T.120

The formal title of the T.120 Recommendation is *Transmission Protocols for Multimedia Data*. T.120 was approved by the ITU-T in Q1, 1996. It contains a conceptual description of the T.120 series of recommendations that define multipoint transport of multimedia data in a conferencing environment. T.120 describes the interrelationships between the constituent standards that make up the Series. It also describes how the T.120 series can be used in support of other ITU-T standards, namely, the H.32X family of videoconferencing specifications. T.120 encapsulates the concepts it expresses in six supporting ITU-T recommendations, arranged in a layered hierarchy. Each layer leverages its preceding layers to define protocols and service definitions.

T.121

The formal title of the T.121 Recommendation is *Generic Application Template*. T.121 describes a conceptual model of a T.120 application and defines those operations that are common to most T.120 application protocols. This portion of the series aims to facilitate the process of applications development. It does this by providing a common structure of protocols and services that underlie T.120 applications. This structure ensures a consistent approach to the development of T.120-compliant applications. The Generic Application Template does not impose rules on the structure of application software but rather defines what that applications that software might be called on to support.

T.122

The formal title of the T.122 Recommendation is *Multipoint Communication Service for Audiographics Conferencing—Service Definition*. The T.122 MCS is a generic connection-oriented service that collects point-to-point transport connections and combines them to form a Multipoint Domain.

T.122 supports highly interactive multimedia conferencing applications. It supports full-duplex multipoint conferencing among an arbitrary number of connected application entities over a variety of networks (as specified in T.123). T.122 assumes error-free transport connections with flow control. It uses these connections to provide broadcasts and to support multipoint addressing (one to all, one to sub-group, one-to-one). It also ensures selection of the shortest path to reach the receiver and uniform data sequencing. T.122 uses tokens to resolve resource contention or channel availability.

T.122 also works in tandem with T.125, a multipoint communication service protocol. Together, the two form Multipoint Communication Services, the regulative portion of T.120 conferences.

T.123

The formal title of the T.123 Recommendation is *Protocol Stacks for Audiographic and Audiovisual Teleconference Applications*. T.123 presents a uniform OSI Transport interface and services to the MCS layer above it. Essentially, it is comprised of numerous network-specific transport protocol stacks. Network categories are addressed in T.123 profiles and include ISDN B channels, POTS, packet-switched networks (IP), and others. Each profile may extend as high as layer seven in the OSI reference model, depending on the mode selected.

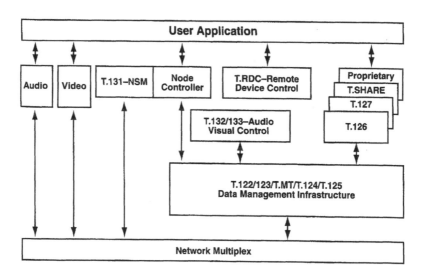

Figure 8-2. The T-120 Family of Recommendations (Laura Long).

T.124

The formal title of the T.124 Recommendation is *Generic Conference Control*. T.124 provides a high-level framework for conference management and control. Generic conference control functions provide tools for the establishment and termination of

144

sessions and the coordination of conference management. They also allow oversight of the nodes that are participating in a conference, and provide a registration directory to track them. A portion of the specification is devoted to managing the roster of application capabilities and the Application Protocol Entities that support them.

The services addressed by T.124 include:

- Conference establishment and termination
- Managing the conference roster
- Managing the application roster
- Application registry services
- Conference conductorship
- Miscellaneous functions

T.125

The formal title of the T.125 Recommendation is *Multipoint Communication Service Protocol.* T.125 defines a protocol that operates across layers of multipoint communication hierarchy. It specifies the format of messages that are passed between T.125 entities, and procedures that govern their exchange over a set of transport connections. The purpose of T.125 is to implement the services defined by ITU-T Rec. T.122.

T.126

The formal title of the T.126 Recommendation is *Multipoint Still Image and Annotation Protocol.* T.126 defines the protocol to be used by a wide array of user applications that require the exchange of graphical information in a heterogeneous (multi-vendor) environment. Applications include the annotation and exchange of still images, simple whiteboarding, screen sharing, and remote computer application piloting. T.126 uses services provided by T.122 and T.124.

The T.126 protocol permits conference-wide synchronization of multi-plane/multi-view graphical *workspaces.* An extensible set of bitmap, pointer and parametric drawing primitives can be directed to these workspaces. The protocol also supports advanced options (keyboard and pointing device signaling), and can be used in remote application piloting and screen sharing. T.126 also supports in-band facsimile exchanges. T.126 is deliberately extensible; it allows any new or extended capabilities that are not defined in the original specification to be added later. Information about these extended capabilities is passed in-band.

T.127

The formal title of the T.127 Recommendation is *Multipoint Binary File Transfer.* T.127 defines multipoint binary file transfer within an interactive conference. It provides mechanisms that facilitate simultaneous file distribution (broadcast) and retrieval. It also provides for private distribution of files to a selected subgroup. T.127 permits file distribution control by a chairperson.

T.127 protocols are simple and versatile. They provide core functionality with the ability to extend that core to meet the demands of more sophisticated applications.

T.SHARE

The formal title of the T.SHARE Recommendation is *Multipoint Computer Application Sharing Protocol*. T.SHARE has not been adopted, but the work of its authors is worthy of reconsideration later.

T.SHARE defines a cross-platform, multipoint computer application-sharing model. It allows a computer application hosted at one site to be viewed at all sites within a session. It includes provisions that arbitrate which site is currently in control of the session's pointing device and keyboard stream. The result of T.SHARE is that multiple sites can concurrently host one or more shared applications.

T.SHARE has two modes of operation. One, the *Legacy Mode* was designed to be compatible with Microsoft's NetMeeting and PictureTel's Group Share products. This was included to achieve industry compatibility in a market dominated by some very large companies. The Legacy Mode coexists harmoniously with all other T.120 application protocols, but no future enhancements will be made to it.

The Base Mode is modeled after the Legacy Mode but it makes more extensive use of generic conference control for capabilities negotiation and session management. Physical Data Unit organization and encoding has been brought in line with other T.120 application protocols. The Base Mode will provide the basis for future extensions to T.120's application sharing support.

T.SHARE does the following:

- Provides a protocol for font matching and text exchange
- Provides a protocol for basic window management
- Defines a protocol to manage shared pointers
- Supports compressed bitmap exchanges
- Supports rectangle, line and frame operations
- Defines caching policy for bitmaps, pointers, color tables and desktop save areas
- Defines protocol for the communication of keyboard & pointing device events.

T-SHARE has been available on the Macintosh platform, UNIX, and various Microsoft Windows platforms (Microsoft's Remote Desktop Protocol, RDP, is a single-session extension of T.SHARE).

T.120 FUTURE ENHANCEMENTS

T.RES

The T.RES is concerned with Reservation Systems. This T.120-oriented set of specifications was designed to simplify the process of reserving a conference across product platforms. Like T.SHARE, T.RES deployment has suffered at the expense of manufacturers who do not necessarily consider cross platform interoperability to be in their interests. It is quite unfortunate.

There are four components to T.RES—T.RES.0, T.RES.1, T.RES.2 and T.RES.3. T.RES.0 provides a reservation system overview. T.RES.1 defines the user-to-reservation system interface. This protocol specifies how user terminals access conference reservation systems. T.RES.2 defines a reservation system-to-MCU interface. T.RES.3 defines a reservation system-to-reservation system interface.

T.MT

T.MT is a set of specifications that are designed to allow Multicast Transport—e.g., to enable the multipoint conference service (MCS) to ride on the top of IP-based networks. To accomplish this, T.MT relies on several different methods to transport data across IP-based networks. The first is UDP (see Chapter 9). The second is OSI end-to-end services on top of the Internet suite of protocols (TCP/IP). OSI over TCP relies on RFC 1006, an important IETF document ("ISO Transport Arrives on Top of the TCP/IP") co-authored by Dwight Cass and Marshall T. Rose. Additional transport techniques include Multicast Transport Protocol-2 (MTP-2), which is a revised version of MTP (originally introduced as RFC 1301) and Real-Time Multicast Protocol (RMP); T.MT also includes an *other* category for IP-based transport protocols yet unspecified.

T.MT's deployment has been limited because Internet service providers do not want their already over-subscribed networks to have to support the additional traffic that Multicast introduces. As bandwidth becomes cheaper and, subsequently, as the backbones become more robust, T.MT may play an important role.

T.130

T.130 (along with T.132) was determined in March 1997 and ratified in 1998. Like T.120, it is an overview document that incorporates subordinate recommendations. These provide a network-independent control service for Real-Time audio and video streams. Implementation of the completed standard will allow users to coordinate the audio and video components of a conference across diverse network boundaries. At the IMTC's Fall Forum in Sunriver, Oregon, Pat Romano, Director of Advanced Development, Polycom, Inc., delivered an update on the ITU-T's progress as it relates to T.130. Our discussion of T.130 is primarily based on this presentation. Even at the end of 1999, little exists on T.130 because it simply has not been deployed. Vendors all state that they are interested in T.130 and that they are following the standard.

The T.130 specifications are:

- T.130   Real Time Audio-Visual Control for Multimedia Conferencing
- T.131   Network Specific Mappings
- T.132   AVC: Infrastructure Management
- T.133   AVC: Conference Service Management
- T.134   Multimedia Application Text Conversation protocol (T.CHAT)
- T.135   Secure T.120 conferencing
- T.RDC   Remote Device Control
- T.TUD   Transportation of user-defined bitstreams between endpoints

T.130

The formal title of the T.130 specification is *Real Time Audio-Visual Control for Multimedia Conferencing*. T.130 and T.132 were determine by the ITU-T in March of 1997 and ratified in 1998. T.130 provides a description of how terminals with diverse capabilities can manage audio and video transcoding, continuous presence and other features in a conference. T.130, where possible, leverages existing ITU-T control protocols (H.242/3, H.245). It addresses virtual networks and virtual and distributed multipoint conferencing units—a concept that is new to the H.32X and T.120 family. In addition, T.130 allows a terminal to request and measure the network quality of service (QoS) with respect to Real-Time streams.

The T.130 series defines an architecture, a management and control protocol, and a combined set of services that constitute an Audio-Visual Control system (AVC). T.130 supports the use of Real-Time streams and services in a multimedia conferencing environment. T.132, outlines the protocol and services in terms of management and control. T.130 enables devices (e.g., conference server, gateway, or MCU) to provide endpoints with T.132 video and audio services. Such services include stream identification, on-air identification, video switching, audio mixing, continuous presence, and remote device control.

The T.130 series leverages H.320 and T.120. It is compatible with systems in which audio and video are transmitted independently of T.120. However, it is also compatible with systems that are capable of transmitting multiple media types within a common multiplex. T.130 is useful for any conferencing scenario that requires multipoint audio or video. Because of the transmission requirements of Real-Time dataflows, the audio and video streams are transported in independent logical channels. However, T.130 relies upon the services of GCC and MCS to transmit control data.

Although the T.130 suite addresses manipulation of real time streams within an established conference topology, it does not define mechanisms to set up that

148

topology. Terminals are not required to implement the protocols referenced in T.130 to participate in the audio and video portions of a T.130-based conference. However, terminals that do not may have very limited options as they relate to audio and video control services.

T.131

The formal title of the T.131 is *Network Specific Mappings*. Each network requires unique control mechanisms that manage audio and video streams. T.131 specifies the mapping between the network-independent control mechanisms of T.132/3 audio video control (AVC) and that of the active network. Network specific mapping mechanisms reside in a node controller. Given that AVC depends on the data services provided by T.120, it must provide for the inheritance of audio and video streams that are activated prior to the start-up of T.120.

PSTN	Defines interaction between AVC and H.245 in H.324 systems
ISDN	Defines interaction between AVC and H.242/3 in H.320 systems
LAN	Defines interaction between AVC and H.245 in H.323 systems
ATM	Defines interaction between AVC and H.245 in H.310 systems
DSM-CC	Defines interaction between AVC and DSM-CC DAVIC systems

Figure 8-3. Networks supported by T.131

T.132

The formal title of the T.132 is *Audio Video Control: Infrastructure Management*. The T.132 protocol performs operations that are needed by T.133, Audio-Video Control: Conference Services. T.132 specifies mechanisms for multimedia capabilities exchange (whereby terminals handshake and exchange information about the services that they support/require). It also defines procedures used to arbitrate access to the various channels and support mechanisms that, collectively, sustain a conference. T.132 characterizes the procedures used to configure and control the audio-visual infrastructure.

The T.132 specification introduces the concept of zones, whereby a zone is a group of terminals that share a common network type. For instance, videoconferencing terminals connected in a switched digital arrangement (e.g., via ISDN) might comprise Zone 1 while terminals connected to an H.323 LAN might constitute Zone 2. Zone 3 might contain terminals connected via POTS. A zone manager would represent each zone on behalf of the entire bridged connection (where all three zones are linked). T.132 defines procedures that arbitrate the access to audio-visual infrastructure across zones.

T.133

The formal title of the T.133 Recommendation is *Audio Video Control: Conference Services*. In the T.130 family, T.133 defines audio video control conference services. These include:

- Real Time Channel Management
- Video switching
- Video processing and transcoding (including continuous presence)
- Audio mixing
- On air indication
- Source identification and selection
- Floor control
- Privacy

T.133 relies on the services provided by T.132, which it uses to convey service-oriented capabilities across zones and between terminals.

T.RDC - Remote Device Control

The formal title of T.RDC is *Remote Device Control*. The T.RDC specification provides for remote control of cameras, VCRs, microphones and other peripherals. It defines a mechanism to select and configure audio and video sources. Using the T.RDC specification, T.130-compliant terminals can define the control of audio mixing, video switching and continuous presence facilities.

T.RDC is defined in the spirit of a T.120 application protocol that additionally uses the services of AVC (T.132/3). It attempts to define a number of standard device types (cameras, microphones, VCRs) with standard control attributes. This model accommodates network equipment and unmanned audio/video device paradigms in addition to conventional audio/video terminals. The technical direction of T.RDC is to endow T.130-compliant nodes with the ability to advertise attached devices and their associated attributes. A T.RDC protocol will be defined to control, configure, query status and receive event notifications from remote peripherals.

Development of T.120 Compliant Products

With the T.120 standards complete, third-party developers must now implement them. Several development environments are available for doing so. Perhaps the oldest and most widely used is DataBeam's T.120 toolkit for developing multimedia data-sharing applications. The toolkit encapsulates the complex T.120 specification into a development environment that can be used to create complex, standards-based point-to-point and multipoint applications. Others have followed in DataBeam's footsteps. DataBeam deserves special mention because it has been involved in the T.120 process since inception, and because it has displayed admirable restraint in avoiding any temptation to put its own particular spin on things. DataBeam has acted in the interests of the T.120 family of standards and, therefore, in the interests of users, vendors, and the market in general.

Using T.120 mature development environments accelerates the development of software applications and network infrastructure products (i.e., PBXs, bridges, routers, network switches, and LAN servers). T.120-based products can deliver industry-wide interoperability in the areas of data collaboration and document conferencing. For a complete list of companies that supply these products, refer to Appendix H, Interactive Multimedia Suppliers (by Category).

T.120 / T.130: A Tribute to the ITU-T and IMTC

To respond to proprietary audio video system, standards-based products must provide roughly equal capabilities. The argument that proprietary is better is only difficult to counter when standards bodies stop short of complete specifications that address real-world needs.

The ITU-T and the IMTC perform a tremendous service for the video communication and data collaboration industry— and for the users of industry products. Their accomplishments, to date, reflect their pragmatism. Equally important is their ability to work quickly and thereby keep abreast of proprietary advances. Because of their diligence, these two groups have essentially discouraged even the largest players from *going it alone*.

All the goals that T.120 initially set out to accomplish are completed. Moreover, goals were added throughout the standards-setting process. Not only are most of the core standards stable and complete, they are already being revised after years of exhaustive effort.

The ITU-T will continue to work on the T.120 and T.130 protocol suites; years will pass before they consider those standards complete, if they ever do. Technology marches along and those who draft standards must lock step. On the horizon are plans to extend the infrastructure standards to leverage new communications technologies as they emerge (e.g., ATM, wireless networks, and cable television networks). The IMTC will also work to expand the set of application protocols (e.g., chat, generic object management, and remote procedure invocation).

CHAPTER 8 REFERENCES

"A Primer on the T.120 Series Standard," *DataBeam*. December 1999. http://www.databeam.com/ccts/t120primer.html

"Ratification Schedule for the T.120 and H.323 Series Standards," *DataBeam*. December 1999. http://www.databeam.com/standards/schedule.html

Adams, Chris. "An Introduction to Digital Storage media—Command and Control," *Nortel*. 1996.

"T.120-Based Conferencing To Explode," *Newsbytes* September 25 1996 PNEW09250010.

Brown, Dave. "Bytes. Camera. Action!" *Network Computing* March 1, 1996 v7. n3 pp. 46-58.

Grigonis, Richard. "H.324 Video and T.120 Data Will Change Your Life," *Computer Telephony* October, 1996 pp. 122-131.

Halhed, Basil R. "Standard Extend Videoconferencing's Reach," *Business Communications Review* September, 1995 p52.

Knell, Philip D. "The Multipoint Revolution: Where We're Headed," *Telecon XVI Show Issue* October 1996 v15 n5 p. 13.

Labriola, Don. "The Next Best Thing," *PC Magazine* December 1995 v14 n21 pNE 1-9.

Lee, Yvonne L. "Videoconferencing Takes Off As Vendors Standardize Wares," *InfoWorld* June 3 1996.

Masud, Sam. "Conference server beams Internet users in real time," *Government Computing News* June 10, 1996 v15 n12 p. 55.

Romano, Pat. "T.120 Update," *IMTC Fall Forum*. Sunriver, Oregon, October 1996 (Courtesy of the IMTC and Polycom. Inc.)

Rose, Marshall T. *The Open Book*, The Wollongong Group Inc. Prentice Hall 1990.

Sullivan, Joe. "How T.120 Standards Support Multipoint Document Conferencing," *Business Communications Review* December, 1995 p. 42.

9

MOBILE VIDEOCONFERENCING AND INTERACTIVE MULTIMEDIA

They used to say years ago, we are reading
about you in science class. Now they say, we are
reading about you in history class.

Neil Armstrong

A current discussion of video communication would be incomplete without addressing the dynamic reality and immense potential presented by mobile communications. Modern wireless networks and powerful mobile handsets play an ever-growing role in the world of tomorrow, offering newly enabled services such as Mobile Streaming Media, Mobile TV, and of course the pinnacle of mobile multimedia; Mobile Interactive Voice and Video—commonly known as mobile videoconferencing.

There is little doubt that wireless technologies in general—and specifically cellular technologies—have allowed modern and emerging societies to experience an unprecedented level of freedom in the way we live, interact, and communicate. Cellular technology is pervasive, and with over 1 Billion subscribers globally and a whopping 2 billion subscribers expected by 2008 it is only going to grow more pervasive in the future. Cellular industries make-up a significant portion of the global economy, with 2004 cellular telecommunications sector revenues exceeding $100 Billion in the US alone! Technology investments in Asia and Europe over the past decade have been even more substantial, and are projected to yield unprecedented revenues in the years ahead by enabling new and powerful communications services that consumers are demanding. Africa and the Middle East are seeing huge investments in wireless infrastructure, with many areas of developing countries receiving cell-phone service before they even have running water! Cell phones are a global phenomenon, crossing all demographics and societal strata and providing amazing new opportunities for the people of the world to connect and communicate.

Not surprisingly, mobile technologies are evolving at a breakneck pace—though still *not fast enough* for most consumers! With widespread adoption of this technology less than a decade old, many consider it to be relatively immature and facing a great deal of growth in the years ahead. In many ways, the mobile communications industry is going through the same growth process that the PC industry went through in the 1990's, where the most "bleeding-edge" technology that can be commercially purchased is completely eclipsed only a few months later by something better, faster, and oftentimes less expensive. As a result, the most tech-savvy consumers are changing handsets at least twice annually, and sometimes every 3-4 months. Devices are being replaced not because they no longer function properly, but rather they lack the capabilities enabled by newly improved networks. This cycle is beginning to slow as newer handsets become more modular, mobile handset software becomes more mature and enabling technologies become more standardized.

It is difficult to conceive that only a few decades in the past all of these technologies were either drawing-board concepts or theoretical ideas of futuristic "Science Fiction" writers and dreamers.

Synoptic History Of Mobile Communications

Even thought the "industry" of consumer cellular phones can be dated to the early 1990's, the roots of the technology go much deeper into our history. Arguably the first cell phone user was the 1930's comic book hero Dick Tracy whose "wrist radio" sparked the idea of roving, wireless, mobile communications. Nobody knew (and who really cared) how it all worked—in the post-depression world, something that was "Dick Tracy" became synonymous with incredible and fanciful technology.

Similarly, Jules Verne's post-civil war stories of amazing technology more than a half-century earlier depicted unheard-of wonders that allowed men to travel to the moon and enjoy deep-sea adventures on the Nautilus. Verne imagined a world of electricity and a host of other high-tech Science Fiction gadgets that were yet to be invented! It is interesting to note that now, early in the 21st century, Verne's picture of a high tech world seem oddly prophetic, with real-world examples of his "futuristic" technologies now decades old, rusting and aging in modern museums. It took less than a century for the wildest fictional dreams of the one of the worlds most imaginative minds to become eclipsed with even more amazing science and technology.

In the middle of the 20th Century, post World-War II dreamers cast their vision of a high-tech future, ever pushing the boundaries of possibility and inspiring scientists and inventors to solve "how" one might actually accomplish such incredible things. For example, Gene Roddenberry's Sci-Fi masterpiece about futuristic voyages into "Space, the Final Frontier" detailed a host of non-existent "technologies" that scientists are successfully creating decades later. Capt. Kirk's communicator (the first flip-phone) may have been bigger that Dick Tracy's wrist watch, but hey—it could talk to a space ship orbiting the planet! Detailed descriptions of phasers, tricorders, even the "Transporter beam" are now available in "Technical Manuals" that are becoming less fiction and more science with each passing year. Roddenberry's stories inspired following generations to dream even bigger dreams.

It is not difficult to identify the inspiration of Verne, Roddenberry and countless others who imagined new and never-before considered technologies in the epic Star Wars Sept-ilogy. Concepts of droids, light-sabers, hologram communicators, interactive 3-D games, floating speeders, hyperspace travel (certainly done before, but the streaking stars effect in-and-out of the "jump" was AMAZING) and cities in the clouds are but a few of the notable technologies George Lucas generously provided to an unsuspecting and gape-mouthed audience. I was personally inspired by the handheld Jedi PDA / communicator that Kwi-Gon Jin used in Episode 1, the one that could communicate with the ship, access the computer thru voice commands and display a hologram of the sleek Royal Naboo Spaceship that was used as collateral in a risky bet to... well you know the story. It caused me to think that... given current technologies, that device is not only plausible; I think we will see it in the next decade.

Which brings us back to Dick Tracy's now not-so-special single-function wrist-radio, which today would certainly need to be much more than a "simple" 2-way radio. 75 years of technology advances have shown us that we can do better. It would be a multi-modal, voice-activated, multimedia-enabled, broadband internet-connected, interactive voice-video-data "super communicator." But, believe-it-or-not, we already have devices and communications networks that CAN do all that! Right now there are tens of millions of mobile, handheld devices that can offer multi-modal communications (WiFi, High-Speed Cellular Voice and Data, Bluetooth), support voice activated programming, are "always-on" with instant access to the World-Wide-Web an back-end applications, AND support video communications. The form factor is surprisingly similar to Capt. Kirk's communicator, but today is generically referred to as a "Smartphone." And thanks to PalmOne and Microsoft you can accessorize "a' la Dick Tracy" with an internet-connected wristwatch!

Imagine how much more impressive Capt. Kirk's communicator could have been if it supported all the features of a modern Smartphone: it could have been an MP-3 player with removable media options, an AM/FM radio & satellite TV receiver, a Blood Sugar monitor, heart-rate monitor, sporting a keyboard, Global positioning system... the list is ever growing. There are in fact so many mobile phone options that all the individual components sometimes don't all fit into a single device. This too has been planned-for, with the introduction of short range "Personal Area Network" (PAN) communications, enabling multiple different devices to be interconnected and their functionalities combined. Technologies like Bluetooth allow for combining the best features of several different wearable, pocket-able, handheld and/or laptop devices and create a super-net of personal devices. Alternatively, a PAN could support short range Person-to-person or Mobile-to-Mobile communications for things like interactive gaming, file transfers, ad-hoc networking and security functions (digital car or house keys) and / or monetary transactions (using NFC).

Admittedlty, "Beam me up Scotty" is still a few decades away—but we're working on it, and on a whole host of new and interesting mobile technologies...

UDP

Overview of UDP

The User Datagram Protocol (UDP) is a streaming IP protocol that is being substituted for TCP in Real-Time multimedia applications. UDP is unidirectional; no mechanism is available at the sender to know if a packet has been successfully received. Thus, UDP does not wait for transmission acknowledgements. It carries the packets presented to it but, if those packets are dropped or duplicated, or arrive out of sequence, UDP takes no corrective action.

On the other hand, UDP increases the throughput of an IP network. Since there is no need to wait for acknowledgement by receivers, the UDP is an efficient mechanism for delivering multiple streams of information. TCP, on the other hand, has an inherent need to track and retransmit dropped or corrupted packets. The service that TCP provides has, unfortunately, no value for Real-Time applications. Video communication sequence and timing does not allow for a lost frame of video or packet of audio, to be re-sent and re-inserted in the data stream.

Today, UDP provides the foundation for the other IETF protocols that are being developed to support video communication over the Internet. These include the RTP, the RTCP, and RSVP. UDP is also being used for IP multicasting. Applications developed in accordance with the IETF protocols rely on codecs to detect and recover from packet loss, in conjunction with RTP.

Wireless Technology Evolution

The technologies that make wireless and mobile communications possible are varied; most are standardized, but many of these "standards" are competitive and/or conflicting. Many standards are regional or national and greatly impacted by socio-political-economical drivers. In fact, there are SO MANY competing wireless communications standards that entire volumes are dedicated their definition, with countless more volumes describing and detailing the uniqueness of each. One could not due them just ice in a single volume (or a hundred volumes, for that matter) so we will focus only on those that are most pervasive. To keep everything in perspective, I will first offer an abbreviated history of the development of wireless communications.

In the beginning...we had 2-way radios, HAM, Citizen's band, all of them analog half-duplex solutions. This was the spawning ground for the First Generation—or "1G"—cellular communications period that began in the late 1970s and lasted through the 1980s. 1G systems were the first true mobile phone systems, known at first as "cellular mobile radio telephones" or more commonly as "bag phones" because the apparatus, connections, battery, and transceiver unit were collectively so large they required a suitcase-sized bag to hold them all. 1G cellular networks used the analog Advanced Mobile Phone System (AMPS) in the 800MHz frequency range and were little more sophisticated than repeater networks used by amateur radio operators. They were also prohibitively expensive and lacked national coverage.

It was the mid 1990's when 1G analog cellular phones made way for Second Generation (2G) technologies that introduced digital voice encoding and the first major branch in competing international mobile standards. The three major competing 2G standards were:

- Time Division Multiple Access (TDMA), also known as "Digital AMPS" (D-AMPS) or ANSI-136, which delivered services in the 800 MHz and 1900 MHz frequency bands.

- Code Division Multiple Access (CDMA) also called IS-95. The code division technology was originally developed for military use decades earlier, using code sequences as traffic channels within common radio channels deployed in 1.25 MHz increments.

- Global Services Mobile (GSM) was introduced as the mobile phone platform used in Europe and much of the rest of the world -- though not as widely deployed in the USA. GSM networks operate in the 900Mhz and 1800Mhz ranges in Europe and in the 800MHz and 1900MHz ranges in the US.

Since their inception, 2G technologies have improved and evolved to offer increased bandwidth, enhanced packet routing, and an infrastructure capable of supporting limited mobile multimedia applications. The present state of mobile wireless communications is often called "2.5G", which implies something better than 2G but not yet good enough to be Third Generation (3G). Some examples of 2.5G technologies are GSM/GPRS and CDMA 1xRTT digital networks. These networks support mobile digital data as well as voice calls and (among other things) have given to the phenomenon of digital messaging, also known as the text messaging, Short Message Service (SMS) and mobile Instant Messaging (IM). Core upgrades in 2.5G networks opened the door to mobile applications like email, limited web surfing, and other entertainment services. Although theoretically capable of data rates in excess of 100KB, typical 2.5G connection rates are less than 56Kb, and often less than 20Kb in congested areas.

Which brings us to the Third Generation of cellular communications, more commonly know as 3G. (3G) as defined by the International Telecommunications Union (ITU) specification as providing data bandwidth up to 384 kbps for stationary devices and up to 2 Mbps in fixed applications. Per the specification, 3G services may be delivered over wireless interfaces including but not limited to GSM, TDMA, and CDMA. Additionally, while 3G is generally considered applicable mainly to mobile wireless, it is also relevant to fixed wireless and portable wireless. 3G specifications have been architected to support access from any location on, or over, the earth's surface, including use in homes, businesses, government offices, medical establishments, the military, personal and commercial land vehicles, private and commercial watercraft and marine craft, private and commercial aircraft (except where passenger use restrictions apply), portable users (pedestrians, hikers, cyclists, campers), and—with a nod to Captain Kirk—even space stations and spacecraft. 3G technologies are seen to hold the promise of keeping people connected at all times and in all places around—or above—the globe.

There are numerous competing 3G "standards", but each is required to offer standard minimum capabilities and features, including:

- Enhanced multimedia (voice, data, video, and remote control)
- Multi-modality (telephony, e-mail, SMS, videoconferencing, apps, browsing)
- Broad bandwidth and high speed (upwards of 2 Mbps)
- Routing flexibility (repeater, satellite, LAN)
- Global roaming capability throughout Europe, Asia, and America.

It is germane to note that both initial 3G cellular networks and earlier 2.5G cellular networks are mainly circuit-switched, with connections always dependent on circuit availability. This is in contrast to newer 3G technologies that rely on Packet-Switching technologies. One of the benefits of packet-switching connections using the Internet Protocol (IP) is the concept of a constant virtual connection is always available to any other end point in the network. Packet switching is of great interest to consumers and carriers in that is supports provisioning of new services such as flexible billing methods enabling pay-per-bit, pay-per-session, flat rate, asymmetric bandwidth, and others.

As you can see, the variety of carrier services, their global coverage (or lack of it), and the raw capabilities of each solution will vary greatly, and each technology continues to evolve and "leapfrog" its competition. This bloom of competing technologies actually benefits consumers and carriers alike by enforcing a basic Darwinian theory: survival of the fittest, or in this case, survival of the fastest, most reliable, most extensible and—ultimately—least expensive protocols and solutions. Which technology will ultimately "win" is a guess at best and individual technology strengths ensure that there will be even more options in the years to come.

In theory, 3G is a wireless broadband connection to your mobile phone, opening up a wealth of data-rich services on the move, including (but certainly not limited to) video calls—so you can actually see who you're calling instead of just hearing them. 3G service coverage is still limited to large metropolitan areas in the US, whereas in Japan and Europe 3G coverage is nearly ubiquitous. In those markets with reliable 3G coverage, there has been a steady rise in the usage of premium wireless services, and carriers are seeing a noteworthy rise in their revenues as users begin to enjoy the value and benefits of premium services.

In the US, competing CDMA and GSM solutions have slowed 3G rollouts significantly. The competition has been fierce, and shows every sign of continuing for years to come. Even so, there has been significant progress in US-based mobile solutions.

On the CDMA side, Verizon Wireless offered 3G EV-DO services at the end of 2004 to only 32 cities—including New York, Los Angeles, Dallas, Chicago and Atlanta—covering about 75 million people, or about a quarter of the US population. Not to be outdone, Sprint Corp. plans to make 3G available to about 129 million people in 39 greater metropolitan areas by mid-2005, and across its entire national network by early 2006.

On the GSM side, Cingular Wireless offered 3G service to six cities at the end of 2004, including San Francisco, Detroit, Phoenix and Seattle. But Cingular expects to have most major markets covered by the end of 2006. And because the Cingular network will be UMTS-based, US subscribers to Cingular's service should be able to use their equipment overseas.

Other recent 3G Milestones include:

- A music concert recently streamed to 3G users in the UK.

- A live news report aired on the BBC using a tripod-mounted 3G phone.

- Forget cash or credit cards—you can now use your phone to pay for a variety of services, from restaurant bills to drinks cans.

Despite all the challenges, mobile video is already a reality in the US and beyond. Established 2.5G and early 3G carriers are reporting considerable interest from subscribers in current partnerships with content companies. For example, subscribers on the Sprint PCS Vision network in the United States have sent more than 100 million pictures and 15-second videos since November 2002. In 2003, there were 5.1 million video clip message users, mostly in Japan. By 2008, the ARC Group predicts that over 250 million users worldwide will spend more than $5.4 billion USD annually on mobile video. In short, the best is yet to come!

These are just a few examples of what 3G offers today. Let's look at how these services evolved and what 3G means specifically to Mobile videoconferencing and Mobile Multimedia.

CAN YOU SEE ME NOW?

Mobile videoconferencing is widely regarded as one of the signature services to be provided over 3G mobile networks. Interactive bi-directional multimedia applications like videoconferencing require two to three times more bandwidth than is required by single connection calls. Not surprisingly, mobile operators have invested heavily in 3G licenses and infrastructure—over $100 Billion globally—in order to offer new IP-based services that support voice, video and data. Mobile video will be one of many bandwidth intensive applications that are planned to become a core part of the mobile communications infrastructure.

Figure 9-1: Wireless and Cellular Radio Technologies

Terminology	Definition
1xRTT	Single Carrier (1x) Radio Transmission Technology, a 2.5G developed by Qualcomm referred to as CDMA2000-1x, or errantly as 3G or 3G1x, the current technology used by Sprint and Verizon in the US, offering mobile data rates up to 144Kbps
CDMA 2000	IMT-CDMA, the CDMA version of the IMT-2000 standard developed by the ITU can support mobile data communications at speeds ranging from 144 Kbps to 2 Mbps.
EDGE	Enhanced Data for Global Evolution, broadband wireless technology used or planned by AT&T Wireless, Cingular and T-Mobile, interim solution for GSM networks between GPRS and UMTS.
EV-DO	Evolution-Data Only (*1xEV-DO*), part of a family of cdma2000 1x digital wireless standards, it is a 3G standard based on a technology initially known as "HDR" (High Data Rate) or "HRPD" (High Rate Packet Data), developed by Qualcomm. The international standard is known as IS-856. Unlike other "1x" standards, EV-DO only addresses data - not voice. It requires a dedicated slice of spectrum, separate from voice networks using standards such as 1xRTT. 1xEV-DO offers very high data rates - up to 2.4 mbps - averaging 300-600 kbps in the real world. It is much faster than the 50-80 kbps typically offered by 1xRTT technology. Currently available.
EV-DV	Evolution, Data and Voice, addresses both data and voice, unlike 1xEV-DO, which only addresses data. Combines the high-speed HDR technology from 1xEV-DO with the widely deployed 1xRTT standard. It integrates seamlessly with 1xRTT, providing full backward-compatibility and simultaneous voice and data.
FOMA	Freedom of Mobile Multimedia Access, DoCoMo's brand name for 3G services based on the W-CDMA format. Introductory FOMA services for a limited number of users is to begin at the end of May, with full commercial services due in October.
GPRS	General Packet Radio Service, a 2.5G packet switching upgrade to GSM offering theoretical data speeds of up to 115kbps; introduces the packet switched core required for UMTS; used globally in Europe and Asia, in the US by AT&T Wireless / Cingular and T-Mobile.
GSM	Global Service Mobile; platforms include today's GSM, GPRS, EDGE and 3GSM; developed for operation in 900MHz band and modified for 850, 1800 and 1900MHz bands.
IDEN	integrated Digital Enhanced Network, technology developed by Motorola using 800, 900 or 1.5 GHz spectrum and used by Nextel. This technology will likely disappear in the next 3-4 years.
IMODE	Proprietary mobile data solution successfully used by NTT DoCoMo with the main advantage over WAP being that it is packet-switched network. Thus, it actually runs at the same speed of connection as WAP, but because the network is packet switched, it works considerably faster
HSDPA	(High Speed Downlink Packet Access) is an upgrade for WCDMA (UMTS) networks. It doubles network capacity and increases download data speeds five-fold
OFDM	Orthogonal Frequency Division Multiplexing, broadband wireless technology available today using. Flarion Technologies FLASH-OFDM® wireless broadband system (using 1.9 spectrum band) in trials with Nextel in U.S. and Vodafone KK in Japan.
UMTS	Universal Mobile Telecommunications System, supports W-CDMA radio access technology with deployments in Asia-Pacific, Europe, Middle East and Africa. Offers broadband, packet-based transmission of text, digitized voice, video, and multimedia at up to 2Mbps and offers a consistent set of services to mobile computer and phone users no matter where they are located in the world. Based on GSM, it is the next generation step for most GSM carriers around the world. Deployments for this technology started in 2004 in some countries. Cingular has announced plans to migrate its US networks to UMTS by 2006.
W-CDMA	Wideband Code-Division Multiple Access, deployed in 5 MHz increments; AT&T launched in US in 2004.
WiFi	Wireless Fidelity, refers to 802.11x networks; Although an Ethernet LAN protocol, it functionally provides 3G services in localized "hotspots." Most major carriers offer WiFi options with their cellular data service plans. 802.11b is the most common, offering up to 11MB data rates. Other 802.11 standards (a,g,n) offer enhanced data rates, more reliable and broader coverage.
WiMax	Worldwide Interoperability Microwave Access, deployed on licensed and unlicensed spectrum. Provides high-throughput broadband connections over long distances; implementation of the IEEE 802.16 standard, provides metropolitan area network connectivity at speeds up to 75 Mbps. Can transmit signal 30 miles.

The 3G mobile phones of the 21st century are just beginning to take advantage of sophisticated broadband capabilities and will pave the way for the imminent development of a new breed of personal communicators. With speeds reaching 384K bps, nearly three times faster than basic rate ISDN, for fast moving mobile terminals and even higher speeds for users who are stationary or moving at walking speed, 3G systems are changing the way we communicate, access information, work, and interact socially and personally.

To get perspective, consider that global 3G sales and services revenues are expected to exceed $9 billion in 2005. Other research concludes that by the year 2008 there will two billion mobile users worldwide and that 3G networks will carry traffic from more than one billion of these and account for two-thirds of global cellular service revenues.

In the early 1990's, consumer mobile phones were designed to enable and enhance Person-to-Person communications and—despite their initial size, cost and limited coverage—have forever changed the way people interact. In just over a decade, mobile phones went from a nice accessory to THE critical mode of personal communication and an indispensable business productivity tool. As they have continued to evolve, mobile networks, handsets, accessories (yes, accessories DO matter) and software have matured from simple voice solutions to fully capable broadband IP infrastructures that can support mobile applications—including Voice over IP (VoIP), Multimedia Messaging (MM), Instant Messaging (IM) and interactive, real-time streaming video communication with the same utility and convenience consumers presently enjoy voice, voicemail and SMS.

In order to take advantage of all the 3G-enable carrier services, mobile devices from a growing array of manufacturers and various OS-es support powerful new mobile software (PacketVideo, Goodmood, Extended Systems, etc) that will doubtless impact the way business is done—and will drive the need for reliable, hi bandwidth connections (3G, WiFi, WiMax) services around the globe. Mobile phone-based videoconferencing products include numerous video-ready 3G Smartphones and Mediaphones that utilize wireless broadband connections. Complete mobile videoconferencing solutions are already widespread in Japan and areas of Asia and have been introduced into Europe and most recently the US market. Japanese carrier NTT DoCoMo has been rolling-out innovative technologies over the past two years that supported live video chatting from mobile handsets and reports that estimated five million of its subscribers are already in possession of Smartphones with video-calling functions.

Mobile Device Considerations

The year 2004 saw the emergence of a new class of mobile cellular handheld device—the Smartphone. It is a unique mobile device that is equal parts PDA and business-class cell phone with "always-on" connectivity that offers presence, location-based services, and more. As the ultimately utilitarian mobile device, the Smartphone can also serve as an external high-speed modem for the Laptop, Tablet PC or PDA device using infrared, Bluetooth or some other personal area connectivity.

There are certainly devices other than a Smartphone that can be used for mobile videoconferencing. For example, there are many laptops, Tablet PC's, WiFi-enabled PDA's and a host of other options each offering sufficient connectivity and display features to support a mobile videoconference. Certainly, there are successful examples of camera and broadband-connected laptops being used to participate in videoconferences. Normally, however, these are nomadic configurations, not truly "Mobile" in that they require a bagful of accessories, multiple cables, clip-on cameras, and usually require external power to be used for any length of time. Add to that the fact that even the slimmest laptop is too large to fit into your pocket, that laptops are not designed for "always-on" communications required for dynamically scheduled, impromptu mobile videoconferencing.

The device features that enable truly mobile videoconferencing can vary and should be considered individually based on the application requirements. Some basic qualification guidelines for selecting an appropriate mobile videoconferencing device is one that:

1) Doesn't need to be re-charged during a regular business day

2) Offers a usable screen (large and clear enough to be useful)

3) Has a camera and video screen that both point the same direction

4) Supports some form of user input (touch screen, thumb-board, qwerty keyboard, handwriting recognition, etc)

5) Includes at least 2 modes of communication (cell, wifi, bluetooth, ir)

6) Supports add-on applications (java, c, brew, symbian, ppc)

7) Provides high-quality voice communications

8) Fits into a jacket pocket

Let's face it—a mobile device that needs to be plugged-in, lacks a usable screen or input device, doesn't make great voice-only phone calls and doesn't fit into a jacket pocket isn't really all that mobile. Devices that do meet or exceed these qualifications include selected Symbian models from Nokia, Fujitsu, Samsung, SonyEricsson, Motorola (and others); MS PPC solutions from HP, Dell, etc; various solutions from Qualcomm, NTT, LG and a host of others; and of course, venerable Palm-OS solutions from PalmOne.

The cost of videoconference-capable mobile devices is higher in comparison to standard cellphones, and the premium-service cost of making video calls reflects the enormous investment carriers have made in 3G technology. The market for video-calling devices, whether mobile or fixed wire-line also has an inherent hurdle to overcome in that people can make video calls only to other people who also possess video-capable devices. The growth of the market has been slow initially, but has grown to the point that it is now worthwhile for many more people to invest in new video-enabled mobile devices. To further encourage the purchase of newer and more powerful multimedia handsets, rapidly evolving technology offers devices with improved battery life, better video, enhanced audio and lower costs. Add to this the fact that carrier data services seem to become more affordable every month, the likelihood of calling another phone capable of making video calls can becomes more likely with each video-capable device sold.

In an interesting market evolution, 2004 was the first year in which the sale of Smartphones exceeded Palm-Based PDA's and also the first year in which the number of internet-enabled mobile devices exceeded the number of internet-connected PC's. 2005 and beyond Smartphone sales are projected to be in the 100's of millions per year, making Smartphones the largest and fastest growing market segment of all the mobile video options by a huge margin. And, smart as they are, Smartphones are getting even smarter, even more feature-packed with on-board MP3 players, FM Radios, video applications and continually enhanced multi-media capture and manipulation capabilities.

With the help of new carrier-based Session Initiation Protocol (SIP) features, Smartphones portend to pave a broad avenue for consumers and businesses to gain access to the benefits of interactive mobile applications and services. In many ways, SIP enabled networks are the key to universally-available mobile multimedia services, offering the standards-based infrastructure that will allow many latent Smartphone features to be activated and integrated into a host of consumer and business applications.

Session Initiation Protocol

As discussed earlier in Chapter 7, SIP (Session Initiation Protocol) is a signaling protocol for Internet delivered applications including Internet voice and video conferencing, Push-To-Talk (PTT) and PTT over Cellular (PoC) telephony, presence, events notification, and Instant Messaging (IM). In the most basic terms, SIP is a set of standard Internet protocols designed to connect any fixed or mobile SIP-enabled client devices to any Web-delivered application server any other remote SIP-enabled client. From the perspective of mobile solutions, SIP ensures an integrated IP environment for mobile devices to securely and reliably communicate with each and any IP-enabled application.

As the IETF's standard for multimedia communications over IP, SIP is a text-based, application-layer control protocol (defined in RFC 2543) that can be used to establish, maintain, and terminate sessions between two or more end points. SIP is in some respects comparable to HTTP, although service creation methods based on SIP have been defined to provide enhanced addressing and extensibility features for delivering applications throughout the entire mobile 3G architecture.

SIP can also be used in a wide host of application areas but is uniquely suited for setting up enriched communication sessions such as a real-time video/voice calls (sounds like videoconferencing to me!) between mobile devices. SIP has been globally selected as the call set-up protocol for all IP networks that will be standardized by the Third Generation Partnership Project (3GPP). This means that 3G All-IP terminals and networks will support SIP.

SIP and Java—Refreshing Technology for the Mobile World

Mobile applications delivery is significantly enhanced by Java and specifically by the interaction of Java and SIP solutions. Java is a software technology developed to provide a common base for running applications across a wide variety of computing platforms and has become widely available in mobile terminals. Java provides a standard platform for application development on a wide range of mobile terminals, making it a desirable technology platform to implement SIP-based applications.

Java consists of a programming language and a run-time environment that allows Java applications (like SIP, for example) to run on servers, personal computers, mobile phones and other devices. The Java Community Process in JSR-180 specifies the SIP API for the Java 2 Platform, Micro Edition (J2ME™). Java developers now have a standard and open tool kit to develop SIP-based Java applications that addresses the differing capabilities of various computing platforms vary by using the following 3 editions:

- Java 2 Enterprise Edition (J2EE) - for high-end business machines

- Java 2 Standard Edition (J2SE) - for personal computers

- Java 2 Micro Edition (J2ME) - for smaller handheld devices such as mobile phones and PDAs. The Mobile Information Device Profile (MIDP) is a key element of the J2ME, defining a platform for dynamic networked applications. Developers using MIDP can write applications once and then deploy them to a wide variety of mobile devices.

Since software written in Java is independent of operating systems, each computer or device that runs a Java application must have a Java Virtual Machine (JVM) that adapts the application code for the specific device and OS before executing it. Leaders in the mobile communications industry share the vision for an open mobile computing platform architecture based on Java, and they contribute to the continually evolving specifications for both the MIDP and CDC based Java platforms. Today, MIDP and CDC based Java platforms are available for deploying enterprise applications on a wide variety of business phones from many leading handset manufacturers.

Java was originally developed by Sun Microsystems, Inc. Enhancements and standardization of the Java platform are conducted within the Java Community

Process (JCP). Java™ and all Java-based marks are trademarks or registered trademarks of Sun Microsystems, Inc. Other product and company names mentioned herein may be trademarks or trade names of their respective owners.

A FINAL WORD ON SIP

By joining the fixed and mobile worlds, SIP will enable true service mobility and access independence. It will also support the rich call concept, or concurrent access to images, data, or other value-added information during a single session. This will add a "see what I mean" capability for users through a combination of voice, real-time video sharing, messaging, content sharing, and other services.

The primary advantage of the SIP protocol is the way in which it can integrate voice, video, and other interactive communications services with instant messaging and presence. SIP will greatly enhance person-to-person communications by combining multiple media types and communication methods with presence and community information while making user interaction simple and secure.

Vendor Highlight—PacketVideo

One company leading the way in Mobile Videoconferencing software and services is PacketVideo, headquartered in San Diego, CA. PacketVideo's unique software enables live, real-time, person-to-person, video calls with images of both callers on the screens of their respective phones. At a recent CTIA Wireless Entertainment and IT show (www.wirelessit.com/), company representatives demonstrated PacketVideo's two-way real-time video telephony on the Nokia 6630 Smartphone and other PacketVideo-powered handsets from the leading device OEMs.

PacketVideo's 2Way software is a 324M 3GPP-compliant solution for real-time two-way voice and video conversations and video conferencing. Based on PacketVideo's fourth-generation multimedia platform, the solutions has been deployed globally in mobile networks in Japan, Korea, Italy and the UK.

The basic solution consists of client software that gets installed on the Mobile device and Carrier network-based services.

Released in 2005, PacketVideo's SIP-based client provides enhance functionality that ties into presence-aware network components to offer optimized services as the mobile user moves thru various network segments, each with varying bandwidth, coverage, and capacity.

PacketVideo is the number one supplier of embedded multimedia communications software for mobile phones. PacketVideo software enables mobile phones to take digital pictures, record home movies, play back digital music and videos, and make 2-way videophone calls. Their global leadership is proven by unrivaled relationships with mobile operators, dominance in design wins, and the millions of PacketVideo-powered multimedia phones in the market. PacketVideo was founded in 1998 and is headquartered in San Diego, California. More information can be found at http://www.pv.com.

CHAPTER 9 REFERENCES

"Wikipedia, English Version" en.wikipedia.org/wiki/, January 2004.

Kerner, Sean Michael. "Converged Devices Lead Mobile Demand." November 1, 2004 www.clickz.com/stats/sectors/hardware/article.php/3429541.

"Bell Labs Technical Journal 9(3), 15–37 (2004)", Lucent Technologies Inc. Wiley InterScience (www.interscience.wiley.com).

"Ollila: Smartphones To Spur Mobile Sector" Global News Wire—Asia Africa Intelligence Wire, *Business Daily Update* November 22, 2004.

Mannion, Patrick. "Wi-Fi Alliance targets Wi-Fi-cellular convergence" *EE Times*, Oct 21, 2004.

"VoWLAN usage to grow four-fold by 2006" *FierceWiFi*, November 03, 2004.

Gold, Elliot. "Teleconferencing industry to hit $3.7 billion in revenues for 2004." *PR Newswire Association*, September 9, 2004.

Garfield, Larry. "Report expects big things for mobile video" *Infosync World*, www.infosyncworld.com/news/n/4978.html, 27 May 2004.

"Session Initiation Protocol", www.cs.columbia.edu/sip/, Jan 30 2005 .

"Session Initiation Protocol (sip)" www.ietf.org/html.charters/sip-charter.html, Jan 18, 2005.

10

STANDARDS FOR CONFERENCING OVER IP NETWORKS

*The standards that have come out of the IETF
have been impressive. They are produced
quickly, and they work.*

Carl Malamud

THE INTERNET AND INTERACTIVE MULTIMEDIA

Today's enterprise seeks out networks capable of transmitting all media—voice, still-image moving-image, and data. The goal is to find a single solution for all corporate information flows, to front-end it with a Web browser, and endow it with enough bandwidth so that the average corporate user never has to give the network a second thought.

With the rise in popularity of the Internet and its enterprise-adapted equivalent, the intranet, the direction of the enterprise network is now clear. Corporate America is migrating to TCP/IP (Transmission Control Protocol/Internet Protocol) and abandoning proprietary networking protocols. While open-systems standards are great, it is also true that they evolve in response to a particular need or application set. This holds true for TCP/IP, a set of Internet Engineering Task Force (IETF) protocols that are defined in a diverse group of Request For Comments (RFC) documents. RFCs are a series of notes that contain Internet-oriented protocols. RFC documents also set forth ideas, implementation techniques, observations, studies, measurements, and clarifications related to the diverse set of specifications that collectively run on Internet Protocol (IP) based networks.

Before we can discuss video communication over networks, we must understand how certain IETF protocols function, and how those functions must be modified to meet the unique needs of real-time media streams.

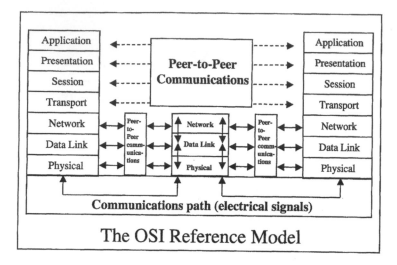

Figure 10-1. The OSI Model.

IP-Based Networks

TCP/IP is a protocol suite. A protocol suite (also known as a "stack") is made up of a set of protocols that permit physical networks to interoperate. Protocol suites are layered in such a way that they break up the interoperability task into a set of services. These layers are often discussed in relation to the ISO's OSI Reference Model. The OSI model describes a conceptual seven-layer architecture that divvies up interoperability tasks. It does this by starting at the top, with an application (conferencing, for instance) working its way downward all the way to where the application's bits are placed onto a physical network.

In a layered protocol, higher-level protocols require services from protocols below them. In return, they provide services to lower-layer protocols. Each protocol in a "stack" also provides services to its peer on the other side of the connection.

TCP/IP conforms to the OSI reference model conceptually, but not precisely. It is has four conceptual layers. There is the Application Layer, which passes messages or streams between devices. There is the Transport Layer, whose primary duty it is to provide reliable, end-to-end communication across applications. There is the Internet Layer, which handles communications from one machine to another. This layer moves packets or datagrams, as they are known in IP networking terminology. Lastly, there is the Network Interface Layer. It is responsible for accepting IP datagrams and transmitting them over a specific network. Using these four layers, TCP/IP forms a virtual network in which individual hosts are identified not by physical network addresses but by IP addresses.

The networks that comprise an inter-network (whether the Internet or an intranet) are physically connected by devices called routers. A router is a special computer that is designed to transfer datagrams from one network to another. The IP protocol routes datagrams by hopping from network to network. The router acts as a connection point between networks, or a *gateway*. Software hides the underlying routing process from the user and thereby makes the process transparent.

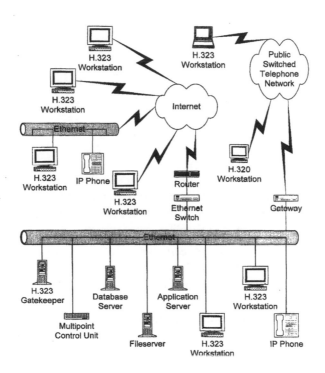

Figure 10-2. An H.323 Network.

Routers perform their task by using IP addresses. IP addresses are actually 32-bit strings that identify specific network devices. To make them easier to remember and understand, people express IP addresses in dotted decimal notation, (e.g., 134.13.9.6). When conveyed across a network, these addresses are converted to hexadecimal notation (hence, the 32 bits). Special database servers, or DNS (Domain Name System, RFC 1034, 1035) servers, translate IP addresses into more recognizable Internet addresses. An example of this is name@company.com.

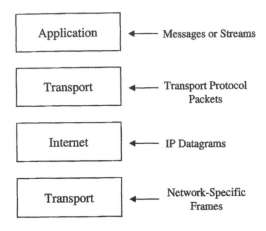

Figure 10-3. Layering in the TCP/IP Protocol Stack (Laura Long).

The component of the address to the left of the @ character is the local part of an address, and identifies a user. The component to the right side of the @ character identifies the *domain*, or the name of the system (whether that person's company or an Internet service provider) that provides services for the user.

A domain name such as *company.com* refers to the company, *company*, in the commercial (*com*) sector. Other suffixes include *org* for non-profit organizations, *edu* for educational institutions, *gov* for government agencies, and *net* for network services providers. The domain *company.com* can be further subdivided in to sub-domains such as *eng.company.com* and *mkt.company.com*. The subdivision will distribute the site in to smaller sections to avoid network congestion and to simplify network management. The term *sub-networks*, or *subnets*, refers to these subdivisions or smaller network segments.

Through a global network of routers and DNS servers, computers anywhere in the world can communicate with any other host, as long as they have that host's Internet address and they have been granted privileges to do so. After DNS does its job at an Internet-wide level, a smaller version of DNS is applied to convert a human-friendly Internet address to a computer-friendly IP address. Lower in the stack, the network interface layer converts an IP address into a machine address (e.g., a 48-bit Ethernet ID).

Internetworking, specifically TCP/IP internetworking, was designed for the exchange of data and documents. It was never intended to meet the rigors of time-sensitive media streams. This becomes glaringly apparent when people try to conduct videoconferences over the Internet. Some desktop conferencing products have tried to deliver synchronized audio/video over the Net. However, most reviewers share the same opinion: these products deliver two results. First, they tend to wreak havoc on the performance of other applications. Second, they produce only minimally acceptable frame rates, with unstable, garbled, and poorly synchronized sound and pictures.

The biggest obstacle to deploying Real-Time multimedia over the Internet is the lack of an appropriate infrastructure and interoperability between different schemes. TCP/IP protocols do what they are supposed to do: transmit data across a network reliably, and in a proper sequence. In theory, when network traffic is light, TCP/IP is capable of reliably transporting Real-Time, synchronized multimedia. However, in the real world, most networks suffer from periodic congestion. Faced with contention, TCP/IP is not able to guarantee that Real-Time media will arrive properly timed, and with all audio and image packets intact.

The Internet Engineering Task Force's (IETF) Audio-Video Transport Working Group is adapting the IP protocol stack to the unique needs of interactive media streams. Several Internet standards have been the result, with more on the way. This chapter examines the most widely-implemented, the User Datagram Protocol (UDP), the Real-Time Transport Protocol (RTP), the Real-Time Transport Control Protocol (RTCP) and the Resource Reservation Protocol (RSVP).

We will also discuss the Real-Time Streaming Protocol (RTSP), IP multicasting.

UDP

Overview of UDP

The User Datagram Protocol (UDP) is a streaming IP protocol that is being substituted for TCP in Real-Time multimedia applications. UDP is unidirectional; no mechanism is available at the sender to know if a packet has been successfully received. Thus, UDP does not wait for transmission acknowledgements. It carries the packets presented to it but, if those packets are dropped or duplicated, or arrive out of sequence, UDP takes no corrective action.

On the other hand, UDP increases the throughput of an IP network. Since there is no need to wait for acknowledgement by receivers, the UDP is an efficient mechanism for delivering multiple streams of information. TCP, on the other hand, has an inherent need to track and retransmit dropped or corrupted packets. The service that TCP provides has, unfortunately, no value for Real-Time applications. Video communication sequence and timing does not allow for a lost frame of video or packet of audio, to be re-sent and re-inserted in the data stream.

Today, UDP provides the foundation for the other IETF protocols that are being developed to support video communication over the Internet. These include the RTP, the RTCP, and RSVP. UDP is also being used for IP multicasting. Applications developed in accordance with the IETF protocols rely on codecs to detect and recover from packet loss, in conjunction with RTP.

UDP and Firewalls

Most organizations that deploy Internet audio and video technologies to conduct business and communicate with their global counterparts demand that these communications be secured through corporate firewalls. Firewalls are usually placed between internal LANs and WANs and external networks such as the Internet (they usually reside directly behind the router). Although it may provide reporting functions, a firewall's primary purpose is access control. Firewalls require configuration; the forfeiture in allowing a service (e.g., UDP) through the firewall is that one creates another door through which a hacker may intrude. Setting up firewall access rules on routers (what can pass through, what will be denied) is a time-consuming, error-prone process. However, it is generally accepted that static filters are not adequate for security, and firewalls are being designed to be simpler to configure. Still, network administrators generally adopt a "whatever is not expressly permitted is forbidden" approach to incoming Internet traffic.

Most Internet (and intranet) applications are delivered as connectionless streams over UDP sockets. Since firewalls approach security in connection-oriented terms, they do not permit packets that are associated with connectionless streams to penetrate the corporate network.

Different types of firewalls handle security differently, of course. There are four general categories: packet-level firewalls, circuit-level firewalls, proxy servers, and dynamic or intelligent packet filters (a well-known term for what this type of firewall performs is *stateful inspection*). Today most categories are combined into hybrid products that offer multiple security levels. At the lowest security level are packet-filtering routers; they are not truly firewalls, but they do examine each (or at least some) incoming IP packets. The router checks the packet's source, destination addresses, or services and, based on what it learns, decides whether to allow the packet on the corporate network. Packet filters are complex to maintain. They are also easy to penetrate.

Circuit-level gateways operate at the session level (as does UDP). They are host-based, and as such, act as a relay program. They receive packets from one host and hand them off to another. This design is more secure than that of packet-filtering routers, but less secure than application-level gateways. Circuit-level gateways are not often used as standalone firewalls, but are combined with application gateways to deliver a full suite of features for a robust firewall system.

Proxy-based firewalls have a detailed understanding of specific IP protocols and are extremely secure because they allow packet analysis at the application layer; packets claiming to be of one type but carrying another type of data will be rejected. They are very good for security because they do not actually allow an outside host to connect to an inside host but, rather, relay information from one to the other. Unfortunately, a proxy firewall does not deal well with connectionless protocols such as UDP. Proxy servers can also add delay to connections if the computers on which they are running are slow or if the firewall code is not efficient (this increasingly becoming less of an issue).

Dynamic filtering firewalls offer sophisticated, customizable, granular access-control schemes that allow network managers to determine who (and under what conditions) they will allow onto corporate networks. Firewalls in this category deal with packets at the applications-level. They are sometimes called *stateful* firewalls because they enhance packet-filtering with historical context to keep a history of all IP traffic that moves through the firewall. If a packet has gone from an intranet to someone on the Internet, it is generally assumed that it is acceptable to reverse the process, and thereby allow the host on the outside to return packets through the firewall. Application-level firewalls adapt best to UDP-supported media streams.

Protocols such as RTP (which attaches a header to UDP packets that can be used to detect packet loss), RTCP (which guarantees quality of service) and RSVP (which establishes service classes) are bringing streaming data to TCP/IP, as supporting router infrastructure is being deployed.

RTP

Overview of RTP

In January 1996, the IETF's Audio-Video Transport Working Group published RFC 1889, entitled, "RTP: A Transport Protocol for Real-Time Applications" and with it, a companion standard, RFC 1890, "RTP Profile for Audio and Video Conferences with Minimal Control." RTP and RTCP are part of the IETF's IIS (Internet Integrated Services) model, along with RSVP (described below). A second version of RTP and RTCP became available shortly thereafter. RTP bridges the gap between the lower-level UDP transport and a Real-Time multimedia application.

The RTP is a streams-oriented protocol. It provides end-to-end network transport services for Real-Time multimedia applications over IP based networks. It can be used in multicast (multipoint) or unicast (point-to-point) applications.

How RTP Works

RTP applications can compensate, to a degree, for the instability of UDP/IP networks by appending a 10-byte header onto each UDP packet. The header is designed with Real-Time transport in mind. It contains a time stamp and a sequence number to keep multimedia packets in the proper order during transmission. The time stamp is used to pace frame display and synchronize multiple streams (e.g., voice, video and

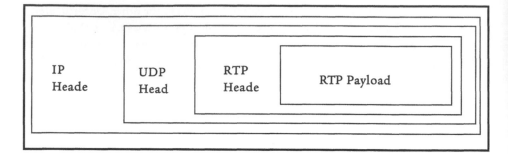

Figure 10-4. IP Packet Containing RTP Data

data) while the sequence number is used to reassemble the data stream at the receiving end. The header also carries information about the information payload (in other words, the types of compression used to generate the media stream.)

Unlike the point-to-point TCP, only end-points (and not network devices) are required to support RTP. An interactive audiovisual application in an RTP-based conference must buffer data and examine the header's timing and sequence information. Using this information the application performs sequencing, synchronization, and duplicate packet elimination. The result is a coherent media stream within which varying degrees of network latency are compensated. The application may also use RTP payload information (also included in the header) to select a codec to be used for the exchange (e.g., MPEG or H.261).

The Future of RTP

RTP is being integrated into many video communication and audio products. One of its shortcomings is that it is limited to IP networks. On the other hand, IP seems to be the direction in which the networking world is headed. Still, no single Real-Time protocol has emerged as the industry standard.

Nevertheless, RTP has gained a tremendous amount of attention as is demonstrated by the proliferation of RTP based IETF drafts:

RFC 2032: Describes a scheme to packetize an H.261 video stream for transport using RTP with any of the underlying protocols that carry RTP.

RFC 2190: Specifies the payload format for encapsulating an H.263 bitstream in the RTP. Three modes are defined for the H.263 payload header. An RTP packet can use one of the three modes for H.263 video streams depending on the desired network packet size and H.263 encoding options employed. The shortest H.263 payload header (mode A) supports fragmentation at Group of Block (GOB) boundaries. The long H.263 payload headers (mode B and C) support fragmentation at Macroblock boundaries.

RFC 2198: Describes a payload format for use with RTP, version 2, for encoding redundant audio data. The primary motivation for the scheme is the development of audio conferencing tools for use with lossy packet networks such as the Internet MBONE, although this scheme is not limited to such applications.

RFC 2250: Describes a packetization scheme for MPEG video and audio streams. The scheme proposed can be used to transport such a video or audio flow over the transport protocols supported by RTP. Two approaches are described. The first is designed to support maximum interoperability with MPEG System environments. The second is designed to provide maximum compatibility with other RTP-encapsulated media streams and future IETF conference-control work.

RFC 2343: Describes a payload type for bundled, MPEG-2 encoded video and audio data that may be used with RTP, version 2. Bundling has advantages for this payload type particularly in video-on-demand applications. This payload type may be used when its advantages are important enough to sacrifice the modularity of having separate audio and video streams.

RFC 2354: Summarizes a range of possible techniques for the repair of continuous media streams subject to packet loss. The techniques discussed include redundant transmission, retransmission, interleaving, and forward error correction. The range of applicability of these techniques is noted, together with the protocol requirements and dependencies.

RFC 2429: Specifies an RTP payload header format applicable to the transmission of video streams generated based on the 1998 version of ITU-T Recommendation H.263. Because the 1998 version of H.263 is a superset of the 1996 syntax, this format can also be used with the 1996 version of H.263, and is recommended for this use by new implementations. This format does not replace RFC 2190, which continues to be used by existing implementations, and may be required for backward compatibility in new implementations.

RFC 2431: Specifies the RTP payload format for encapsulating ITU Recommendation BT.656-3 video streams in RTP. Each RTP packet contains all or a portion of one scan line as defined by ITU Recommendation BT.601-5, and includes fragmentation, decoding, and positioning information.

RFC 2435: Describes the RTP payload format for JPEG video streams. The packet format is optimized for Real-Time video streams where codec parameters rarely change from frame to frame.

RFC 2508: Describes a method for compressing the headers of IP/UDP/RTP datagrams to reduce overhead on low-speed serial links. In many cases, all three headers can be compressed to 2–4 bytes.

RTCP

Overview of RTCP

The Real-Time Transport Control Protocol is a companion protocol to RTP. Like the RTP, it is also standardized in RFC 1889 and RFC 1890. The primary function of the RTCP is to provide feedback on the quality of a connection.

RTCP allows monitoring of multimedia data delivery in a manner scalable to large multicast networks. It provides minimal control and identification functionality to senders and recipients of a media stream. RTCP, like RTP, is simply data from the point of view of routers that handle packets on their way between senders and recipients. However, conferencing end-points can use RTCP to provide any-to-any feedback on current network conditions. Knowing something about network performance, and its impact on a conference, gives participants some options. The sender might reduce the video frame rate to conserve bandwidth. Another choice might be to switch from IP to ISDN transmission. If video is not critical to the collaborative process, the conference manager might elect to eliminate the picture portion of the conference altogether and rely strictly on voice and data.

How RTCP Works

RTCP works by periodically transmitting control packets to all participants in a session, using the same distribution method as the packets themselves. The underlying protocol must provide multiplexing of the data and control packets. One way of doing this is to use separate UDP port numbers.

RTCP carries a persistent transport-level identifier for an RTP source called the canonical name or CNAME. Receivers use the CNAME to keep track of each participant in a conference and to associate multiple data streams from a given participant in a set of related RTP sessions.

All participants in an RTP multicast must send RTCP packets to each other. Therefore, the rate of the RTCP packet flow must be controllable to allow for a very large number of participants. RFC 1889 describes the RTCP and includes a discussion of the RTCP Transmission Interval. It includes an algorithm that calculates the interval between sending compound RTCP packets to divide the protocol's allowed control traffic bandwidth among participants.

RTCP can also be used to display the name of a participant at the user interface. This is most useful in loosely controlled sessions in which participants come and go at will, or when membership control features are not part of an application.

There are five types of RTCP packets. Type 1 is Sender Report or SR. It is used for transmitting and receiving reception statistics from all participants that are active senders. Type 2 is Receiver Report or RR. This type of packet is used to convey reception statistics from participants that are not active senders. Source-Destination packets comprise type three. CNAME is included in such a packet. Type 4 is used to indicate the end of a session and is aptly named BYE. There is also a fifth type, known as APP. It is used to carry application-specific functions.

RTCP is being implemented, along with RTP (whose services it uses), by a number of Internet videophone providers. It is becoming a popular and widely supported protocol, and is included as part of many Internet/intranet products.

RSVP: IETF RFC 2205

Overview of RSVP

The Resource Reservation Protocol (RSVP) relies on its companion protocols (UDP, RTP and RTCP) to promise Real-Time or near Real-Time transport over IP networks. In creating bandwidth-on-demand, it is RSVP's job to request dynamically that network bandwidth be set aside. Such bandwidth is required for building a Real-Time multimedia session between one or more devices.

RSVP acts on behalf of applications to request necessary bandwidth from devices throughout the network in accordance with priority. It allows a receiver to request a specific amount of bandwidth for a particular data stream, and to receive a confirmation or denial of that request. This protocol is the one necessary for guaranteeing QoS (quality of service) on the Internet. RSVP requests the bandwidth needed for a session; it does not have the authority to command that bandwidth and, therefore, cannot guarantee that resource availability.

Standards Approval

RSVP was first outlined in 1994 in IETF RFC 1633, "Proposed extension to the Internet architecture and protocols to provide integrated services, i.e., to support real-time as well as the current non-Real-Time service of IP." It was approved as an IETF standard in October of 1996, and was revised again in 1997 as RFC 2205. Several key concerns remain even after the first version was approved.

How RSVP Works

RSVP includes two components—a client and a network device. The RSVP client is built into the video communication application, which relies on it to request service priority levels from the RSVP network devices that support the transmission infrastructure. These network devices must, in return, recognize and respond to client requests. Upon receiving a request, an RSVP-enabled device considers network conditions and other factors to determine whether the request can be granted. If the situation looks promising, the router signals the application that it can proceed with its transmission.

RSVP requests theoretically result in resources being reserved in each node along the data path, but only in one direction (two-way traffic requires two RSVP flows). RSVP operates on top of IP, as a transport protocol in the protocol stack. Like ICMP, IGMP, or routing protocols, RSVP is a control protocol and does not transport application data. RSVP is designed to operate with current and future unicast and multicast routing protocols and is not a routing protocol. Routing protocols determine where packets go, and RSVP attempts to ensure the QoS of those packets. RSVP makes receivers responsible for requesting a specific QoS (RSVP93). RSVP's weakness lies in the fact that receivers tend to grant the highest QoS to every application that they use. The RSVP protocol carries the user's request to all the routers and hosts along the reverse data path to the data source, as far as the router from which the receiver's data path joins the multicast distribution tree.

RSVP Classes of Service

RSVP has four service classes, although only two are aimed at Real-Time multimedia. The two most common are Guaranteed Delay and Controlled Load.

As the name implies, guaranteed delivery specifies the maximum allowable transmission delay. Controlled load allows priorities to be associated with different streams of data and, in response to those priorities, tries to provide consistent service, even if network usage increases.

For applications that can tolerate the late arrival of some packets, two additional RSVP classes, Controlled Delay and Predictive Service are offered. They allow for specified bounds on delay.

Deploying RSVP on the Client

RSVP sits on top of IP in the protocol stack. Client applications will have to add it to use the services of RSVP network devices. Early implementations of RSVP shipped as middleware (Precept Software has a well-publicized one) but RSVP soon became an inherent part of the desktop operating system (e.g., Windows 98 and Windows 2000).

RSVP's Impact on Network Infrastructure

While RSVP does not require a wholesale overhaul of network infrastructure, it does require support by the switches and/or routers that make up a transmission path. If any device along the path fails to support RSVP, it can circumvent the efforts of the rest that do.

For switches and routers to become RSVP-enabled, they must be able to provide multiple queues. These queues buffer lower-priority data when higher-priority data is being transmitted. Most modern routers offer multiple queuing capabilities. Consequently, they must deal with traffic on a first-in, first-out (FIFO) basis. Here, RSVP faces a challenge. The frame format of Ethernet varies, with no frame smaller than 64 octets (bytes) and no frame larger than 1518 octets. If two bandwidth-reserved frames arrive at a single-queue router simultaneously, the first one to enter the buffer is the first in line for transmission. Because the whole frame must be sent, the second stream is held in queue. If the first frame comes close to Ethernet's maximum 1518 octet frame size, the second application's payload will be noticeably delayed. Because of this, RSVP-enhanced IP networks fare poorly in comparison to

ATM, which uses very small cells (53 bytes), and thereby makes QoS guarantees more realistic.

The real-world effect is that many routers must be upgraded or replaced for widespread deployment of RSVP. The two largest router/switch manufacturers, Cisco (San Jose, California) and Bay Networks (Billerica, Massachusetts), have long since made their products capable of responding to RSVP requests. Cisco was first to market, with its IOS software that began shipping in September of 1996. The two groups most likely to buy and install those upgrades are private corporations and Internet service providers. The Internet service provider (ISP) BBN Planet (now GTE), Cambridge, Massachusetts offered the first commercial implementation.

In 1996, RSVP was largely a test-bed implementation. Vendors, working collaboratively, tried out many different implementations. Most early RSVP implementations left it up to the user to set load and delay parameters. This may work in an experiment, but when RSVP is widely deployed, this approach will quickly fall apart. Users, perennially bandwidth-starved, would promote their application to the front of the service line. In response, vendors are developing offerings in such a way that network administrators (often the decision makers when it comes to selecting products) have the ability to charge for higher-priority service and therefore, to limit congestion.

The IETF is developing specifications to allow users to assign priorities to applications according to their business needs. In the years that it has been perfecting RSVP, competing QoS standards such as MPLS (multiprotocol label switching) and Diffserv (differentiated services) have evolved.

MPLS

In 2004, Sprint joined AT&T, Equant and MCI (legally known as WorldCom) by offering Multiprotocol Label Switching (MPLS) services to its customers, and seemingly sealed the destiny of the standard as the protocol for next-generation IP networks.

MPLS is an IETF initiative that integrates Layer 2 information about bandwidth, utilization, and latency, into Layer 3 (IP) within an autonomous system to enhance IP-packet transmission. MPLS enables network operators to more effectively route traffic around link bottlenecks and failures. It also enables them to apply priorities and preferences in managing diverse data streams, and thereby provide QoS. With MPLS, network operators can sell multiple classes of service with various corresponding price tags.

When a packet enters a MPLS network, a label edge router (LER) attaches a label to it that contains routing information such as destination, delay, and bandwidth. The label (often called an *identifier*) comes with information on the source IP address, Layer 4 socket number information, and differentiated services. Upon completing and mapping this classification, the router assigns packets to appropriate labeled switch paths (LSPs), in which label switch routers (LSRs) label the outgoing packets. LSPs enable network operators to route and divert traffic based on data-stream type and customer.

RTSP

The Real-Time Streaming Protocol (RTSP) is a streaming-media client/server-oriented protocol that was submitted to the IETF in draft form on October 9, 1996, and was defined by the IETF in RFC 2326 in April 1998. RTSP is an application-level protocol for control over the delivery of data with Real-Time properties. It provides an extensible framework to enable controlled, on-demand delivery of audio and video, whether live or stored. RTSP is intended to control multiple data delivery sessions, provide a means for choosing delivery channels such as UDP, multicast UDP, or TCP, and provide a means for choosing delivery mechanisms in conjunction with RTP.

The RTSP allows interoperability between client/server multimedia products from multiple vendors. It can be implemented across a broad array of client-side operating systems such as Macintosh and various flavors of Windows and Unix. It is supported by an equally large number of server-side platforms including Macintosh, Windows NT, and variations of UNIX.

Using RTSP, application developers can create products that will interoperate within the heterogeneous environment of the Internet and corporate intranets. The protocol defines how a server can connect to a client-side receiver (or multiple receivers) to stream multimedia content. Streaming multimedia can be anything that includes time-based information, such as on-demand multimedia playback, live Real-Time feeds, or stored non-Real-Time programming. RTSP is closely associated with the ITU-T H.323 protocol stack, and facilitates merging of the separate spheres of telephony, conferencing, and multimedia broadcasting.

RTSP allows non-traditional broadcasters to leverage IP networks to compete with television and radio programmers. The protocol allows data to be displayed as it is received, rather after all packets are stored and processed. This makes RTSP ideal for developing network computer applications in which hard disks are either non-existent or very small, or when rights management does not favor distributing files.

Until 1999, vendors developed applications independently or in collaborative, but proprietary efforts. This did not meet the needs of the Internet, which relies upon open protocols. Progressive Networks, Inc. (RealNetworks) and Netscape Communications spearheaded the collaborative efforts of forty companies on the development of RTSP.

RTSP allows content providers to create and manage the delivery of programs in a commercial environment. Content providers want to be compensated for their efforts, and RTSP facilitates delivery control and efficiency, quality of service, security, usage measurement, and rights management.

RTSP provides a mechanism for choosing IP-based delivery channels (UDP, IP multicast and TCP) and supports delivery methods based upon the RTP standard.

RealNetworks, one of the two primary RTSP sponsors, offers software developments kits for creating RTSP-compliant applications. In only four years, from 1995 to 1999, RealNetworks software systems became the most pervasive method of streaming media on the Internet. By 2000, the company claimed that more than 80 million unique users had been registered, its RealPlayer download rate had exceeded 175,000 per day, more than 145,000 hours of live entertainment had been broadcast over the Internet, and hundreds of thousands of hours of content were available on-demand. Of course, it must now compete with Microsoft's Windows Media product.

IP MULTICASTING AND ITS PROTOCOLS

IP multicasting (for that matter, multicasting in general) calls for the simultaneous transmission of data to a designated subset of network users. It exists somewhere between unicasting (sending a stream of data to a single user) and broadcasting (sending a data stream to everyone on a network.) Most frame-based networks (e.g., Ethernet) can communicate in all three modes.

Most applications incorporate unicast. Examples of a unicast might include a client contacting a server or two users exchanging messages and files across the network. Broadcasts allow a single station to communicate simultaneously with all other devices on the network. Broadcasts must reach across subnets to reach all machines on an IP network. Broadcasts have their place in the corporate intranet, where they are useful for company meetings, product announcements and addresses by the chair of the board. Broadcasts are not as useful when interactivity is facilitative.

IP multicasting is similar to a pay-per-view cable television channel. A viewer can tune if they are authorized to do so. To handle the invitation process, a sending computer specifies a group of recipients by transmitting a single copy of its multicast to a special IP address. Special protocols are employed to translate from the group address to the IP addresses or group members, and from there, to the MAC addresses of individual hosts. Multicasting also requires a special protocol that is used to manage group membership. Lastly, it requires routers that can interpret multicast protocols and use them to distribute packets only to the proper networks. These three special multicast-specific functions—address translation, dynamic management of group membership, and multicast-aware routing—are discussed in the next section.

Layer-3-to-Layer-2 Address Translation

The hosts that combine to make up a multicast group are assembled under a special class of IP address. This address category is known as a Class D address. There are also Class A, B and C addresses (used for unicasts and broadcasts) and Class E addresses (unassigned at present but reserved for the future).

Class D addresses exist at the network-layer and are used exclusively for multicast traffic. These addresses are allocated dynamically; each represents a select group of hosts interested in receiving a specific multicast. The network-layer (Class D IP address) that is used to communicate with a group of receivers must be mapped over to the data-link layer (typically Ethernet MAC) addresses associated with each group member.

Translation between an IP multicast address and a data-link layer address is accomplished by dropping the lowest-order 23 bits of the IP address into the low-order 23 bits of the MAC address. In order for a host to participate in a multicast, its TCP/IP stack must be IP-multicast aware— most, but not all, are. Network adapters must be multicast-enabled, too. Good ones allow several MAC addresses to be programmed in by the TCP/IP stack. This allows a host to monitor several multicast groups at once without receiving an entire general range of MAC addresses. Network adapters in this category use both hardware and software (drivers) to deliver their functionality.

Dynamic Management of Group Membership

IP multicasting works thanks to an extension of the IP service interface. This extension allows upper layer protocol modules to request that their hosts create, join, confirm ongoing participation in, or leave a multicast group. This process is known as dynamic registration. It is served within the IP module by the Internet Group Management Protocol (IGMP), which was defined by the IETF in RFC 1112.

IGMP provides the mechanism by which users can dynamically subscribe and unsubscribe to multicasts. It ensures that multicast traffic is present only on those subnets where one or more hosts are actively requesting it. Using IGMP, hosts inform their network or subnet that they are members of a particular multicast group. Routers use this registration information to transmit, or abstain from transmitting, multicast packets that are associated with specific groups.

There are three versions of IGMP. IGMPv2 lets a host tell a router that it no longer wants to receive traffic. The router will stop retransmission of traffic, based on a time-out parameter. This is not possible with IGMPv1. IGMPv3 adds source filtering ("only some" or "all but some" sources) to the host membership protocol. Routers merge requests from different hosts so that everyone gets at least what they want (but maybe more). IGMP versions one and two attempted to reduce the group-membership protocol traffic by having hosts suppress group memberships that had already been reported. IGMPv3 transfers the burden of merging multiple requests from the hosts to the routers by eliminating the need for suppression by sending reports to the all-routers multicast group. There are three types of reports: *current state, filter mode change*, and *source list change*. The current state report is used in response to the router's periodic queries, and the filter mode change and source list change (collectively known as *state change records*) are used when the host's state changes in response to an application request. Queries are backwards compatible, so that an IGMPv3 router may send a single message and get responses back from IGMPv3, IGMPv2 and IGMPv1 hosts.

IGMP client support is offered in the TCP/IP stacks of most hosts, including most flavors of Windows. IGMP must also be supported at the router level; CISCO, Bay Networks and others offer routers that can cope with it.

Multicast-Aware Routing

There are several standards available for routing IP Multicast traffic. These are described in three separate IETF documents. RFC 1075 defines the Distance Vector Multicast Routing Protocol (DVMRP). RFC 1584 defines the Multicast Open Shortest Path First (MOSPF) protocol—an IP Multicast extension to the Internet's OSPF. The Protocol-Independent Multicast (PIM) Sparse Mode was defined by the IETF in RFC 2117; PIM-Dense Mode (PIM-DM) is, at the time of this writing, still a draft. PIM is a multicast protocol that can be used in conjunction with all unicast IP routing protocols.

DVMRP uses a technique known as Reverse Path Forwarding to flood a multicast packet out of all paths except the one that leads back to the packet's source. Doing so permits a data stream to reach all LANs. If a router is attached to a set of LANs that do not wish to receive a particular multicast group, it can send a *prune* message back up the *distribution tree*. The prune message stops subsequent packets from flowing to LANs that do not support group members. DVMRP refloods on a periodic basis in order to absorb hosts who are just tuning in to the multicast. The more often that DVMRP floods, the shorter the time that a new recipient will have to wait to be added to a multicast. Flooding frequently, on the other hand, greatly increases network traffic.

Early versions of DVMRP did not support *pruning*. Customers should not implement DVMRP for IP multicasting unless their routers support prune.

DVMRP is used in the Internet's multicast backbone (MBONE) which has a following in the academic community. The MBONE's primary application is the transmission of conference proceedings although a variation of it (CU-SeeMe) also supports desktop conferencing.

MOSPF was defined as an extension to the Open Shortest Path First (OSPF) unicast routing protocol. It will only work over networks that use OSPF. OSPF requires that every router in a network be aware of all of the network's available links. Each router calculates routes from itself to all possible destinations. MOSPF works by including multicast information in OSPF link state advertisements. An MOSPF router learns which multicast groups are active on which LANs, and then builds a distribution tree for each source/group pair. The tree state is cached and recomputed every time that a link state change occurs (or when the cache times out).

MOSPF works best when there are only a few simultaneous multicasts at any given time. It should be avoided by environments that have many, simultaneous multicasts or by environments plagued by link failures.

PIM works with all existing unicast routing protocols, and is usually more popular from both a customer and a vendor perspective because it is simpler to implement and to manage. PIM supports two different types of multipoint traffic distribution patterns: dense and sparse. PIM-Dense mode (PIM-DM), a proper subset of PIM-Sparse Mode (PIM-SM) performs best when senders and receivers are in close proximity to one another. It is also best adapted to situations where there are few senders and many receivers. PIM handles high volumes of multicast traffic and it does well when the streams of multicast traffic are continual. PIM-DM uses Reverse Path Forwarding and resembles DVMRP. The most significant difference between DVMRP and dense-mode PIM is that PIM works with whatever unicast protocol is being used; PIM does not require any particular unicast protocol.

Sparse-mode PIM is optimized for environments with intermittent multicasts in with many multipoint data streams, and for senders and receivers that WAN segments separate. It works by defining a Rendezvous Point. When a sender wants to send data, it first sends to the Rendezvous Point. When a receiver wants to receive data, it registers with the Rendezvous Point. Once the data stream begins to flow from sender to Rendezvous Point, the routers in the path will optimize the path automatically to eliminate superfluous hops. PM-DM is simpler because it does not require the control overhead that is necessary for a Rendezvous Point, which is often not needed when leveraging the data-handling capacity in the switch infrastructure, and one is not running a connection oriented network (e.g., ATM). Another difference between PIM-SM and PIM-DM is that the latter does not transmit periodic *joins*, only explicitly triggered grafts/prunes. Although PIM-DM is not yet a RFP, it does work, and it is interoperable.

Sparse-mode PIM assumes that no hosts want the multicast traffic unless they specifically ask for it. It works well in corporate environments where multiple groups may need to multicast across WANs, but only occasionally.

Conferencing and the Internet

Enterprises will continue to seek networks that facilitate the transmission of every type of data, including Real-Time multimedia. A debate has subsided for now, and the enterprise network of choice is a refined set of IP protocols that runs over fat pipes. Some companies circumvented IP's best-effort design limitations by adopting ATM cell switching for LAN and WAN, but ATM's demise finally seems imminent. Deterministic ATM-based networks currently have the advantage in service quality, but IP protocols are widely implemented and are constantly evolving. Moreover, it is very difficult to find skilled resources with an ATM background (or to pay for them once one has found them). In contrast, IP-oriented network management expertise is prevalent. Right or wrong, the current Internet/intranet infatuation only strengthens the position of IP-networking. Still, IP was designed in the 1970s when interactive multimedia over packet-switched networks was hardly conceivable; no matter what tires, engine, and transmission one puts on a '62 Dodge Dart, it is still a '62 Dodge Dart.

ATM would seem to be the future. It apportions bandwidth in a connection-oriented arrangement, and thereby accommodates multiple streaming data flows. IP-based connectionless networks transmit in bursts and force video and audio streams to adapt, if not suffer. ATM has an inherent provision for service quality. IP-based protocols are being retrofitted to meet this requirement. ATM can support TCP/IP traffic and any other type of traffic, both gracefully and with applications-specific adaptation layering. IP-based protocols do their best to adapt, but their best is often contrived and awkward. ATM upgrades are expensive but applications such as Real-Time streaming media are motivating organizations to take a second look.

We will discuss ATM in greater depth in the following chapter as we move downward in the stack to look at physical and data-link layer transmission systems.

CHAPTER 10 REFERENCES

Baker, Richard H., "Fighting fire with firewalls," *InformationWeek* October 21, 1996 (Computer Select).

Comer, Douglas E., *Internetworking with TCP/IP Volume I: Principles, Protocols and Architecture,* 2nd edition, Prentice Hall 1991.

Coy, Peter, Hof, Robert D., Judge, Paul C. "Has the Net finally reached the wall?" *Business Week,* August 26, 1996 n3490 pp. 62-67.

Crotty, Cameron. "The Revolution will be Netcast," *Macworld* October 1996 v13 n10 pp. 153-155.

Frank, Alan. "Multimedia LANs and WANs," *LAN Magazine* July 1996 v11 n7 pp. 81-86.

Fulton, Sean. "Steaming multimedia hits the 'net," *CommunicationsWeek* July 22, 1996 n620 p. 52-58.

Kosiur, Dave. "Internet telephony moves to embrace standards," *PC Week* November 18, 1996 v13 n46 p. N16.

McLean, Michelle Rae. "RSVP: promises and problems," *LAN Times* October 14, 1996 v13 n23 p. 39-42.

Plain, Steve. "Streamlining," *Computer Shopper* December 1996 v16 n12 p 620.

Rogers, Amy. "A blueprint for quality of service," *CommunicationsWeek* November 18, 1996 n638 p. 1.

Schulzrinne, H., Casner, S., Frederick, R., and V. Jacobson, RFC 1889, "RTP: A Transport Protocol for Real-Time Applications," *IETF,* January 1996.

Schulzrinne, Henning and Casner, Stephen, "RTP: A Transport Protocol for Real-Time Applications," Audio-Video Transport Working Group, Internet Engineering Task Force, working draft, October 20, 1993.

Steinke, Steve. "The Internet as your WAN," *LAN Magazine* October 1996 v11 n11 pp. 47-52.

Tannenbaum, Todd. "IP Multicasting: Diving Through The Layers," *Network Computing* November 15, 1996 pp. 156-160.

11

CIRCUIT-SWITCHED NETWORKS

"It's kind of fun to do the impossible."
Walt Disney (1901-1966)

Two types of networks can be used to transport video/data conferencing signals, circuit-switched and packet-switched. In this chapter, we will discuss circuit-switched networks; in the next, we will discuss packet switching.

Until Ethernet really became established, in 1979, the concept of packet switching was scarcely implemented anywhere. Today, packets traverse LANs and WANs all over the world; indeed, many people use the word networking to refer to packet-switched service exclusively, without acknowledging that circuit switching, a different type of networking, is older by almost a century.

Today, the bulk of video conferencing takes place over circuit-switched telephone lines, such as ISDN or Switched 56, or over dedicated lines, such as T-1 or fractional T-1. Even when the conference endpoints are workstations on the same LAN, they typically use telephone lines, not the LAN, for video conferencing traffic. For that reason, we will start our discussion of video/data conferencing networks with a look at the public switched telephone network (PSTN).

THE PUBLIC SWITCHED TELEPHONE NETWORK (PSTN)

Installing a video communication system requires some understanding of the circuit-switched *public switched telephone network* known as the *PSTN*. The PSTN is the set of digital dial-up services offered by carriers. It includes network-based services and network-based switching.

The PSTN consists of analog and digital components. In order to identify what is analog and what is digital, we will divide the PTSN into three broad categories: *access, switching,* and *transport.*

Access is the portion of the circuit that connects the customer's premises to the first network switching point. Another term used to refer to access is *local loop.* A local loop has two ends: the one that connects to the customer's equipment and the one that connects to the telephone company's (*carrier*) switching equipment. The switching equipment is said to reside in the carrier's *central office* or *CO.* Switching allows one node to be connected to another node and is accomplished via addressing (in which an address is a telephone number). A circuit-switched connection can be made directly within the serving central office (if the same CO also serves the called party) or carried across the PSTN to a distant CO.

The local loop is usually conditioned to provide analog transmission service (POTS). Most customers in the United States can buy, as an option, digital local access. The remainder of the PSTN is digital. CO switches are digital and the trunks that connect them are high-capacity digital facilities.

Prior to 1996, only local exchange carriers (LECs), with a few minor exceptions, provided the local loop, in a monopoly arrangement. Although allowing monopolies to exist seems like a misguided practice, at the time that the U.S. PSTN was being constructed, this was really the only alternative.

The Old PSTN

In 1934, Congress passed the Communications Act. It created the Federal Communications Commission (FCC). The Act was intended to regulate the market for telecommunications services, and allow everyone to have a telephone. At the time that the Act passed, the U.S. was building its telecommunications infrastructure. The companies that supplied the funds (primarily AT&T) wanted to be assured of a reasonable return on their investment. Permitting the construction of competitive networks would jeopardize an investor's profitability. Moreover, duplicating infrastructure was wasteful. As long as regulation ensured that a single operator (carrier) did not engage in predatory practices, one network per area would suffice. The funds required to build regional telephone systems could be spread across more customers. Universal Service would be the result.

For more than sixty years, the structure held; the U.S. enjoyed (arguably) the best telephone system in the world. It benefited one organization enormously: AT&T. However, in 1974, the Department of Justice (DoJ) accused AT&T of anti-competitive practices and filed an antitrust suit. After years of maneuvering, AT&T and the DOJ (Department of Justice) reached an out-of-court settlement. Dubbed the Modified Final Judgment, or MFJ, it became effective on January 1, 1984. It separated the Bell Operating Companies (BOCs) from AT&T and grouped them into seven holding companies. Today these holding companies (which now number five, not seven) are loosely referred to as the Regional Bell Operating Companies or RBOCs.

The MFJ permitted the RBOCs to provide local telephone service as monopolies, but precluded them from providing long distance service. AT&T became, exclusively, a long distance carrier; the MFJ banned it from providing any type of local service. To distinguish between long distance and local service, Local Access and Transport Areas (LATAs) were established. A RBOC or independent LEC could carry a call that originated and terminated within a LATA. Once a call crossed a LATA boundary, a long distance carrier (inter-exchange carrier or IXC) was required to haul it.

Because of the MFJ, the RBOCs were highly regulated. Their activities were watched by the FCC, the DOJ and, above all, the state public utility commissions (PUCs are known by other names in some states). The PUCs were tasked with the dual responsibility of keeping phone rates affordable and telephone companies profitable enough to provide high-quality service. The goal of this delicate balancing act continued to be the provision of universal service.

Over the years, telephone service became truly universal. More than 96% of U.S. households have telephones. A new concern arose, however. In the process of making telephone service universally affordable, regulators had effectively discouraged the largest phone companies from investing in network upgrades. The local loop (POTS) was narrowband and analog. POTS service is not compatible with the requirements high-bandwidth applications (e.g., multimedia). Congress, the FCC, and other administrative branches of the U.S. government studied how to introduce competition to the local loop in order to stimulate a complete overhaul.

The New PSTN

On February 6, 1996, Congress passed the Telecommunications Competition and Deregulation Act of 1996. The Deregulation Act was aimed at promoting competition in local loop, cable television, and long distance service. It blurs the distinction between types of carriers (e.g., LEC, IXC, cable television, and electrical utility). Today anyone can file to be a competitive local exchange carrier (CLEC) and, in the future, local exchange carriers (the Regional Bell Operating Companies or RBOCs and independents such as GTE) will be able to provide long distance service. Cable television service is up for grabs since the act overturned restrictions that prohibit RBOC/cable television cross-ownership. Electrical utility (power) companies can also pursue the local access market. One popular method for merging telecommunications and electrical networks is to embed fiber optics technology (which uses optical signaling) into the core of an electrical cable. Electrical and optical signaling can coexist, so a power company can maintain a single, multipurpose infrastructure.

Those who wish to compete in the lucrative local access market have three choices. They can buy from existing local loop services from the LEC at a discount and resell them at a profit; they can build their own up-to-date digital networks, or they can combine these two strategies. The third choice is a popular one. Most companies that are trying to grab a piece of the local access market (e.g., AT&T, Qwest/US West, or MCI WorldCom/Sprint) are pursuing the hybrid approach.

Ramifications of the Telecommunication Act of 1996 (derisorily referred to as the Attorney Employment Act) transcend the scope of this text. Monthly, weekly, and daily publications are the best source for the most current information—many good sources are listed in the reference sections of this book. The Deregulation Act has blurred the distinction between carrier types. For instance, when MCI provides local service, is it a LEC or an IXC (we use terms such as LEC and IXC broadly throughout this text even though references may not be precisely accurate)? The act has fostered competition and perhaps more innovation and development than AT&T introduced throughout its entire monopoly. However, it has also allowed market consolidation and, consequently, control by a few dominant players.

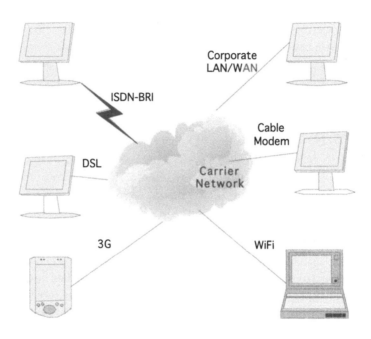

Figure 11-1. Carrier Access Options.

Two broad categories of digital circuit-switched WAN access services exist: dedicated and switched. In the following section, we will address the differences between these two categories.

DEDICATED ACCESS

In the past, customers usually connected a video communication system to a long distance carrier's point-of-presence (POP) through a dedicated digital circuit (sometimes called a private line). A dedicated digital circuit is a permanent configuration that routes traffic exclusively across the network of the carrier to which one is connected. This approach is declining in popularity because it causes problems when a conferencing application involves external companies. Connecting through an IRC's POP is acceptable, however, when all videoconferences will be intra-company or all participants use the same IXC.

A dedicated access connection can be provided over a T-1 or lower speed digital circuits, depending on the application's bandwidth requirement. Most companies install a T-1 for other purposes (long distance service, data communications) and then segregate channels (typically between two and six) from the rest of the pipe for video communication applications.

A survey of 75 telecommunications managers, performed by Personal Technology Research (Waltham, Massachusetts) in 1993, found that 77% of those interviewed used codecs requiring bandwidths of 384 Kbps or lower. An organization that falls into this category may want to consider Fractional T-1 (FT-1). Providers of FT-1 use a digital access cross connect system (DACS) to split a T-1/DS-1 signal into one or more DS-0s. IXCs package FT-1 differently, with the low-end speeds starting at 56 or 64 Kbps and ranging, in 64 Kbps multiples, through 768 Kbps. FT-1 terminates on a specially equipped multiplexer. Prices vary widely, depending on bandwidth.

When dedicated access is chosen over switched, the segregated T-1 approach is usually more cost-effective than the FT-1 approach. Sharing a dedicated access pipe among multiple types of traffic often allows a customer to cost-justify a T-1.

Not only can dedicated access be ordered in several different sizes but customers can also choose from whom to buy that access. The most costly approach is to procure access directly from an IXC, who leases the service used to connect the customer to its closest POP from a LEC. In this scenario, the customer pays the IXC a flat fee to provide end-to-end circuit reliability. The monthly charge covers the LECs' access charge and compensates the IXC for assuming responsibility for the performance of the entire circuit.

Another approach is to go directly to a LEC and request a dedicated connection to an IXC (Base Line Service). In this type of arrangement, a customer pays two bills: one to the LEC, and one to the long distance provider. It is possible that even more than two bills will be rendered: depending on how many LECs' areas the circuit passes through to reach the IXC's closest POP. The total cost for access is lower than with a dedicated connection because, in times of circuit malfunction, *the customer is* responsible for problem resolution.

Probably the least costly way to obtain access is to buy it from a competitor of the local telephone company, an alternative carrier known as a competitive local exchange carrier (CLEC). Historically, these tend to specialize in digital access service. Managers who have the option of using CLECs often find that they are, in addition to being price-competitive, very service-oriented. Their networks tend to be newer, but many gain depth-of-experience by hiring former telephone company employees. CLECs often offer faster response time for circuit installation and repair, and are flexible (and interested) enough to fill atypical customer requests.

Astute managers ensure that its carrier deliver T-1 service over the shortest possible path because billing reflects how far a customer is from the CO. It is also prudent to weigh the discounts awarded for long term-commitments against pricing declines that a deregulated industry promotes. The post-Deregulation Act environment offers other choices, too. Cable television operators, electrical utilities, mobile network operators, and others seek a piece of the local exchange market. Shopping around can be confusing, and rewarding, depending on how the process is managed.

Signaling and Synchronization in the Digital Network

Carriers, when taking orders for T-1 service, always ask what type of signaling/framing should be used. The video communication system provider usually knows what to ask for—that is, if they are to install a *turnkey* solution. There are two choices: Alternate Mark Inversion (AMI) and Extended Super Frame (ESF)/Binary Eight Zero Substitution (B8ZS).

The digital telephone network is synchronized, in order to manage the stream of zeros and ones that run across it. Synchronization relies on clocking, which uses cues from the bit streams to maintain order. Generally, a voltage indicates a one bit and no voltage indicates a zero bit. The network uses the voltages of the 1 bits as clocking pulses. In signaling, no more than eight bit times should pass without a signal. However, it is not usual for a customer's data to have eight or more successive 0s and thus provide no heartbeat for clocking. Without synchronization, the network shuts down.

One solution has been to dedicate one bit out of every eight as a control bit. If all the other bits in an eight-bit string are zeros, then the control bit will be set to one and synchronization will be maintained. The need for this control bit is the reason for switched 56 service (rarely used anymore), which runs over 64 Kbps DS0 circuits, but provides only 56 Kbps service. Synchronization borrows the eighth (least significant) bit to maintain clocking. Thus, only seven out of eight bits are available to users.

An alternate solution to the synchronization problem was developed in the mid-1980s. It involves forcing a deliberate violation (called a bipolar violation or BPV). The BPV is used to maintain network clocking, but networking knows not to interpret BPV pulses as data. This technique, binary eight zero substitution, is called B8ZS. DS0 circuits that support B8ZS from end to end can carry 64Kbps of user data—in other words, offer clear-channel service.

The point of this digression is this: if channels of a T-1 carrier will be used for some of the newer clear-channel digital services offered today (e.g., ISDN), it is best to order ESF/B8ZS. ESF/B8ZS also supports 56 Kbps service. If the T-1 carrier will be used for 56 Kbps service exclusively, AMI signaling is a safe choice.

Later in this chapter, we will further explore network signaling, when we introduce *out-of-band signaling* ESF/B8ZS is not to be confused with out-of-band signaling, which uses an entirely separate packet-switched network overlay to pass network information.

Channel Service Units/Data Service Units

Leased line connections rely on special terminating equipment. When the leased line is a T-1, this equipment is of two types, a channel service unit (CSU) and a data service unit (DSU). In most cases, a CSU includes a DSU.

Customer premises equipment (CPE)—particularly equipment with a data rate lower than 56 Kbps—produces asynchronous bit streams. Asynchronous transmission relies on start and stop bits to distinguish between bytes of data. Furthermore, in asynchronous traffic, the interval between bytes is usually arbitrary. The digital signaling hierarchy, on the other hand, relies on synchronous signaling, in which senders and receivers exchange clocking information in order to identify the boundaries between units of data.

CSUs and DSUs are used to condition asynchronous bit streams, so that they can be passed over the synchronous network. DSUs connect to CPE via RS-232 or V.35 interfaces. They adjust between asynchronous and synchronous timing systems by putting bytes of data onto the network at precise intervals. This activity is known as rate adaptation.

Second, a DSU converts digital pulses from the format used by video communication (or other computerized equipment) into the format used by the network. In computers and codecs, for example, ones are represented by a +5 volt charge and zeros are represented by the absence of voltage. This is called unipolar nonreturn to zero signaling. Carriers use bipolar signaling, in which a zero voltage level represents zeros, and ones alternate between +3 and -3 volts. In networks, bipolar signaling is preferable to unipolar signaling, which tends to build up a DC charge on the line.

The CSU ensures that a digital signal enters a communications channel in a format that is properly shaped into square pulses and precisely timed. In addition, it manages other OSI layer-one signal characteristics such as longitudinal balance, voltage isolation, and equalization. Nearly all CSUs allow the carrier to test the line by performing a loopback test.

The customer is always responsible for buying, installing and maintaining both the CSU and the DSU (or a combined version of the two).

Carrier Gateways

Each carrier structures the signaling and management portion of its digital networks a little differently. These networks, after all, were developed during a time when competition was fierce. Carriers did not exchange network information because it was considered proprietary. The legacy of the 1980s lives on today, when it comes to digital network interconnection. If dedicated access service will be used for video communication, bear this in mind: While the IXCs' analog networks are universally interconnected, their digital networks are not. This presents a problem when two different companies want to set up a videoconference when both use dedicated digital access, but each uses a different IXC.

Placing a call through a carrier gateway is still the most reliable way to bridge between carrier's digital networks when dissimilar dedicated access causes connection problems. The cost for using these gateways varies widely, based on speed, location, type of service, and carriers. All the major long distance carriers offer these gateways (AT&T, MCI WorldCom).

Some gateways are one-way. In other words, carrier A's customers may be able to gateway to the network of carrier B, but customers of the carrier B may not be able to gateway to the network of carrier A. When a customer of carrier B needs a videoconference with a customer of carrier A, he must arrange for that party to initiate the call. Since the customer of carrier B initiated the call, they are billed for it. Moreover, carrier A's customer *must* be registered with carrier B's video communication network, even though that carrier does not serve them. Registration requires connectivity tests, a process that can take up to a week. In short, first time, spontaneous connections between dedicated access customers served by different carriers are almost impossible to arrange. Explaining this to a room full of executives who are waiting for a videoconference with an important customer may be stressful.

SWITCHED ACCESS

The way to avoid this problem is to access an IXC's digital network using switched access service provided by the LEC. Switched access service is service that allows a customer to flexibly route calls across different long distance carrier's networks. To achieve this flexibility, a customer must generally buy their access service from a company that is *not* their long distance carrier. Since the distinction is blurring between long distance carrier and local access providers, it is important to ask the access provider if the switched connection being installed can be used to call anyone, on any other network, by simply dialing their telephone number.

Switched digital service choices for video and data conferencing include Switched 56, ISDN Basic Rate Interface (BRI), ISDN Primary Rate Interface (PRI), and asynchronous transfer mode (ATM). Switched access should not be confused with dedicated access service, which may also rely on ISDN PRI or ATM. In a switched access arrangement, channels can be assigned flexibly, to any long distance carrier *and* a customer can override the long distance carrier assignment by dialing a carrier access code.

Long Distance Carrier Access Arrangements

When ordering local service, customers are required to select a primary inter-exchange carrier (PIC). The LEC indicates the PIC choice by entering the selected carrier's access code in a routing table or database. When a "1+" long distance call is made—in other words, a long distance call that is not specifically directed to a different long distance carrier—it defaults to the PIC. In a switched-to-switched connection, it does not matter what PIC the called party uses.

In a switched-to-dedicated access connection, an extra step may be required, but only if the called party uses a different IXC than the caller. When a difference exists, the caller dials a five-digit carrier access code, which forces the call over the long distance network that serves the called party. Examples of carrier access codes include 10288 for AT&T, 10222 for MCI, and 10333 for Sprint.

Switched access is easy to use but it has two cost-related drawbacks. First, IXCs typically offer lower rates for calls conducted over dedicated access arrangements than for switched access calls. The reason for this is simple. An IXC must pay the LEC a per-minute fee for long distance calls that pass over the local network. This fee does not apply when a call originates over a dedicated access connection—in that case, the LEC gets its compensation in the form of a *private line monthly rental charge.* IXCs pass the additional cost of switched access on to their customers. Consequently, the cost for switched-to-switched connections is higher than dedicated-to-dedicated connections.

The second reason that switched access might be more costly is duplication. Most companies that are in the position to install video communication have at least one, if not more, T-1 connections to an IXC. Often there are channels standing idle. It seems logical to use these channels for video communication, and thereby eliminate a separate monthly cost for switched access. However, an IXC often charges a sub-multiplexing fee to split out video communication channels from the rest of a T-1 pipe. That charge can, in some cases, equal or exceed the cost of separate switched digital access channels.

Several categories of switched digital access service exist. Before we continue, we should further examine network signaling and control methods. Differences between signaling systems may affect the type of switched digital service offered.

Signaling: In-Band Versus Out-Of-Band

Network signaling is used to control the operations in a carrier's network and to establish connections between communication endpoints. The two categories of network signaling are *in-band* and *out-of-band.* The most noticeable difference between the two is the fast call setup times that out-of-band signaling delivers—often less than one second, compared to the 30 seconds (or more) setup time required to establish an in-band signaled connection.

Out-of-band signaling involves the separation of network signaling from customer data. This arrangement delivers the entire bandwidth of a digital channel for use by the customer. When signaling is separated from the transmission path, the result is clear channels, which are, by definition, 64 Kbps wide.

In-band signaling is an older method of establishing network connections. It requires that a customer relinquish some channel bandwidth to the carrier, who uses it for signaling. This technique of deriving bandwidth for network overhead is known as *bit-robbing.* A bit-robbed 64 Kbps line will only carry 56 Kbps of customer data.

In-band signaling steals valuable bandwidth from the customer's channel and is inefficient. Call set-up requests are routed progressively, from switching center to another, using the same circuit that the call will travel over after the connection is established. The method is slow and inefficient because each switching point introduces processing delays. A call could travel nearly all the way to its destination only to be blocked at a busy CO. During the time it took to reach its point of failure, the doomed call will have tied up circuits that other calls could have used.

Inside their own networks, carriers use out-of-band signaling methods to establish connections. Requests for call set-up and teardown are conveyed separately from the call itself, as are special messages that indicate caller-ID, class of service status, and other customer-specific data. The signaling portion of the network is packet-switched while the part that carries customer data is circuit-switched. Out-of-band signaling can be extended all the way to the customer's premise, but only if a special path is provided for its transit. If it is not, the carrier must use a portion of the line for signaling. The customer, who is able to transmit only 56 Kbps of data, pays the price.

Out-of-band signaling was first introduced by AT&T in 1976, when long distance competition forced it to become more efficient. Even in the '70s, AT&T's network was shared by millions of calls on any given day. Slow set-up times, and bandwidth wasted on failed calls, necessitated additional circuits. To streamline the process, AT&T devised a way to signal network status on a call-by-call basis. In order to identify available facilities, AT&T developed several huge databases of circuit inventories. These were accessed using a system known as Common Channel Interoffice Signaling (CCIS). CCIS circuit status requests were replicated on databases across a separate, analog packet-switched network, operating at speeds between 2.4 and 4.8 Kbps. Multiple routing-request packets streamed along the common highway, one after another.

In 1980, the CCITT approved an updated out-of-band signaling system. Known as Signaling System 7 (SS7), it also relies on a separate, packet-switched network. SS7 differs from SS6 in that it is digital and operates at either 56 or 64 Kbps. Call set-up times are significantly reduced because the network is so fast.

The SS7 protocol has gradually developed over the last decade to become the "nervous system" of the PSTN. Almost all modern carriers have migrated the *trunk side* of their networks to SS7 but the *line side* (the part that provides local access service to customers) sometimes uses bit-robbed signaling. Central offices must be retrofitted to accommodate clear channel service, and upgrades can be expensive. Thus, bit-robbed (switched 56 and T-1) service is sometimes all that is available to customers wanting digital service.

Switched 56 Service

Although switched 56 service has become a relic, we will discuss it because it is still used and it was one of the original enabling telecommunication technologies for videoconferencing. The IXCs and LECs offer *switched 56* dial-up digital service. It is a non-standards-based digital adaptation of a regular telephone line, which means it uses in-band signaling. As its name implies, it allows customers to call up and transmit digital information at speeds up to 56 Kbps, using regular dialing techniques. Switched 56 service is billed like a voice line—e.g., a monthly charge plus a cost for each minute of usage.

AT&T was first to offer switched 56, in 1985, when LECs offered no comparable form of switched digital access. Most carriers, including MCI, and Sprint, developed a switched 56 offering. Even though it is not standards-based, any two customers who use LEC-provided switched 56 service can interconnect, even when an IXC crosses a LATA line.

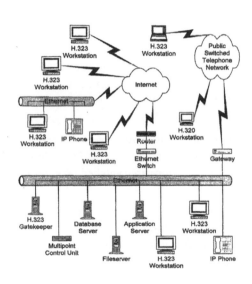

Figure 11-2. In-Band Vs. Out-Of-Band Signaling (Laura Long).

In most switched 56 videoconferencing applications, two channels are aggregated to yield one 112 Kbps circuit. This is bare-minimum bandwidth for group-system videoconferencing but many organizations get by with it.

Organizations that want the freedom to call anyone, regardless of which long distance carrier serves them, use the LEC's switched 56 lines (or, more commonly now, ISDN BRI) to reach an IXC's switched 56 service. Users of switched 56 (using either switched or dedicated access) can place calls to ISDN customers, as long as the ISDN equipment can accept switched 56 calls, as most can. Placing a video call to a customer using switched 56 service is much like making a long distance phone call. The caller dials two phone numbers, one for each channel. Placing a call using switched 56 is just as easy. One dials the telephone numbers associated with the channels that connected to the video communication system on the other end. Because signaling is in-band, the connection may take a minute or more to establish. Part of the delay is usually associated with the codec handshaking process, however.

Three different serving arrangements exist for switched 56: the four-wire variety (INC USDC) that AT&T developed, a two-wire version that Northern Telecom developed (Datapath), and Circuit Switched Digital Capability (CSDC), which is yet another two-wire version that few LEC's offer. Two-wire switched 56 is transported at rates of 144 or 160 Kbps, and multiplexed using ping-pong modulation. Also called time-compression multiplexing (TCM), this technique supports transmissions from only one end of the connection at any given time, and simulates bi-directional (full-duplex) transmission by rapidly accepting signals from both ends on an alternating basis. Four-wire switched 56 provides a communications path for each end of the connection, with each pair of wires operating at only 56 Kbps. Two- and four-wire implementations of switched 56 are compatible. From a subscriber's point of view, there is really not much difference between the two except that four-wire service permits loops between the subscriber and the CO to span a greater distance.

Switched 56 service terminates in a DSU, which connects to CPE via an RS-232 or V.35 interface. Customers must buy the proper type of DSU; they differ depending on whether the carrier provides two- or four-wire switched 56 service. The carrier should specify which type is required.

In switched 56 applications, DSUs not only manage clocking and bipolar conversions, they also signal the destination of the call to the telephone company switch. DSU dialing conforms to the Electronics Industries Association (EIA) RS-366 dialing standard. The RS-366 interface uses the AT command set that was originally employed by Hayes modems. The AT command specifies an auto-call unit (ACU) that takes a line off-hook, detects dial tone, dials a number, detects ringing, and recognizes call completion or failure.

Switched 56 service is reliable, universally available, and easy. Its drawback is that local access is expensive. Only in unusual circumstances would one buy switched 56 services today when ISDN or, better yet, IP is a choice.

ISDN BRI

ISDN is an all-digital subscriber service—a single transport medium that can carry any type of information: voice, data, or video. The Basic Rate Interface, one of the two primary ISDN interfaces, relies on a single twisted copper pair of wires— the same two wires on which digital POTS service rely. Using this wired-pair, it delivers two 64 Kbps circuit-switched B channels and one 16 Kbps packet-switched *delta* (*data* or D) channel. The B channels can be aggregated. For instance, a videoconference can be carried over the 128 Kbps data stream derived by combining two B channels. Low-end video communication applications (in which the need for motion handling and picture clarity is moderate) make extensive use of ISDN BRI service. Where it is offered, it is usually priced competitively, particularly in comparison to Switched 56 service. Not only are monthly rates low, installation charges are also reasonable. The biggest problem with BRI is that is not universally available.

If ISDN can deliver 144 Kbps of combined bandwidth, it would seem reasonable to expect POTS lines to do the same. Yet, even when POTS service is equipped with very fast modems, it can generate a maximum data transfer rate of only 33.6 Kbps (56 Kbps modems—that, at the time of this writing, require a number of caveats— eliminate digital-to-analog conversion and, therefore, do not provide analog service). The reason that ISDN is more than three times as fast as analog POTS is filtering. The PSTN uses filters to keep unwanted frequencies out of the 4 kHz analog voice pass band. Digital signals are not impeded by spurious frequencies, so the same twisted pair cable can, using digital modulation techniques, deliver much greater throughput.

Speed, reliability, and lack of an alternative account for ISDN BRI's popularity. Approximately one million ISDN BRI lines existed at year-end 1995. This represents a 300% increase from the preceding year. It is estimated that, in 1996, this number more than doubled to 2.5 million. Although video communication applications accounted for only a small portion of the installations (5-10%) ISDN BRI suits interactive streaming media applications.

ISDN BRI service charges vary greatly, both within the United States and abroad. Rates fluctuate not only due to tariff differences, but also in relation to the distance between the customer and central office. On average, ISDN lines are priced about 1.5 to 2 times higher than POTS lines.

ISDN service has many advantages over POTS. ISDN data connections are set up in less than a second compared with over 30 seconds for POTS connections. ISDN can also be configured to provide enhanced Caller ID and other custom network features. There are many B and D channel configuration choices (although choices narrow when videoconferencing is passed over the B channel).

ISDN BRI can terminate up to 18,000 feet from a CO (in some areas, 24,000 feet), without signal reinforcement. COs have to be equipped for ISDN; if they do not have this capability, or if the customer is too far away from the CO, a telephone company may be able to provide ISDN BRI through the use of a Basic Rate Interface Terminal Extender (BRITE) card. BRITE technology uses T-1 carrier to transport up to 24 ISDN BRI channels. Since each BRI has three channels—two B and one D—this means that a BRITE card can support eight total ISDN BRI connections.

The ISDN NT-1

Eventually, ISDN BRI ends up at the customer's premises as a single twisted pair. These two wires terminate on a building block or protector and are then connected to an NT-1 (network termination, type one, device). As originally conceived by the CCITT, the NT is service-provider-owned equipment installed on the customer's premises. It establishes the point of telecommunications service delivery by a carrier. In the U.S. only, provision of the NT is the customer's responsibility (as mandated by the FCC's Network Channel Termination Equipment Order). On one side of the NT-1, circuit integrity is the responsibility of the carrier. On the other side, it is the responsibility of the customer. As such, the NT-1 is just like any other network demarcation (demarc) point.

The NT-1 is about the size of a stand-alone modem and requires a power supply. The power supply plugs into a commercial power outlet in the customer's equipment room. This is a drawback of ISDN compared to POTS. POTS is powered by batteries in the CO. ISDN BRI requires more power than can be provided over the local loop. If an ISDN BRI-supported videoconferencing system must remain operational even during times of commercial power failure, a customer must provide a battery or uninterruptible power supply.

NT1 units allow up to eight different devices to be daisy-chained to the ISDN circuit, and thereby enable all eight to use it simultaneously or to contend for the circuit. Thus, as many as eight videoconferencing devices could share the same ISDN BRI line and contend with one other for the B channels as needed.

NT-1s and their associated power supplies vary widely in price; the cost difference correlates to capability. More expensive NT-1s can terminate ISDN offered by both AT&T and Northern Telecom. The cheaper ones are of switch-specific design. Some equipment vendors build the NTI-1 into CPE. The disadvantage to a built-in NT-1 is that it cannot be shared by other systems.

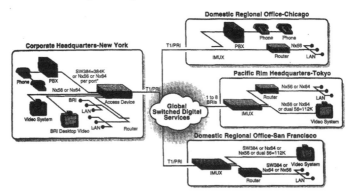

Figure 11-3. Global Switched Video Network (courtesy of Ascend).

The ISDN Terminal Adapter

The ISDN specification also requires a terminal adapter or TA. ISDN TAs place calls, make connections, and transfer digital data across the ISDN circuit. They consist of a call-control module and a B channel and D channel transport module. The call control module manages the resources of the ISDN line, using the Q.931 protocol to perform signaling functions. Q.931 is discussed below.

TAs can be internal (operate from the computer's local bus) or external (in which they are limited to the speed of the computer serial port—typically 115 Kbps. In videoconferencing applications, the TA is usually built into software.

Configuring a TA can be tricky. For instance, the person doing the set-up must know the type of CO with which the TA is communicating. Between North America and Europe, there are many different versions of the D-channel protocol. Some of these are based on standards (see National ISDN, below). Others are unique to a type of central office (e.g., AT&T 5ESS Custom, Northern Telecom DMS-100, European 1TR6, NET3, etc.).

 One of the biggest configuration challenges in setting up the TA had to do with entering the proper Service Profile Identifier (SPID). The customer must enter the SPID into the ISDN TA before a device can, using the D channel, initialize at layer three of the OSI Reference Model.

A TA sends out its SPID each time it communicates with the CO—and each TA has a unique SPID. The SPID is associated with a set of ISDN features and capabilities. SPIDs can be assigned *only* after a customer has notified the LEC about what ISDN service and features are required. The carrier then translates this configuration into two or more SPIDs. Since eight TAs can share one ISDN NT-1, it is possible that eight SPIDs might be entered during an ISDN BRI configuration. Microsoft and others have tried to address the confusion over SPIDs in software.

Some products offer TAs that support both POTS and ISDN. This allows a videoconferencing application to use ISDN B channels when the far-end can accommodate ISDN and POTS when it cannot. A good TA will adhere to the Multilink Point-to-Point Protocol (PPP), which defines a common method of negotiating a PPP session over two B channels. This is more important for data conferencing than for video communication. Small routers provide more bandwidth, but configuring them can be somewhat complex.

ISDN PRI

ISDN PRI consists of 23 B channels and a single 64 Kbps D channel. PRI is packaged much like North American T-1 carrier (24 total channels). ISDN, however, differs from T-1 in its signaling and synchronization methods. T-1 uses bit-robbing, and thereby results in a pipe made up of 24 channels, each of which offers 56 Kbps of bandwidth. PRI uses out-of-band signaling, and thereby results in a pipe with 23 B channels (each of which offers 64 Kbps of bandwidth) and one 64 Kbps D channel.

An ISDN PRI customer has exclusive use of the 23 B channels, which can be used separately or aggregated for data transport. Although PRI B channels are used for video and data conferencing, it is not common to dedicate an entire ISDN PRI pipe to a single videoconferencing system.

In both PRI and BRI arrangements, the ISDN D channel extends the intelligence of a carrier's SS7 network beyond historic boundaries, and allows it to communicate with customer premise equipment. The customer and the carrier can share the D channel, (e.g., the customer can transmit packet-switched data over it) however this is very rare.

PRI terminates in a network terminal device called an NT-2. The NT-2 provides the same functions for ISDN PRI service as the NT-1 provides for BRI lines. In addition, the NT-2 performs switching (to route individual B channels) and bonding (to aggregate them into Nx64 pipes). Although it is not always possible to access BRI from a LEC, ISDN PRI is generally available from an IXC. PRI is used for many things, including inbound and outbound long distance, remote LAN access, and videoconferencing. A PRI pipe typically terminates in a PBX or an ISDN hub.

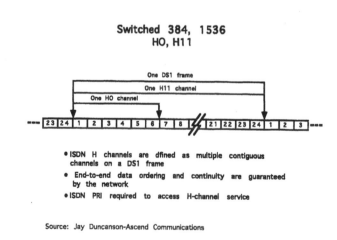

Figure 11-4. Switched 284, 1536 (HO, H11) Framing (Michael Bayard).

In an ISDN-enabled PBX environment, the PRI comes in to the trunk side of the switch (the side leading to the telephone company) and video communication CPE hangs off the station side. To make a video call, a station goes off-hook (perhaps using software) and dials a number. If there are ISDN B channels available on the PRI trunk, one or more are allocated to the station. Using the term station, we include both desktop video and data conferencing systems and group-systems.

Issues Related to ISDN

When the first set of ISDN standards was ratified in 1984, the CCITT (now the ITU-T), intended that it provide a migration path to move the historically analog PSTN to an end-to-end digital system. While the ITU-T was busy planning an evolution, customers were envisioning an immediate revolution to digital service. This did not account for the cost of upgrading a carrier networks for ISDN. The carriers that did deploy ISDN in the 1980s found inadequate demand for high-bandwidth digital service. ISDN, it was said, should stand for "innovation subscribers do not need."

A decade after ISDN's introduction, several *killer applications* almost simultaneously emerged. Remote LAN access, telecommuting, Internet access, and low-bandwidth video communication proved that the CCITT had been right—and visionary in their thinking. The problem today is this: ISDN deployment is by no means seamless. In the present competitive environment, telephone companies face revenue pressures; budgets are much tighter as a result. Customer demand can lead to upgrades or work-arounds (back-hauling from an ISDN-capable CO, for instance), but implementation still takes time.

Even when ISDN is available, its clear channel capabilities may not extend beyond the LEC's regional boundaries. This is due to an oversight in the original CCITT specification. ISDN defines two interfaces, the network-to-user (N-interface) and the user-to-user (U-interface). It does not now, nor will it ever, define a network-to-network interface. This means that extending ISDN's clear-channel capabilities between carrier's networks requires negotiated agreements and special engineering. When the LECs and IXCs work cooperatively, ISDN can be arranged to offer end-to-end digital service on a clear-channel (64 Kbps) basis. The component that makes ISDN capable of offering end-to-end digital service is SS7; it *is* the network-to-network interface. However, D channel signaling has been implemented in various proprietary ways and one carrier's SS7 system might not be compatible with another.

The trouble with SS7 interoperability can be traced to a rather sluggish standards-setting process. In 1984, the ITU-T published the first ISDN specifications. Unfortunately, it was incomplete as it related to signaling over the D channel. Without a comprehensive standard, the switch manufacturers were left to their own devices to fill the holes in the D-channel protocol. AT&T and Northern Telecom, the two dominant CO manufacturers in North America, turned out proprietary interpretations. Since switching systems could not interpret each other's D-channel protocols, inter-network signaling had to be passed in-band. Even today, it is very difficult to arrange for 64 Kbps clear channel endpoint-to-endpoint, using LEC access. Resolving this haphazard arrangement was the goal of *National ISDN*.

National ISDN

In 1991, the Corporation for Open Systems (COS) and the North American ISDN User's Forum (NIUF) developed an initiative to standardize the implementation of ISDN within the United States. In November 1992, Bellcore, the COS, and the NIUF introduced National ISDN-1 (NI-1), a standard that aims to provide a consistent interface among LECs, IXCs, and equipment manufacturers. NI-1 was introduced at a week-long event called the Transcontinental ISDN project, or TRIP '92. The NI-1 standard, which precisely defines ISDN Basic Rate Interface, has been widely implemented.

National ISDN-2 got underway in late 1993. In the NI-2 standard, Bellcore strengthened NI-1, and thereby provided for service uniformity and facilitating operations and maintenance. NI-2 also defines various PRI features, including circuit-switched voice and data call control, packet-switched data call control, switched T-1, fractional T-1 service, and D-channel backup. NI-2 did not address Non-Facility Associated Signaling (NFAS)—the ability for one D channel to provide signaling services for multiple PRIs. NFAS allows a customer with multiple PRI connections (typically 11, maximum) to allocate only one D channel to handle signaling for all of them. Each additional PRI, after the first, has 24 B channels.

The third iteration of National ISDN was NI-95. NI-95 improved PRI and defined how ISDN is used for access to personal communications services (PCS) networks. NI-96 listed new ISDN features and capabilities and gave references to Bellcore requirements documents for these services. Current information on National ISDN is available at www.nationalisdncouncil.com.

In May 1996, the U.S.'s largest LECs, under the aegis of the National ISDN Council (Gaithersburg, Maryland), agreed to implement a standard procedure for a customer's establishment of ISDN service. These companies included all regional Bell companies, Cincinnati Bell Telephone Co., GTE Corporation, and Southern New England Telephone Co. The ISDN ordering standard includes codes, known as EZ-ISDN codes, which cover the two most commonly ordered ISDN BRI configurations. Neither of the two EZ ISDN versions (EZ ISDN 1 and EZ ISDN 1a) are aimed at easing the ordering process for videoconferencing configurations (two circuit switched B channels configured for voice/data and a single packet-switched D channel).

The standard also includes a uniform SPID format. Under the new definition, a SPID is exactly fourteen digits in length whereas, in the past, SPIDs could be anywhere from 11 to 20 digits.

The video codec connects to the station side of the PBX with a BRI connection. Only a handful of switches offer a National ISDN-1 (NI-1) BRI interface (two that do are the NEC ICS and the Northern Telecom Meridian with Release 18 or newer software). Most switches require some type of adapter to be plugged into their regular digital station ports to support NI-1 connections. At any rate, the videoconferencing equipment, when placed behind a PBX, is assigned a DID (direct inward dialing) number, that makes it accessible through the PBX. The advantage to this scenario is that the videoconferencing system shares PRI channels on a call-by-call basis with other applications. The disadvantage is that the connection is essentially dedicated and, therefore, results in the same problems with inter-IXC calling that are encountered with other dedicated access methods.

ISDN H0, H1 AND MULTIRATE

ISDN is designed to allow customers to select different bandwidths easily by combining, on a flexible basis, the number of ISDN B channels necessary to deliver a desired level of motion handling and picture resolution. Today a North American customer can set up a videoconference at 384 Kbps or 1.536 Mbps using a long-distance carrier's H0- and H11-based products.

H0 and H11 service use several ITU-T protocols for call set-up. These are Q.920, Q.921 and Q.931. Link Access Protocol-D (LAPD) is formally specified in ITU-T Q.920 and ITU-T Q.921. LAPD is an OSI layer-2 protocol. The connection-oriented service establishes a point-to-point D channel link between a customer and a carrier's switch. Data is sent over the link as Q.921 packets. Whereas the D channel is packet-switched, B channel connections are circuit-switched and are established, maintained, and torn down using the Q.931 interface (also known as ITU-T I.451). In Q.931, the ITU-T specified how 'H' channels—multiple contiguous channels on T-1 or E-1/CEPT frames—can be bound together and switched across an ISDN network.

In addition to the switched 384 H0 standard and the switched 1.536 H11 standard, the ITU-T specifies H12 (European switched 1.920. All are part of Q.931, and all make it possible for high-bandwidth circuits to be created on-demand. The network guarantees end-to-end synchronization and clear-channel access. However, the customer's premise equipment must bond the channels and pass them to the network over PRI access.

H0 dialing calls for six contiguous DS0s to be combined into a single 384 Kbps channel. At 1.544, PRI can carry 4 H0 channels or 3 H0 + D; at 2.048 it can carry 5 H0 + D. H11 dialing combines 24 DS0s into a channel-less arrangement with an aggregate bandwidth of 1.536 Mbps. H12 combines 30 channels into a channel-less arrangement.

In 1991, AT&T became the first U.S. carrier to offer commercial H0 service, by making it available on a dedicated-access (both ends) basis. AT&T offers H11 service, as well. MCI WorldCom also offers H0 and H11 service.

An even more recent development, ISDN multirate, also known as Nx64 or switched fractional T-1 service, lets users combine, on a call-by-call basis, between 2 and 24 B channels.

INVERSE MULTIPLEXERS

Multirate ISDN has never become ubiquitous. An alternative is to use hardware to achieve variable-rate bandwidth, or *bandwidth on demand*. To dial up multiple 56 or 64 Kbps channels, a customer can use an inverse multiplexer (I-MUX). An I-MUX breaks a high-bandwidth videoconference into multiple lower-speed channels for transport across a switched digital network.

I-muxes were introduced in 1990. They break a high-speed data stream (such as that produced by a codec) into N (where 'N' typically indicates some number between 4 and 24) lower-speed switched digital channels. I-muxes use the EIA RS-366 standard to provide dialing commands to the network. They keep a signal synchronized so that a compatible device on the other end can rebuild a single high-speed signal from the multiple lower-speed signals it receives.

Synchronization is necessary because a switched digital network is a mesh of circuits. Many different paths can be taken to get between two points, and calls are routed across this mesh independently. If a high-speed videoconference were subdivided into lower-speed fragments, each fragment could take a different route to its destination. Routing disparities can cause data stream synchronization problems.

I-muxes perform *delay synchronization*. In other words, they hold a frame open until all its fragments reach their destination. Once received, the I-MUX reconstructs the frame and presents it to the codec (or other high-bandwidth device). Depending on the configuration, I-muxes can perform additional tasks. Some are switches (hubs) that patch incoming channels to outgoing channels. Others receive ISDN PRI signals and convert them to T-1 or E-1/CEPT. An I-MUX may, among other things, convert protocols, perform overflow routing, and provide private line back-up service.

When I-muxes were first introduced, they could not interoperate—identical devices had to be installed on both ends. This presented another obstacle for inter-company videoconferencing. To address the problem, a consortium, which included almost 40 vendors, developed the BOnDInG standard; the first draft was approved on August 17, 1992.

BOnDInG stands for Bandwidth On Demand Interoperability Group. The BOnDInG specification describes how I-muxes can dial up channels across a network and combine them to create a single higher-speed connection. BOnDInG is discussed in more depth in Chapter 13.

FOLLOW THE MEGABIT ROAD

Today, organizations can get as much bandwidth as they need. Internet Protocol (IP) networks offer all-in-one support for voice, data, image, and motion video and have, therefore, become the norm.

At the bottom of the network, however, are more choices than ever before. Cable modems and DSL modems have become the standard for connecting homes and even backup connections for small businesses. However, the real news is that wireless is displacing wired access. At offices, DS-1, DS-3, and a technology known as Synchronous Optical NETwork, or SONET have become the common physical network connections. Above the physical layer (but below the networking functions of the upper layer transport protocols) lies another technology known as asynchronous transfer mode (ATM). Above the ATM layer are two basic services: those that are designed for variable bit-rate (data-oriented) transmission and those that are aimed at constant bit-rate (streaming media—voice and video).

SONET is closely associated with ATM, but is not necessarily dependent upon it. Both require the speed that metallic media cannot reliably support over long distances. SONET can carry DS-0 and its multiples, and thereby provide a bridge between the older, slower digital signaling hierarchy and ATM's faster (minimum 155 Mbps) data delivery rates. With its smallest increment of provisioning VT-1.5 (the next increment, at 51.84 Mbps, is STS-1), SONET provides 1.7 Mbps of bandwidth using time division multiplexing to simultaneously transmit multiple dedicated data streams. Bellcore's GR-2837 standard maps ATM cells onto SONET, and turns a SONET network into a cell-switched (packet-switched) pipe that uses the full bandwidth and mitigates ATM's inherent waste. High-bandwidth (e.g., carrier, corporate WAN) lines are increasingly carrying IP traffic directly over SONET without ATM. DWDM fiber is projected to soon carry IP with no need for either ATM or SONET.

The North American digital hierarchy specification was developed for copper media. The aggressive deployment of fiber optic cable makes the digital signaling hierarchy (based on DS-0 multiples) highly inefficient. Optical networks can transport data at speeds far in excess of 41.84 Mbps (DS-3). Synchronous optical network (SONET), an optical transmission interface originally proposed by Bellcore—and later standardized by ANSI—was conceived to replace the copper-oriented digital signaling (DS) interface over time, yet remain compatible with it throughout the upgrade process.

SONET resumes where the digital signaling hierarchy concludes. The SONET and SDH standards establish a standard multiplexing format that uses 51.84 Mbps channel multiples as building blocks. Lower level ANSI and ITU-T digital signaling hierarchies used a 64 Kbps (DSO, EO) building block. The 51.84 Mbps rate was specified because the overhead involved in framing precluded higher speeds from being exact multiples of the 64Kbps base speed. A 51.84 Mbps signal can carry a 44.84 Mbps DS-3 signal, with room left for overhead.

SONET line rates are referred to as Optical Carrier (OC) speeds. The electrical signal interfaces that correspond with these speeds (known as STS) have also been identified. The 51.84 Mbps base speed for SONET is known as OC-1/STS-1. Higher speeds are multiples of this rate. OC-3/STS-3 has a speed of 155.52 Mbps. It is expected to be an important speed level for establishing multimedia transmissions and, of the specified optical standards, it is the slowest one capable of 100 Mbps LAN speeds. SONET also includes a 622.08 Mbps, (or OC-12/STS-12), specification, and a 1.244 Gbps (OC/24/STS-24) specification. SONET's highest speed is the 2.48 Gbps OC-48/STS-48 specifications.

SONET specifies not only line speeds, but also transmission encoding and multiplexing methods and establishes an optical signaling standard for interconnecting equipment from diverse suppliers. A synchronous multiplexing format carries DS-level traffic. The ITU-T developed a European-oriented version of SONET known as synchronous digital hierarchy (SDH). In short, SONET provides the physical infrastructure for a next-generation PSTN switching architecture. That architecture is known as B-ISDN and its most widely-known implementation is ATM.

ATM

B-ISDN was developed by the CCITT to be a 155 or 622 Mbps public switched network service. In specifying B-ISDN, the goal of the ITU-T was to create a public network that is ideally suited to carrying the signals that are intrinsic to multimedia, which have a wide range of traffic characteristics.

The CCITT's "Task Group on Broadband Aspects of ISDN" looked at a variety of interactive broadband services that would be required in the public network. Of course, they considered video telephony and video and data conferencing, HDTV, and high-speed file transfer—all topics that we cover in this book. Indeed, the B-ISDN protocol stack, and its ATM layer, would not have been necessary were it not for Real-Time interactive data types that needed to share a public network with less time-sensitive traffic.

To reiterate, ATM is a layer in the B-ISDN model. Below ATM in the stack lies SONET. Above it lie various layers that adapt the ATM transport structure to meet the needs of different media types and applications. An ATM network switches data of all types in fixed-length cells (for that reason, its also called cell relay). The cell size is 53 bytes; of this, 48 bytes are filled with customer data and 5 bytes are used for cell header information. Cell relay works because information is "chopped" so finely that the network can easily fit many cells across one high-speed link. Although ATM has a great deal of overhead (more than 10%) it works well in fast network (155 Mbps and up) environments.

ATM user equipment connects to the ATM network via a user-network interface or UNI. Connections between ATM-provider networks are made via a network-network interface (NNI). Data contained in a cell header are used by the UNI, the NNI, and ATM switches during transport. Specifically, the first two fields in the ATM header at the NNI are the 12-bit virtual path identifier (VPI) and the 16-bit virtual circuit identifier. The 28-bit address formed by the VPI/VCI is used in switching a call, hop-by-hop, across the network. These 28-bits change as a call passes from ATM switch to another. The forwarding switch and the receiving switch track the VPI/VCI pairs (sender and receiver) locally, but only for a brief interlude, then the cell moves on.

Cells are arranged in logical connections called channels. This confuses some people who think ATM is designed to support both connection-oriented and connectionless services. Although it does support connectionless services, ATM is, over all, a *connection-oriented process.* A cell may not be sent until a channel has been established to carry it. ATM packages data into cells according to its type—e.g., datagram, voice sample, video frame. It does so in accordance with the layer above ATM in the B-ISDN protocol stack, the ATM adaptation layer or AAL. Five types of AALs have been identified in the B-ISDN model. The key ones are packet switching service adaptation and circuit switching service adaptation. In adapting non time-sensitive traffic for transport, it is possible to segment large packets into smaller ones and route them to be reassembled sequentially. However, time-sensitive traffic (voice and video) is processed differently, using buffers to keep data flowing in a smoothly paced stream.

When one ATM end-node needs to send data to another end-node, it requests a connection by transmitting a signaling request to the network. The network passes the *requested path* to the destination. If the destination node and the sender can negotiate a connection, a switched virtual circuit (SVC) is completed. When the devices are finished with their exchange, the SVC is *taken down* by all switching points originally engaged in setting it up.

The sender also arranges end-to-end quality of service by specifying call attributes such as type and speed. As discussed previously, audio and video transmissions are time sensitive, but are adequate with some inaccuracy. Data transmissions, in contrast, require absolute accuracy—e.g. no absence of information—but are not time sensitive. Because ATM establishes connectivity through switching rather than through a shared bus, it offers dedicated bandwidth per connection, high aggregate bandwidth, and explicit connection procedures.

ATM is, by definition, capable of providing isochronous service. End-nodes have a way of requesting specific network characteristics when they establish the call. For instance, a node can inform the network of the average bandwidth it will require and whether it will be bursting or sending Real-Time data. The network will condition a circuit based on this information. The network keeps a cumulative inventory of its bandwidth. As new requests for service arise, it decides, based on probabilities, whether it can service the new caller's request. If it cannot, it rejects the request until adequate bandwidth becomes available. For this reason, ATM is an ideal networking technology for interactive streaming media.

Although ATM and ISDN are mistakenly thought of as separate technologies, it is ATM is, in fact, a B-ISDN implementation. Cell switching is the protocol that is used to move packets across B-ISDN and ATM networks. B-ISDN, and its real-world manifestation, ATM, blur distinctions between circuit- and packet-switching.

Digital Subscriber Line (DSL)

As metropolitan cable service operators (MSO) increasingly offer VoIP telephone services, Telcos are striking back by offering video over digital subscriber line (DSL). The H.264/AVC video compression standard is making this practical. At bandwidths of up to eight Mbps, DSL exceeds the requisite two Mbps necessary for broadcast-quality MPEG-2 digital video. However, since most subscribers' bandwidth tops out at about 1.5 Mbps, they are limited to streaming MPEG-4 SP (MPEG-4 Simple Profile). Multimedia Research Group, Inc. expects worldwide IP video services subscriptions to grow to eight million by 2007, and for US revenue to reach about $6 billion in that period. H.264/AVC reduces full-screen DVD-quality digital transmission bandwidth requirements to 700 Kbps. By reflecting the adoption of a video communications standard (H.264) for video broadcast and the desire for Telcos to boost ARPU (average revenue per user) by adding IP services, this vignette illustrates the merging of broadcast video and video communications.

For decades, Telcos have used HDSL to conserve wire pairs without sacrificing T1 speeds. However, in 1989, Bellcore conceived an idea for transmitting video, audio, and data, at rates of more than one Mbps over ordinary twisted-pair telephone lines. The crux of the idea was to transfer information more quickly in one direction than in the other (asymmetrically), with downstream rates of 1.5 Mbps and upstream rates of 16 or 64 Kbps. Such asymmetry would match rate requirements for (one-way) video broadcast and information retrieval over data networks over much greater distances than symmetric means could. They termed this new technology Asymmetric Digital Subscriber Line (ADSL).

ADSL downstream speeds have increased to nearly one Gbps (at the price of distance) and upstream speeds have increased similarly. Major suppliers now mass-produce ADSL chip sets, and the footprint and power consumption of ADSL transceivers has significantly decreased. China has become the clear leader in DSL with almost 14 million lines by April 2004. With about 66% of the world's mass-market broadband subscribers, DSL is the leading broadband technology.

Bellcore conceived ADSL primarily for video at 1.5 Mbps, and early ADSL trials focused on Video on Demand (VOD) applications. However, the primary market is now Internet based interactive multimedia. Cable TV companies and telephone companies are furiously racing to develop infrastructure or to grow by acquisition so that they can win over users who desire high-speed Internet access from residences and remote offices. Data communications, to the Internet or to corporate LANs, is driving demand for ADSL and cable modem based network access. ADSL is only a method of moving bits from one end of copper wires to another; it depends upon access networks to perform network management, protocol conversion, data concentration, routing, multiplexing, and signaling between servers and personal computers.

Sensing an imminent market boom, manufacturers started producing transceivers (modems) in summer 1996 that increased ADSL's downstream and upstream capabilities to as much as 9 Mbps and 640 Kbps, respectively. ADSL customers would connect exponentially faster (downstream) than they were connecting at the time with conventional V.34 modems. Perhaps more importantly, ADSL provided those customers with always-on network access. Since ADSL leverages existing POTS infrastructure, providers could implement it while interoperating with V.90 modems (for those who do not live close enough to a Telco CO—central office—to receive DSL service) and other existing technology.

The absence of standards presented a challenge. AT&T Paradyne defended its proprietary *Carrierless Amplitude and Phase Modulation* (CAP) DSL. The American National Standards Institute (ANSI) advocated Discrete Multitone (DMT) technology—the ASNI Working Group T1E1.4 approved DMT in its T1.413ADSL standard. Discrete Multitone divides the 1 MHz spectrum offered by a phone line into 256 4 kHz channels and varies the bit densities on each channel to overcome noise and interference that may be present in sections of that spectrum. Proponents contend that DMT is better than CAP on noisy lines because it maximizes throughput on good channels and minimizes use of channels that have significant interference. In contrast, CAP uses Quadrature Amplitude Modulation (much like V.90 modems) and relies on a combination of amplitude modulation and phase shifts to increase line capacity over a single line. AT&T integrated a feature of DMT, Rate Adaptive Digital Subscriber Loop (RADSL) to reduce each customer's bandwidth relative to its distance from a central office (CO) and its wire quality—at the cost, of course, of bandwidth.

To address the lack of interoperability between vendors, a consortium of organizations led by Intel, Microsoft, and Compaq announced the "Universal" Asymmetric Digital Subscriber Line (ADSL) Working Group. The group's stated purpose was to accelerate ADSL adoption and availability by proposing a simplified version that would deliver high-speed modem communications over existing phone lines to consumers, and would be based upon an open, interoperable ITU standard. By reducing the complexity of the on-site installation and eliminating the need for new wiring at the user's home, Universal ADSL promised to facilitate cost-effectively DSL deployment. Moreover, by leveraging existing rate adaptive ADSL (RAADSL or RADSL) technology, the standard promised to open the market to additional customers by automatically configuring itself for the fastest rate for any given line. In June 1999, the ITU-T approved the G.992.2 standard, widely known as *G.lite.*

Based on the same underlying technology as standard ADSL, G.lite provides a roadmap for vendor interoperability and eliminates the need to dispatch a technician to install a splitter at the customer premise. Once NSPs provision the central office equipment, the customer connects a G.lite modem as easily as she may have installed a conventional modem. G.lite modems began showing up in retail outlets in fall 1999. G.lite works with Digital Loop Carrier, the local loop infrastructure that connects customers who are located more than 18,000 feet from the central office, and thereby opens the technology to a broader audience.

The new focus on the Internet and corporate LAN access has also forced changes in access network architectures. Because IP dominates the Internet and private networks, NSPs are constructing access networks to support IP to customer premises. However, recognizing ATM's inherent ability to support converged services, and QoS, NSPs are building access networks to relay IP via ATM multiplexing and switching. The first ADSL trials employed a router-based architecture in which ADSL modems perform MAC layer concentration by connecting to an IP router through an Ethernet hub or switch each Central Office (CO). Now, Digital Subscriber Line Access Multiplexers (DSLAM) displace the router by connecting CO modems directly to a shared high-speed backplane.

The multiplexing protocol in this scenario is Ethernet from the PC to the router, and IP for Multiplexing multiple connections into single PCs. NSPs can gradually migrate to this configuration from legacy equipment by integrating various pieces over time. Doing so involves no protocol conversion in the network, and is compatible with Internet and LAN protocols. However, this router-based architecture will not scale for the broad residential broadband market, and cannot be easily adapted for mixed services and switching. For instance, a user who desires to connect alternately to a NSP and to a corporate network requires two separate routers. Moreover, it does not support QoS for video and voice. Consequently, NSPs generally deploy ADSL with ATM layer 2 multiplexing.

Currently, Network Service Providers are offering ADSL networks with Permanent Virtual Circuits (PVCs), and ATM-based paths between user terminals and NSPs or corporate LAN gateways. Users cannot alter PVCs; they require network administrator intervention. IP routing over a PVC provides access to the Internet or to a corporate network. ATM Switched Virtual Circuits (SVC) were envisioned as the upgrade path. A SVC connection is established in Real-Time in response to signaling messages from the customer. SVCs would greatly reduce the effort to provision service to a new customer, and would permit customers to roam freely between Information Service Providers.

Switching traffic through an ATM backbone provides services flexibility and provides the best compromise for convergence. By inserting a DSLAM with integrated ADSL modems for each subscriber at each end office, networks reach subscriber premises over existing copper lines. The ATM cloud resides within the ADSL access provider's access network. SVCs would allow sequential Network Service Provider (NSP) selection from a given client. Such premises may be comprised of an Ethernet connection between the ADSL modem and the PC, an ADSL modem that is integrated into a PC, or a premises network. To serve multiple PCs it might be necessary to install a small router at the premises.

PVCs do not scale as a network grows to millions of customers and thousands of circuit changes a day. SVCs allow users to connect and disconnect just as they make telephone calls, but are difficult and costly to implement. However, PPP over ATM simulates a limited form of customer-controlled switching that enables users to access numerous private networks in sequence.

NSPs have adopted Point-to-Point Protocol (PPP) over ATM because it is easy to integrate with existing protocols and operating methods. They use RADIUS (remote authentication of dial-in user services) servers for authentication, authorization, accounting (AAA) services that are dependent on PPP session establishment to end-users. By adopting the same service access protocol that dial-up users use, NSPs facilitate service convergence and leverage a mature service platform. In lieu of an end-to-end ATM network (or of IPv6), *Layer 2 Tunneling Protocol* (L2TP) Access Concentrators (LAC) map multiple PPP sessions within individual PVCs from the access network to PPP sessions within an L2TP tunnel. Multiple tunnels may exist between the LAC and an NSP where the tunnel used to convey a packet depends upon the priority assigned to the packet.

The NSP-gateway concept can also be deployed via a shared Broadband Access Server (BAS) with PPP Terminated Aggregation (PTA). The same BAS that a NSP would use to terminate PPP can be operated within the ADSL access network. Each AAA server manages sessions to the BAS through a RADIUS proxy interface to the BAS and thereby permits service autonomy. The BAS strips PPP encapsulation from the user packets before forwarding the packet to a NSP. The BAS serves as an IP router and participates in the IP domain of each NSP when IP is the L3 protocol. SVCs or the NSP gateway (LAC or BAS) permit users to alternate between service providers sequentially, and permit the network access provider and Internet service provider to efficiently provision service to large numbers of customers.

ADSL accommodates ATM transport with variable rates and compensation for ATM overhead. The ATM Forum recognizes ASDL as a physical layer transmission protocol. ADSL modems use forward error correction to reduce errors caused by impulse noise, and isolate POTS channels from digital channels to protect them from interruption because of a digital services failure. Virtually all providers offered ADSL by 1998, many with the idea of installing neighborhood Optical Network Units (ONU) to provide virtual fiber to the curb (FTTC).

DSL serves only customers who reside within 12,000–18,000 feet, or 2.25–3.4 miles, from their central office (CO) because signals on copper *attenuate*, or become corrupted over distance (IDSL will travel up to 26,000 feet). Carriers who must install digital repeaters to provide users with ADSL service must charge more for services.

Moreover, carriers who have enabled voice to transmit over longer distances by installing loading coils in the local loop must make expensive modifications to carry ADSL transmissions. Cable modems have eclipsed DSL in the US, but DSL is the wide leader worldwide. Because cable modems are a shared architecture, DSL provides performance and security advantages.

Cable Modems

Cable modems comprised 63% of the US broadband market; with nearly 17 million subscribers by April 2004, the US was the largest cable modem market. South Korea and Japan, respectively, boast five million and four million subscribers.

Nevertheless, cable modem implementation has not been without its tribulations. Cable modems got a slow start compared to DSL because the network infrastructure required massive retrofitting, and because of an absence of standards. In 1997, only 10% of cable plants could accommodate two-way digital connectivity, but cable companies have since invested in hybrid fiber-coaxial networks.

The IEEE *802.14 Cable TV Media Access Control (MAC) and Physical (PHY) Protocol Working Group* was formed in May 1994 to develop international standards for data communications over cable. Their charter was to submit a cable modem MAC and PHY standard to the IEEE in 1995, but the standards were delayed until almost 1998.

In January 1996, the *Multimedia Cable Network System Partners Ltd.* (MCNS), a limited partnership that was comprised of Comcast, Cox, TCI, and Time Warner, issued a RFP for a project management company to swiftly research and publish a set of high-speed cable data services interface specifications. The MCNS Data Over Cable Service Interface Specification (DOCSIS) RFP persuaded MediaOne, Rogers Cablesystems, and CableLabs to join the coalition and thereby constitute a provider organization that serves 85% of U.S. cable subscribers and 70% of Canadian subscribers. Consequently, more than 20 vendors announced plans to build products based on the MCNS DOCSIS standard in March 1997.

To reduce product costs and time to market, MCNS minimized technical complexity and developed a technology solution that, although, not ideal, was adequate. In contrast, the 802.14 group had been seeking to create a standard that would provide excellent quality and stand the test of time. At the physical layer, *modulation formats for digital signals,* the MCNS and IEEE specifications are comparable; both MCNS and the 802.14 upstream modulation standards incorporate QPSK (quadrature phase shift keying) and 16QAM. The 802.14 specification supports the ITU's J.83 Annex A (European DVB/DAVIC), B (MCNS), and C (Japanese) standards for 64/256 QAM modulation, and provides a maximum 36 Mbps of downstream throughput per 6 MHz television channel.

The media access control (MAC), which sets the rules for network access by users, is where the two significantly differ. The 802.14 group specified ATM as its default solution from the headend to the cable modem. MCNS specified a scheme that is based on variable-length packets and that favors the delivery of Internet Protocol (IP) traffic. Both specify a 10Base-T Ethernet connection from the cable modem to the PC. Because ATM is deterministic, it provides superior QoS for integrated delivery of video, voice, and data traffic to cable modem units (as long as the cable modem is an ATM device). The 802.14 group envisioned ATM as a long-term, flexible solution.

Cable operators, whose priority was to provide high-speed Internet services to consumers, believed that ATM would add unnecessary complexity and cost to cable modem systems. MCNS members' intent was to leverage the low costs of ubiquitous Ethernet and IP networking technologies. Standardized DOCSIS cable modems started shipping in limited quantities in the third and fourth quarters of 1998. DOCSIS defines MPEG-2 transport for the downstream transmission convergence sublayer. MPEG-2 allows both video and data signals to be carried in a single multiplexed bit stream that occupies one analog channel, and thereby preserve channel capacity.

The ITU recognition DOCSIS as an international standard in 1998, but the IEEE 802.14 continued its own work. In a token gesture to acknowledge the Working Group's work, MCNS indicated that it would implement IEEE 802.14's advanced PHY specification within DOCSIS. In meeting that commitment, MCNS asked the vendors whom the IEEE had selected to develop an advanced PHY for its own standard, to develop the advanced PHY for DOCSIS 1.2. Many believed that the IEEE 802.14 specification might become a standard for operators outside North America, or that the specification might apply to corporate data services, while DOCSIS would be deployed for residential services. These speculations now seem improbable. Despite good intentions and a better specification than that developed by MCNS, IEEE 802.14 was simply unable to complete a specification in time.

DOCSIS 1.0, first issued on March 1997, offered support for high speed data services. It was a huge success; 30-million DOCSIS modems shipped worldwide as of Q3-2003. DOCSIS 1.1, first issued on March 1999, introduced support for telephony, gaming, and streaming media, and allowed for QoS, service security, and backward compatibility with DOCSIS 1.0. DOCSIS 2.0, first issued on December 2001, added the capacity for symmetric services, increased upstream capacity, improved robustness against interference (A-TDMA and S-CDMA), and backward compatibility with DOCSIS 1.0 and 1.1.

For the downstream physical layer, DOCSIS 1.0 specifies ITU J.83B digital data distribution over existing North American cable TV networks using existing RF equipment. It combines 64 or 256 QAM modulation of the downstream signal, a variable length interleaver, trellis coding, and Reed-Solomon error correction coding. For the downstream transmission convergence sublayer, DOCSIS defines MPEG-2 transport to enable both video and data signals in a single multiplexed bit stream that occupies one analog channel. DOCSIS accommodates most cable network architectures because it uses a range of symbol rates including QPSK or 16 QAM modulation, Reed-Solomon error correction coding, and transmit power automatically adjusted by the MAC. The DOCSIS MAC enables initializing new modems on the network, sharing upstream bandwidth between active modems, periodic ranging and power adjustment to decrease the effects of temperature on the cable plant, and modem management. The DOCSIS standards for Baseline Privacy use 56-bit DES encryption of data traffic. It leverages the RSA public-key encryption algorithm for key management and distribution, with a key size of 768 bits for DOCSIS 1.0 and 1024 bits for DOCSIS 1.1. DOCSIS RF-return cable modems forward packets using IEEE 802.1D forwarding rules, as modified by the DOCSIS specifications. Telco return modems forward packets using IP forwarding rules. In both modes of operation, these modems are capable of layer two and layer three packet-filtering, per operator-defined filter rules.

The DOCSIS specifications were developed to enable hardware and systems interoperability, and automated configuration and management. Cable Modems and Cable Modem Termination Systems (CMTS) from numerous vendors can use the DOCSIS specifications to operate on the same network. In the DOCSIS model, a cable modem is authorized by the CMTS for use on the network, and configures itself according to parameters that it receives from the headend, without user involvement. When the cable modem is powered up and logged on to the network for the first time (or if the data frequency has been changed), it automatically scans the downstream frequency spectrum to locate the data channel. If the cable modem has previously accessed the network, it will have stored the previous session frequency and will immediately tune to the data channel. Once it finds the DOCSIS data signal, it looks for a message that contains the basic parameters for the upstream channel (e.g., frequency, modulation, symbol rate, FEC parameters.). The cable modem then transmits a message to the CMTS with a request for additional information that will enable it to connect to the network. Through a series of messages and interactions, the cable modem establishes IP connectivity using DHCP, and then receives additional configuration parameters via TFTP so that it can configure itself. Once the cable modem is configured, it registers with the CMTS and becomes authorized to use the network. Optionally, the final step of the initialization is establishment of the parameters to use Baseline Privacy. Once configured and authorized, it is a standard Ethernet network device.

The DOCSIS downstream transmission specification is based upon ITU J.83B technology developed for digital TV applications. In 64 QAM mode, the maximum bandwidth available on a 6 MHz channel is approximately 27 Mbps after error correction. When running in 256 QAM mode, the system can support rates up to 38 Mbps. It can be set anywhere in the DOCSIS downstream spectral range of 88-860 MHz. Upstream, the cable modem supports two modulation formats, QPSK and 16 QAM, and 5 different symbol rates. The spectral range for upstream transmission is 5-42 MHz.

The structure of the network is IP over Ethernet. The primary standards incorporated in the DOCSIS system architecture are IEEE Ethernet standards 802, 802.2, 802.3, and RFC-791 IP data standards. Once the transmission protocols have been established, the network operates transparently as an Ethernet network. DOCSIS cable modems are software upgradable during initialization or on command from the network operator via an SNMP command and a TFTP file transfer of the new software image. The upgrade process can fail at virtually any point without leaving the cable modem unusable. The cable modem makes several checks of the new file to ensure that it is intended for that particular make and model of cable modem, since several different makes and models of cable modem may be operating on a single network. There is an MD5 checksum on the new code image that is used to ensure that it has not been corrupted in transmission.

Despite the broad acceptance of DOCSIS in the US, the European community did not adopt it as is. It did adopt what has become known as EuroDOCSIS, which offers a 33% wider downstream channel.

Despite access to 65 million homes, cable providers initially had little entry to business market. Having since dropped fiber lines, they are now also formidable competitors in outside of consumer markets. Moreover, given that DOCSIS 2.0 was designed to accommodate videoconferencing, cable providers may comprise the best platform for providing wireline-based video communication.

HDTV and DTV

The United States' commitment to Digital TV and the strength of its own semiconductor and computer technology stymied it ability to deploy HDTV. The United States, was the first nation to support digital high-definition technology, but was the last to broadcast HDTV. Europe shipped its Digital Video Broadcasting (DVB) standards for satellite, cable and terrestrial broadcasts years before the US. DVB is not high-definition, but the MPEG2-based technology was deployed while the US squabbled about digital HTDV. Meanwhile, Japan was broadcasting 13 hours of its analog (not digital) Hi-Vision programs every day by 1996. Japan significantly lowered the cost of its High-Definition TV equipment and quietly developed a digital replacement. By the end of 1999, Japanese consumers bought more than 746,000 Hi-Vision receivers.

The FCC reserved an extra 6 MHz of spectrum for HDTV and Standard Definition TV (SDTV), and the technology is spawning myriad data services and *datacasting* in the same TV-broadcast spectrum. However, no laws or regulations define how digital-TV licensees might use the new spectrum. Moreover, many people belittle the flexible standard, which the Grand Alliance developed and the Advanced Television Systems Committee (ATSC) recommended to the FCC, as a technology niche that affects TV manufacturers but not the computer industry.

The ATSC, a private-sector organization that develops voluntary standards for the entire spectrum of advanced-television systems, adopted two additional standards in 1996: a "program guide for digital television," and "system information for digital television." These specifications are part of the overall HDTV system on which the United States has standardized. They are important for fostering interoperability between different set-top boxes or TV systems for cable, satellite, multichannel/ multipoint distribution service (MMDS), and satellite master-antenna television systems. The system information defines the transmission parameters that a variety of digital decoders need for accessing and processing digital and analog transmissions. The program guide provides a common format that TV receivers and set-top boxes can decode.

To accelerate the conversion to digital television the FCC ruled, in August 2002, that manufacturers must, by July 2004, include digital tuners in TV sets that have screens that are 36″ or larger. The same requirements would be phased in by 2007 for all other sets.

PBS unveiled the first US HDTV network early in 2004; shortly thereafter, HD-Prime became the nation's first all-HDTV cable service, and VOOM became the first all-HDTV/Digital satellite-to-home service. CBS, ABC and NBC broadcast most of their live programming in full 1080i HDTV in the 2004-2005 broadcast season, including the 2004 Olympics in Greece.

While these broadcasts began in HDTV, not all local stations and cable systems were capable of relaying full HDTV signals. As a result, satellite-to-home services such as DISH, VOOM, and Direct-TV, which dedicated channels to full 1080i HDTV programming, increased in popularity. Multi-path reception causes ghosts in analog TV; it causes complete loss of sound and picture in digital TV. Nevertheless, it is the standard the US chose, and manufacturers seem to be innovating their way out of the problem.

It is unclear how or if US TV broadcasters will ever recoup the billions of dollars they have invested in the conversion to digital TV. However, it seems clear that the strategy includes leveraging the infrastructure with innovative interactive multimedia offerings.

CHAPTER 11 REFERENCES

Butt, Haroon. "World Broadband Growth: DSL Dominates." September 9th 2004. http://www.itudaily.com/new/home.asp?articleid=4090901

Brown, Ralph W. "DOCSIS - 2004 and Beyond." (CableLabs® Media & Financial Analysts Briefing May 2004). *Cable Television Laboratories, Inc.*

"H.264 & IPTV Over DSL." *Intel Corporation.* 2004.

Thompson, Jim. "Move Over Cable, Here Comes G.lite." *ISP News.* June 23, 1999. http://www.internetnews.com/

Starr, Tom, Maxwell, Kim. "ADSL Access Networks," ADSL Forum, May 1998. http://www.adsl.com/

Carr, Jim. "Solving Asynchronous Transfer Mode," *VARbusiness* March 1, 1996 v12 n3 pp. 84-92.

Flanagan, William A. Frames, *Packets and Cells in Broadband Networking* Telecom Library, 1991.

Riggs, Brian. "ISDN ordering: not quite so EZ," *LAN Times* November 11, 1996 v13 n25 pp. 45-47.

Rockwell, Mark. "Carriers agree on ISDN service codes," *CommunicationsWeek* April 1, 1996 n603 p. 45.

Steinke, Steve. "Getting data over the telephone line," *LAN Magazine* April 1996 v11 n4 pp. 27-29. The ATM Forum: Mountain View, California.

12

PACKET-SWITCHED NETWORKS

You could not step twice into the same rivers, for
other waters are ever flowing on to you.
Heraclitus (c. 535–475 B.C)

NEW CHALLENGES FOR OLDER NETWORKS

Today's typical large organization manages multiple independent networks: each optimized to carry unique types of traffic: voice, data, and video. This may not be efficient but, until recently, may have been functional because the characteristics of each traffic type made it incompatible with the others. This is quickly changing. Whereas in the past, applications contained only one type of traffic, today they combine voice, video, and data. Streaming media on already-congested LANs, or worse, WANs can make a network unusable.

Internet Protocol (IP) is not only the clear winner in packet-switched networking, it is the clear winner for networks, period. It is not just the winner in the Local Area Network (LAN), it is also the winner in Wide Area Networks (WAN). Carriers are purging legacy protocols such as frame relay and Asynchronous Transfer Protocol (ATM) as quickly as they can, and moving to pure-IP networks. The reason for this is that organizations are seeking economies and utility by consolidating voice (which, historically, has included video) networks into data networks. The characteristics of video and voice communication strain such networks. Bandwidth allotted for voice or video streams should exceed the video stream itself by about 20%, and the total bandwidth used for those streams should never comprise more than 33% of total network bandwidth.

Reliance upon networks is driving organizations to get serious about infrastructure design and management. What they have is already straining under the weight introduced by changes in the network usage model. Traditional LANs and WANs began bog down when organizations moved from character-oriented host/terminal computing models to more interactive client/server architectures. Client/server *computing,* which splits a user level program into client processes and server processes, requires significant interaction between the two. Powerful CPUs on both sides of the exchange pump out data in a fashion that peppers a conventional LAN with bursts in the hundreds of Mbps.

Along with the shift to client/server (perhaps preceding it in many cases) came seemingly small differences in applications. For instance, in the early 1990s, users began to attach files to electronic mail messages. Large differences came next—organizations added Web browsers to the standard desktop configuration. Suddenly the network was bursting at the seams with media-rich information pulled down from the Internet. As network throughput steadily declined, companies set out to build their own, internally-oriented versions of the Web—intranets became the rage.

In some enterprises, the term network management may be an oxymoron. Yet even as companies struggle to accommodate an overabundance of bits, the usage model is changing again. Early adopters are deploying MPEG-compressed playback and interactive video applications over their networks. This often introduces congestion to a LAN segment. When such dense traffic traverses WAN links, traffic jams are almost assured. Fortunately, most organizations, before launching new applications, adapt their LANs and WANs to cope with the load.

Adapting the network for streaming media means migrating toward infrastructures (hopefully enterprise-wide) that match service quality to traffic requirements. Before we talk about how this is done, we should review the differences between connection-oriented and connectionless networks.

Connection-Oriented vs. Connectionless Networks

Phone calls and circuit-switched videoconferences establish a path between parties for the duration of an exchange. Until the communication concludes and the connection is dropped, no one else can use the circuit. A busy signal is returned to those who try. This type of arrangement is said to be *connection-oriented*. Connection-oriented networks work well for time-dependent transmissions—streaming media—that require a constant bit rate (CBR). Applications in this class not only require a CBR, they need a guarantee that they will receive it. The term used to describe this type of service is *isochronous*. Isochronous service provides predictable, unnoticeable delays, and regular delivery. It is mandatory for any LAN that is to carry video communication.

Packet-switched Ethernet LANs are connectionless. This design is most effective with *bursty* or *transaction-oriented* data in which long lulls may occur between the transmissions of associated bits. For the sake of efficiency, all devices are connected to a common medium. Busy signals are never returned in this scenario. Instead, connectionless networks react to increased traffic by slowing down. When they do, users tend to take corrective action by twitching their cursor, reissuing commands and trying to open new applications. Of course, this only makes things worse. Eventually the congestion clears and throughput increases accordingly.

Even though users become annoyed, delays are usually acceptable in networked text and graphics applications. Real-time delivery is not critical to comprehension. To mediate congestion, frames of data are briefly buffered.

Buffering is not an option for interactive video exchanges. Human protocols, which have evolved over millions of years, are highly sensitive to timing. Regular and sequential delivery is mandatory. Speech must be *synched* to lip movement. Pauses are more than distractions. Delays in excess of 325 milliseconds are debilitating.

Unfortunately, when streaming media moves over Ethernet networks, pauses are the norm, not the exception. Because of Ethernet's resilience and adaptability, however, there are several different approaches to modifying existing Ethernet LANs to make them more suitable for multimedia traffic.

When upgrading LANs, it is important to plan if multimedia communications will not be confined locally (which will usually be the case). The wide area network (which is increasingly also becoming Ethernet) must also be considered. Although intranets are very important, multimedia communications will not be confined to the populace of a single enterprise. Rather, an Internet or telephony (any-to-any) model, that allows almost infinite endpoints, is the goal. With this in mind, we will examine how one might condition an Ethernet LAN for interactive multimedia traffic.

STREAMING MEDIA OVER ETHERNET

Ethernet was developed in the early 1970s to support laser printing. The Institute of Electrical and Electronic Engineers (IEEE), in their 802.3 specification, standardized the technologies that underlie Ethernet. Ethernet is founded on the Carrier Sense Multiple Access with Collision Detection (CSMA/CD) media access control (MAC) protocol. In CSMA/CD networks, nodes broadcast frames of data to all other stations on the same segment. Because the segment is shared, 802.3 LANs employ a listen-before-transmitting scheme. If a sending device hears no oncoming traffic, it transmits its data, which has been addressed to assure that only the intended node listens and responds to it. The system works smoothly unless two stations broadcast at precisely the same time (which is common when subnets support numerous devices). In that case, a collision occurs. Both devices, after hearing a jamming signal warn of the collision, must back off and wait a random period before retransmitting. Contention and collisions limit throughput on Ethernet and preclude any guarantee of timely signal delivery. Therefore, CSMA/CD is quite inadequate in real-time, interactive multimedia applications.

Since its invention, Ethernet deployment methods have continuously evolved. Ethernet began as 10Base-5. The name was descriptive: 10 refers to 10 Mbps, Base pertains to Ethernet's baseband method of transmission and the 5 reflects a network diameter of 500 meters. As time went on, 10Base-5 became known as Thick Wire Ethernet because the coaxial cable it required measured about one-half inch in diameter. After 10Base-5 came 10Base-2 that used coaxial cable about half as thick as

10Base-5. Thin-net cable, as it was also called, offered easier manipulation, and cleaner attachment to a workstation. A signal can be pushed a long way over coaxial cable. Thus, older, 10Base-2/ 10Base-5 Ethernet LANs could daisy chain devices together to support an entire building or a campus. Some older Ethernet LANs supported as many as 300 devices. The problem with this approach (in addition to the obvious congestion implications) was that if any node had a problem (e.g., a bad cable connection or a faulty network adapter) it brought down the entire network.

In 1990, the introduction of 10Base-T Ethernet (also known as IEEE 802.3i) eliminated the need for coaxial cable altogether. Instead, it relies on unshielded twisted pair (UTP), using only two of the four pair of wires found in most UTP cable. 10Base-T is arranged in a star-hub topology, much as voice stations connect to a PBX. Hubs, which reside in floor-level equipment closets, are equipped with port. These ports are wired to individual nodes on a network. Data packets enter a hub through a single port and exit through all ports.

Wireless networking is now doing to 10Base-T (and its faster siblings), what 10Base-T did to coaxial cable. As the various wireless Ethernet technologies get faster and more reliable, wire will increasingly exist only in the network core. PCs will shed the RJ-48 jack, but the distances involved in large campus and carrier networks (not to mention scarcity of spectrum in the unlicensed bands) are likely to ensure that wire has a place for in the near future. However, inclusion of Gigabit Ethernet NICs in PCs is likely to decelerate the adoption of slower wireless networks.

How can one adapt a best-effort network infrastructure (that was designed for printing) to video communication and other bit-intensive multimedia applications? Because there is so much of it installed, the question is an important one, and one with countless answers.

ADAPTING ETHERNET FOR STREAMING MEDIA

LAN Microsegmentation

As applications become more demanding, network managers may divide LANs into smaller segments in an approach known as microsegmentation. Whereas a LAN might have supported 64 devices in the past, hub bandwidth may be divided to serve many segments with only a few ports. A single hub might be capable of supporting as many as 30 different LAN segments. Individual hubs are connected via backbones that conform to one of two designs: collapsed or distributed.

In a collapsed backbone, hubs connect to multi-port bridges (no intelligence), switches (intelligence generally at layer 2), or routers (intelligence at layer 3). Each subnet (sub-network) served by the departmental hub is, typically, bridged to an Ethernet network that connects all devices to a campus or building hub (or series of hubs) and, from there, to other sites and to WAN services.

The packet switching technique that conventional routers use does not scale well. As

226

traffic increases, the router's bus becomes saturated (the term 'router' is largely synonymous with the term 'layer-3 switch'). Consequently, its switching logic (a router must read an entire packet before it can switch that packet to its destination) may fall behind and cause long delays. These delays devastate video applications.

The distributed backbone approach employs a faster backbone network to interconnect campus or building hubs. Hub-based LANs are bridged or routed to the fast backbone. In contrast to the collapsed backbone method of relying on one large router, distributed backbones rely on a number of moderately sized routers. If segmentation is used as a method to increase speed, the network design becomes a collapsed backbone configuration.

Microsegmentation has consequences. Bridge ports and router ports are expensive. Moreover, microsegmentation is hard to administer. When clients and servers are not isolated on the same subnet, they must traverse LAN boundaries. This often causes backbone networks to slow to unacceptable levels. Another drawback of traditional shared-media hubs is that everything connected to a particular hub competes for a slice of the hub's backplane bandwidth.

Microsegmentation offers a short-term solution that is worth consideration only if the objective is to introduce low-speed video communication to the desktop in a limited fashion. Most organizations have abandoned microsegmentation in favor of a new type of arrangement—switching hubs.

LAN Switches

The easiest and least expensive way to increase throughput for high-traffic desktop video communication is to replace conventional Ethernet hubs with LAN switches. Switching allows the efficient interconnection of multiple LAN segments. Ethernet switches are hybrid internetworking devices that combine circuit switching with packet switching (layer-3 Ethernet switches are, in fact, high-performance routers). Network segments (or devices) are connected to switch ports. Traffic between ports is conveyed across the switch backplane. In this manner, large LANs can be constructed from many smaller LAN segments.

In a switched Ethernet environment, users contend for bandwidth only with other users on the same segment. The smaller the segment, the greater the bandwidth available to each station. Because the switching device's backplane speed is faster than individual segments, the aggregate capacity of the LAN is increased.

Switching hub ports have an incoming packet cycle and an outgoing packet cycle, each of which operates at 10, 100, 1000, and 10,000 Mbps. If the switch supports LAN segments, it is said to *segment switch* (most switches do this). Switching hubs can also be arranged to *node switch,* in which case each desktop is provided with its own Ethernet segment. This is cost-effective only in instances in which a workgroup requires high-speed support, such as full-motion video communication.

Whether a switch is configured to segment switch or node switch, it operates as follows: During an incoming cycle a port opens up to accept a packet. As the packet enters the hub, its MAC layerdestination address is examined and it is briefly buffered until the outgoing hub port associated with the destination address opens to accept traffic. If the packet is destined for a node on the same segment as the sending device, it does not cross the switch's backplane. If the packet is addressed to a node on a different segment, it must move across the high-speed backplane and through the silicon-based switching fabric. In this way, Ethernet switching hubs support multiple simultaneous 10/100/1000 Mbps connections between nodes, segments and combinations thereof.

In switching, only the sending and destination port are aware of the packet's existence. Hence, the network is not overwhelmed with extraneous packets. In effect, a switch changes Ethernet from a contention broadcast system to a direct delivery system. In a node-switched arrangement, it creates the illusion that each station has its own private subnet that allows it to send and receive data at Ethernet's full 10 Mbps (for a total throughput of 20 Mbps). This same bandwidth can be sub-divided among a small group of stations in a segment switched arrangement. Thus, isochronous service is delivered to a user or group of users whose real-time applications drive the need for a LAN environment from destructive collisions.

Switches are available for 10, 100, and 1000 Mbps Ethernet. Numerous manufacturers offer products in each category. The 100 Mbps switches are most effective when positioned as traffic control points for the backbone. In this environment, a hub is attached to every port on the Ethernet switch. 100 Mbps Ethernet switches can also extend the diameter of a 100 Mbps Ethernet LAN beyond its 250-meter limit. The per-port price of the 1000 Mbps Ethernet switch limits its deployment to high-end applications. Switched 100 or 1000 Mbps Ethernet would most often apply to video applications in situations where multimedia servers store MPEG-2 encoded material that must traverse the LAN. In a few cases, 1000 Mbps is being extended directly to the desktop but, to obtain maximum value from this approach, supported nodes should be equipped with fast (PCI class or higher) buses.

The Ethernet switch is best suited for resolving throughput problems—for instance, where several nodes on the same subnet all try to deploy real-time audio-video applications at once. Throughput would decline accordingly and performance would be affected for everyone. To resolve the problem, an Ethernet switch might be installed. It would not necessarily have to replace a hub port-for-port (although switches are replacing hubs in most networks). It could be used to isolate a few high-powered desktops from the rest of the workgroup or to create smaller LAN segments with only a few nodes each. In this way, the switch divides the network into smaller collision domains.

Switched Ethernet is inexpensive enough to justify replacing hub ports with switch ports. With no modifications, category 3 cable plant is adequate for 10 Mbps

Ethernet networks, and category 5 cable plant is adequate for 100 Mbps Ethernet networks. Gigabit Ethernet, originally designed to run only over fiber, will run over category 5 cable, but is best over category 7 cable. This allows a company to introduce more bandwidth with almost no downtime and very little cost.

Ethernet switches come either as chassis-based modules that support plug-in modules or as stand-alone stackable units. Chassis-based switches cost more, but they are highly scalable and generally offer redundant power supplies and central management. Stackable switches are more affordable and can be added incrementally. Port densities vary but most stand-alone switches offer between four and 48 ports, and each port will support a separate 10/100 Mbps Ethernet segment (in many cases, each port cannot only switch, but also route). Some switches offer high-speed port (1000 Mbps) options, and most offer virtual LAN (VLAN) capabilities.

Switching hubs use two different techniques to handle frames. Cut-through switching begins to forward a packet to its destination port immediately after decoding its MAC address—before the entire frame is received. This approach reduces the device latency that is incurred when a packet is delivered from one port to another. However, errors are detected through the examination of the CRC (cyclic redundancy check) bits at the end of a packet. Although cut-through switching adds virtually no delay, retransmission of a packet is sometimes necessary after it reaches its destination, which consumes network bandwidth.

Store-and-forward switching receives the entire frame into a buffer before forwarding it to the destination port. In this approach, the entire packet is checked before it is passed to its destination port. Although this method is highly reliable in terms of detecting and dealing with errors, it introduces a small amount of latency. Most store-and-forward switches comply with the Spanning Tree algorithm that discovers the best path to each destination and stores it in a look-up table. Subsequent packets that arrive for the same MAC address are sent out through the port specified in the table. Administrators can also use the table to create a redundant physical network for security.

Kalpana (which means imagination in Hindi) pioneered the concept of Ethernet switching in 1990. According to the results of a 1996 study performed by the Dell'Oro Group, U.S. switching/wiring hub revenues were $3.9 billion in 1994. In 1995, they grew to $6.1 billion, and they have not looked back since.

Given the advent of higher-layer switching (predominantly layer-3), store-and-forward switching is no longer much of an issue. Layer-3 switching is really *wire-speed* routing. So-called non-blocking switches execute routing in silicon rather than in software. By also offering backplanes with bandwidth capabilities that exceed the aggregate bandwidth of all ports, some manufacturers mitigate the non-deterministic nature of Ethernet's collision-detection based architecture.

EIA/TIA 568a

Cable plant plays a big role in allowing Ethernet to run at 100 Mbps speeds. In 1991, the Electronics Industries Association and Telecommunications Industry Association (EIA/TIA) introduced a cable-rating scheme. Knows as the Commercial Building Telecommunications Wiring Standards, EIA/TIA 568 (which has since been revised as EIA/TIA 568a) provides electrical performance specifications for Category 1 through Category 5 UTP. The EIA/TIA categories measure the resistance of the wire to frequency bleeding between pairs (also known as near-end crosstalk) and its ability to pass signals at different frequencies (known as the frequency attenuation curve).

The EIA/TIA 568a categories are as follows:

- Category 1 wiring is basic, low-grade UTP telephone cable that works well for low frequency transmissions such as voice service. By definition, it is not suitable for data transmission.

- Category 2 UTP is UTP that is certified for use as fast as 4 MHz (slower-speed token-ring).

- Category 3 is certified up to speeds of 10 MHz, and is the minimum quality cable required for 10Base-T. UTP in this category must have at least three twists per foot, with every twist conforming to a different pattern. Category 3 cable is the most common cable today, with over 50% of networks using it to support desktop devices.

- Category 4 is the lowest grade UTP suitable for use up to 16 MHz (e.g., for 16Mbps token-ring).

- Category 5 is certified up to 100 MHz. It is acceptable for FDDI over copper and 100Base-T.

- Category 6 offers twice the usable frequency spectrum for signal transmission of a category 5 or 5e system and twice the information carrying capacity. A category 6 channel provides additional signal bandwidth and about 10 dB of near-end crosstalk loss margin over a category 5e channel.

- Category 7 may offer as much as 1 GHz over twisted-pair cabling and thereby enable broadband video (with an upper frequency requirement of 862 MHz) to operate over class F cabling with concurrent connections to other networking applications. Category 7 cables include individually screened twisted-pairs and an overall shield. Pending category 7 specifications require connectors to provide at least 60 dB of crosstalk isolation between all pairs at 600 MHz.

Potential adopters of 100 or 1000 Mbps Ethernet should carefully examine their existing cabling to determine the necessary adjustments and require that the supplier certify that what is in place conforms to the specification in question. When it comes to EIA/TIA 568, every inch of cable plant—from the network adapter in the workstation to the wiring closet, and including punch-down blocks, patch cords and panels—must adhere to the standards. Customers that are not certain about how their cable plant conforms to EIA/TIA 568a should test it, using a cable tester from a company such as Tektronix, Fluke, or Scope. Without an accurate reading on conformance, a manager cannot make an informed decision about what 100 or 1000 Mbps solution will work in his or her environment. When conducting a test, technical personnel (or subcontractors) should test *all* pairs. The reason is that if the plant turns out to conform only to Category 3 cable, all four pairs that make up a UTP bundle will have to be used to support 100VG or 100Base-T4 connections. Category 3 cable will not support 1000 Mbps Ethernet.

Fast Ethernet

In 1994, the Fast Ethernet Alliance completed a wiring specification that provides end users with interoperable 100 Mbps Ethernet products that can exist in a variety of cable plant environments. The completed specification, which includes three specifications, defines how 100Mbps Ethernet should run over Category 3, 4, and 5 UTP and over fiber optic cable.

The three specifications for 100Base-T Ethernet are 100Base-TX, 100Base-FX, and 100Base-T4. A slightly different specification, called 100VG-AnyLAN (IEEE 802.12) competes with 100-Base-T. The key difference among the 100-Mbps products is in the kind of cable they use and, in the case of 802.12, how well they cope with the needs of time-sensitive video packets.

Within the 100Base-T family, the cabling differences are as follows 100Base-TX requires Category 5 UTP, 100Base-FX requires fiber optic cable, and 100Base-T4 uses Category 3, 4 or 5 UTP, but four pairs of wire are required for each node.

Over unshielded twisted pair cable, 100Base-T can support only two repeater hops (compared to 10Base-T's three to five hops) and its theoretical collision domain is limited to only 205 meters—less than one tenth that of 10Base-T. These limits can be overcome by deploying stackable hubs and switches.

Although estimates vary, it is safe to say that Category 3 UTP is installed in more than 50% of the telecommunications networks in North America. That simple fact makes 100Base-T4 a viable option for networks with pervasive Category 3 cable. 100Base-T4 needs four pair. In 100Base-T4 implementations, it is important that every wire and every connection in every cable is good. This is not always the case. Although 100Base-T4 is an easy step to take, the market shows little interest in it.

100Base-TX requires two pairs of Category 5 wires. The Category 5 requirement often forces organizations that select to pay for a new cable system. In various tests, 100Base-TX has been proved to handle increasing traffic loads much more effectively than 10Base-T. For instance, it is possible to run it at loads of up to 70% before any signs of congestion begin to appear.

To lessen the impact of upgrading an entire network to Category 5 cable, many of the

100Base-TX vendors are shipping adapters and hubs that can support both 100Base-TX and 100-Base-T4. Because all 100Base-T solutions use the same Ethernet packets, no extra bridging hardware is necessary to mix 100Base-TX and 100Base-T4 hubs in the same stack. Some 100Base-T4 hubs will even integrate both 100Base-TX and 100Base-T4 ports to accommodate both technologies.

Fast Ethernet adapters, better known as 10/100 adapters, are designed with both 10-Mbps and 100-Mbps chips inside. A majority of Fast Ethernet adapters includes drivers and provides a single port with the capability to switch automatically between 10 Mbps and 100 Mbps without any modification on the client side. The availability of 10/100 adapters make it possible to buy incrementally and upgrade the LANs at some future date by merely replacing the hub or switch with one that supports 100Base-T. With pricing comparable to 10-Mbps adapters, buying Fast Ethernet adapters makes a great deal of sense.

With their double speed capability and competitive pricing, Fast Ethernet adapters are a no-brainer purchase for your network. The trick is to buy the right adapter. Look for one that provides a single port for automatic switching and make sure that it does not rely too heavily on the CPU. When it comes time to switch to Fast Ethernet, having this groundwork in place will make the transition easier. The fiber-based 100Base-FX systems are not vulnerable to radio frequency interference and can extend connections out to over a mile of cable. FAST Ethernet and Gigabit Ethernet are the clear winners in high-bandwidth LAN infrastructure.

GIGABIT ETHERNET

A backbone's bandwidth must be greater than that of the nodes that connect to it, and greater even than the aggregate of all links that feed into it. Therefore, as 100 Mbps Ethernet extends to the desktop, faster standards such as Gigabit (1000 Mbps) Ethernet will become increasingly common on backbones. One need only consider the plummeting cost of 100 Mbps Ethernet network interface cards (NIC) to understand just how fast this migration is occurring.

Gigabit Ethernet (also known as 802.3z) was initially developed by the Gigabit Ethernet Alliance (GEA), which consisted of about 50 vendors including 3COM, Bay Networks (now Nortel Networks), Cisco Systems, Compaq, FORE Systems, Intel, Network Peripherals, Silicon Graphics and Sun Microsystems. The GEA's goal was to provide a low-cost and simple upgrade path for networks that already use Ethernet technology. The GEA presented their work to the IEEE, which formed the 802.3z Task Group to study the preliminary specification. On November 18, 1996, the 802.3z Task Group finalized a set of core proposals as the basis for writing the first draft of the standard. The first draft of the 802.3z specification was distributed in late January of 1997, and the standard was adopted in early 1998.

Gigabit Ethernet is based on the 10Mbps and 100Mbps Ethernet standards outlined in the 802.3 specification. It includes support for CSMA/CD (modifications must be made to maintain a half-duplex mode of operation) and uses the same frame format as Ethernet and Fast Ethernet. The first draft of the specification focuses primarily on physical media requirements. Gigabit Ethernet's physical (PHY) layer is a high-speed adaptation of Fiber Channel. An "Independent Interface" is also included in the draft that will allow the MAC protocol to be decoupled from the specifics of the lower layers—and thereby enable the independent development of other PHY types. Work is being done on adapting Gigabit Ethernet to operate over distances of up to 100 meters on unshielded twisted pair (UTP) cabling. Using LX (longwave length transceiver) GBICs and single-mode fiber, Gigabit Ethernet can span 5 kilometers with Gigabit Ethernet in 1999.

Modifications to CSMA/CD include two new features that enable efficient operation over a practical collision domain diameter at 1000 Mbps. These features are carrier extension and packet bursting and they will only affect operation in half-duplex mode. When Gigabit Ethernet is operating in full-duplex mode, it is identical to 802.3u (100Base-T) Ethernet, only faster. The 802.3z specification includes minor MAC changes, but frame length minimums and maximums and frame formats remain the same. Overall, 802.3z is very similar to its lower-speed predecessors.

Because of raw speed and of cost reductions that companies have achieved by adopting ASIC-based models, Gigabit Ethernet is dominating the backbone using switch-to-switch, switch-to-router or switch-to-server connections. Conventional router manufacturers have struggled to adapt their software-based architectures to achieve the high speed of low price of ASIC-based architectures. In a switch-to-switch scenario 1,000 Mbps pipes could be used to connect 100/1000 switches. Most such switches would deliver, in turn, 100Base-T to the desktop. In another scenario, a video or other high-performance server might be equipped with a Gigabit Ethernet NIC that connects it at 1,000 Mbps to a switch and, through the switch, to desktop video applications supported by 100Base-T. In yet another scenario, a shared FDDI backbone might be connected via a FDDI hub or an Ethernet-to-FDDI router with Gigabit Ethernet switches. In any case, Gigabit Ethernet has relegated FDDI to being a legacy technology; while many must accommodate FDDI, few are installing it.

The emerging specification will allow users or nodes to retain their existing NOS, NIC drivers and applications. Nodes will not be affected since Gigabit Ethernet is emerging primarily as a backbone technology for aggregating 10-Mbps and 100-Mbps endpoints at 1000 Mbps or for connecting fast switches or servers.

On July 28, 1998, the Gigabit Ethernet Alliance announced that balloting for the Gigabit Ethernet over copper proposed standard (1000BASE-T) had formally begun, and that the IEEE 802.3ab Task Force (that developed the draft standard) had resolved all outstanding technical issues. The Task Force in conjunction with the IEEE 802.3 Working Group (which oversees Ethernet Standards) voted unanimously to initiate a Working Group ballot for 1000BASE-T. The 1000BASE-T draft standard was intended to enable Gigabit Ethernet to extend to distances of up to 100 meters over Unshielded Twisted Pair Category 5 (Cat 5) copper wiring, the primary cabling medium inside buildings. Representatives from approximately 120 networking, computer, component and test equipment companies participated in the Alliance.

Concurrently, the group began work on an enhanced version of Category 5 wiring and Category 6 and Category 7 wiring standards. For most applications, Cat 5 will be adequate for the foreseeable future because it easily handles 10 Mbps Ethernet, and even 100 Mbps Ethernet. However, 1000B-T will require all four pairs of the Cat 5 cable rather than the two pairs that previous LAN technologies required. Enhanced Cat 5 was created to address this and to remedy crosstalk and line noise issues that are related to Gigabit Ethernet. Category 6 cabling can run at 200 MHz, twice as fast as Cat 5, and Category 7 would run at 600 MHz. These robust cables will have their places, but businesses that require a network faster than Gigabit Ethernet will adopt fiber optics.

ATM—Not Just a Place to Get Cash

Enterprise networks are undergoing profound changes. Even as the bridged or routed LAN-based internetworking model is being assimilated, a cell-switched architecture, that is ideally suited to carrying multimedia traffic, is being overlooked. Of course, we are referring to ATM.

Specifically designed to carry different types of traffic, ATM leverages the gigabit speeds offered by fiber-optic cable. It was designed to provide scalable, broadband, low-latency, cell-switched services that fuse the LAN into the WAN and computer networks into telephone networks. ATM easily supports high loads of converged data, voice, and video traffic.

Standards for managing service quality, bandwidth allocation, network latency, and data flows have long been established for ATM. These issues are still dogging Ethernet. Nevertheless, Ethernet is what network administrators know, and it is the clear winner.

ATM technology has not entered the mainstream of enterprise networking, despite its obvious advantages. A big reason is simply that it is different in concept and management from the packet-switched and routed networks that form the basis for current-day LANs and WANs. ATM is different and complex (perhaps the most complex networking scheme ever developed). Hardware-intensive switches require the overlay of a highly sophisticated, software intensive, protocol infrastructure. This infrastructure is required to both allow individual ATM switches to be linked into a network, and for such networks to internetwork with the vast installed base of extant LANs and WANs.

234

How ATM Works

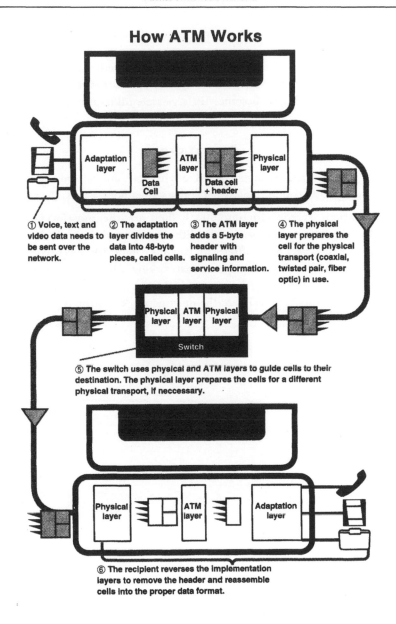

① Voice, text and video data needs to be sent over the network.

② The adaptation layer divides the data into 48-byte pieces, called cells.

③ The ATM layer adds a 5-byte header with signaling and service information.

④ The physical layer prepares the cell for the physical transport (coaxial, twisted pair, fiber optic) in use.

⑤ The switch uses physical and ATM layers to guide cells to their destination. The physical layer prepares the cells for a different physical transport, if neccessary.

⑥ The recipient reverses the implementation layers to remove the header and reassemble cells into the proper data format.

Figure 12-1. How ATM Works (Laura Long).

Cost and complexity have prevented organizations from implementing ATM in the LAN, but the WAN (which is largely managed by carriers) is different. Carriers have not aggressively sold ATM, and are now purging it altogether as they transition to running IP directly over SONET or directly over optical networks. As virtually all traffic (voice, video, and data) becomes IP, large-scale MPLS networks are displacing both ATM and SONET.

CONCLUSION

The connection-oriented nature of ATM might have guaranteed it a permanent quality advantage over connectionless networks, were it not for mass IP adoption.

Ethernet's various speeds (10/100/1000/10,000) are the most widely deployed architecture. However, to compensate for Ethernet networks' inability to guarantee service, network designers must offset with copious bandwidth. It is beyond the scope of this book to predict the future of internetworking. All that is certain is that, as bandwidth demands continue to grow and services are converged, the market will favor the arrangements that facilitate the most network control. ATM was designed for converged networks.

CHAPTER 12 REFERENCES

Dornan, Andy. "Which Wi-Fi?" *Network Magazine*. January 1, 2005. http://www.networkmagazine.com/shared/article/showArticle.jhtml?articleId=55301823

Caruso, Jeff. "Beyond Category 5," *Network World*. July 1999. http://www.nwfusion.com/

Callaghan, Marilyn. "Proposed Standard for Gigabit Ethernet Over Copper Sent to Official Ballot." *Gigabit Ethernet Alliance,* July 1998. http://www.gigabit-ethernet.org/

Beck, Peter. "Twisted pairs challenge broadband signals," *Electronic Engineering Times* September 23, 1996 n920 pp. 54-60.

Cali, Paul D. and Deans, John D. "Installing fast Ethernet backbones," *UNIX Review* October 1996 v14 n11 pp. 46-50.

Karve, Anita, "Gigabit Ethernet prepares for the backbone," *LAN Magazine* October 1996.

Korzeniowski, Paul. "No end in sight to internetworking boom," *Computer Reseller News* October 21, 1996 n706 pS39-40.

Lipschutz, Robert P. "Power to the desktop," *PC Magazine* October 8, 1996 v15 n17 pNE1-6.

Merenbloom, Paul. "Boost your network with Fast Ethernet, fiber, switched hubs," *InfoWorld* October 28, 1996 v18 n44 p. 59.

Schnaidt, Patricia. "Plug in at 100," *LAN Magazine* March 1994 p. 12-30.

Sellers, Philip. "Gigabit Ethernet Flexes Its Muscles," *Computing Canada* October 10, 1996 v22 n21 pp. 1-2.

PART FOUR

EVALUATING AND BUYING VIDEO COMMUNICATION SYSTEMS

13

GROUP SYSTEMS

The engineering is secondary to the vision.
Cynthia Ozick

INTRODUCTION TO GROUP-SYSTEM VIDEO COMMUNICATION

With the excitement over personal video communication, group systems are easy to overlook. As personal conferencing systems accommodate single users at multiple locations, group systems accommodate multiple viewers in a single location. Personal conferees operate from a desk a mobile device; group-system viewers convene in a conference room.

Set-top appliances and PC-based conference room systems have matured to dominate the video communication industry. Room-based video communication system sales are exploding at the departmental, small office, and branch location level. The reason is that they offer innovative technologies, useful operator features, ease-of-set-up and ease-of-use, reduced support needs, and lower price points. A Wainhouse Research survey found 75% of respondents to use appliance-based group systems, and less than 30% to use PC-based group systems. Appliance-based systems also edged out software-based systems at the desktop. For a few thousand dollars, one can buy a high performance, integrated, easy-to use group system. Systems commonly include powerful embedded web capabilities and integrated web-based presentation and collaboration systems. This is significant since most users say that they used presentations graphics software during a video conference.

At the heart of a group system is the *electronics module* that supports the codec. This module also includes audio and video plug-in jacks, and provides an interface to the network. It is usually integrated with other components (a monitor, camera, speakers and a microphone) and either cabinet- or cart-mounted. In the past, systems integrators bought the codec and applied their expertise to tie in monitors, cameras, audio, PC-graphics, document presentation, and projection systems. Today, manufacturers largely integrate that into systems.

Figure 13-1. Roll-About System (courtesy of VCON).

Cabinet- or cart-oriented systems with wheels are known as roll-abouts. Most group systems today fall in this class. A roll-about's codec used to be mounted mid-cabinet and concealed behind doors but, in today's systems, it is integrated into one small unit with a camera (and sometimes speakers) that sits on top of one or two monitors. Such systems are referred to as *set-tops*.

Group-oriented video communication systems are broken into two categories: high-end/boardroom and low-end/small group, but the differences are becoming subtle. High-end systems are feature-rich and use top-of-the line components. It is typical to find them equipped with multiple microphones, a high-performance audio sub-system (with echo cancellation), and stereo plus sub-woofer or surround-sound.

With two monitors, participants can view graphics on one while they talk face-to-face with conferees on another. In the absence of graphics, they can dedicate one monitor to showing them what the far-end is seeing.

Figure 13-2. Set-top System (courtesy of Polycom).

Today, small-group systems are almost exclusively set-top systems. They ship with a codec, a monitor, a fixed camera and all necessary electronics. Features are limited; to reduce costs, manufacturers take a *one-size fits all* approach. Often small-group system installation is as simple as plugging in phone lines and power cables. Network service jacks can be extended to multiple conference rooms if systems are truly mobile. However, roll-abouts are not always moved; jostling of the electronics or the connections can lead to problems.

GROUP-SYSTEM VIDEO

Until 1996, PictureTel, VTEL, and CLI dominated the videoconferencing market, and from 1994–1996, PictureTel Corporation of Danvers, Massachusetts assumed the lead. This dramatically changed in 1996 when VTEL acquired CLI, and Polycom expanded out of the audioconferencing market with its ViewStation set-top products. By perfecting the genre of video communication products that PictureTel made popular with its SwiftSite line, Polycom shot to the forefront and

then acquire PictureTel in October 2001. Tandberg has become the standout in high-end room systems. Other competitors fight for a small fraction of the market.

Founded in 1976, CLI was the first of the U.S. codec manufacturers to introduce a group system product: CLI's VTS 1.5 reached the market in September 1982. It provided good picture quality, particularly considering that a single T-1 provided the signal. The VTS 1.5 was the first product to make heavy use of intraframe coding. In 1984, CLI tightened up its breakthrough codec: the VTS 1.5E.

It provided very good picture quality, particularly considering that a single T-1 pipe was used to transport the signal. The VTS 1.5 was the first product to make heavy use of intraframe coding. In 1984, CLI tightened up the algorithm and rolled out a new codec: the VTS 1.5E. It compressed the signal down to bandwidths of 512 Kbps. The VTS 1.5E featured a new algorithm, Differential Transform Coding (DXC™), which

243

offered picture quality improvements so dramatic that it became the cornerstone of CLI's product line. A variation of DXC called discrete cosine transform went on to form the basis of three important image compression standards: JPEG, MPEG and H.261/H.263.

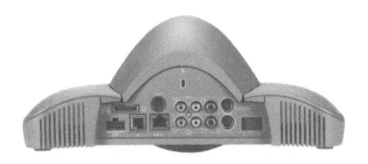

Figure 13-3. Rear View of Set-top System (courtesy of Polycom).

In 1984, PictureTel was formed. Its research in low-bandwidth digital video signal compression led to the introduction, in 1986, of a compression technique dubbed Motion Compensated Transform (MCT). MCT became the basis of PictureTel's C-2000 codec that compressed video and audio signals down to transmission rates of 224 Kbps. In 1988, PictureTel introduced its C-3000 codec based on a new algorithm called Hierarchical Vector Quantization (HVQ) that compressed the signal to 112 Kbps with acceptable quality. Switched 56 service was making its debut in the U.S. and ISDN BRI was emerging in Europe, so HVQ was well-timed.

In 1988, PictureTel shipped 150 C-3000 systems. In 1989, they sold more than 600 systems. Over the next five years, the majority of room-system market growth came from the low-bandwidth sector and, in the early 1990s, PictureTel assumed the lead in the global videoconferencing marketplace.

In 1985, VTEL (formerly VideoTelecom Corporation) was founded in Austin, Texas. In 1986, it developed a line of PC-based systems for the LAN environment.

Customers were not ready for LAN-based video communication applications, and the company shifted its attention to conferencing over the circuit-switched WAN. In 1989, VTEL began shipping the Conference System 200 and 300. Competing in the low-bandwidth end of the market (56–768 Kbps), these offered a new twist. VTEL used a 386-PC platform and based its products on DOS. For the first time, some videoconference system upgrading could be achieved by simply shipping customers a new set of floppy disks.

Through the years, VTEL wisely focused on vertical markets such as telemedicine, distance learning, legal/judicial, and finance. VTEL has done particularly well in telemedicine and distance learning markets and sells more systems there than all other manufacturers combined.

In the early 1990s, the video communication industry experienced upheaval. Standards were approved and codecs were mass-produced. The popularity of low-end roll-abouts resulted in deep margin cuts. Intel announced ProShare in 1995 and began to move the market toward the desktop. The room-system market did not dry up, but many customers began rethinking strategies. With desktop-focused contenders such as Microsoft, the mix of suppliers has vastly changed. This is not to say that Polycom and VTEL do not also sell into the desktop market, it is to say that the dominant room-system codec suppliers' commodious days are done. While market growth is soaring, margins are being squeezed.

Desktop-orientation fueled the standards-effort, which extends across all classes of product. The plug-and-play approach simplifies a network administrator's life and protects the corporate investment. It is unheard of today to buy a group-oriented system that does not support H.323, much less H.320, specifications. Moreover, packet-switched (H.323 and SIP) gateways are now replacing circuit-switched products. What has made this possible is interoperability. Standards have been good for the industry, specifically the H.32x standards. Approved by the United Nations' ITU-T in the neutral environment of Geneva, Switzerland, H.320 fostered peaceful coexistence in the video communication market.

H.320's Impact On Group Video communication Systems

The ITU-T approved its H.320 standard on December 14, 1990. Beginning in 1991, second tier videoconferencing product manufacturers (e.g., BT, Hitachi, Mitsubishi, NEC, Panasonic, and Toshiba) began building H.320-compliant products. The first major U.S. codec manufacturer to incorporate H.320 into its product was VTEL. It provided free software upgrades to users of its newer systems.

In 1991, American video communication systems manufacturers sold about 3,500 systems. Of that number, fewer than 5% were shipped with Px64 (the heart of H.320). Once the standard became well known in the U.S., customers began demanding that the two main players, PictureTel and CLI, make it available. Early in 1992, CLI, announced H.261 upgrades at a cost between $6,000 and $20,000.

PictureTel was next, and offered its standards-based Link-64 upgrades for $9,000. By 1992, more than 70% of the 6,500 codecs sold globally included the standard. Generally, products were shipped in a multi-algorithm arrangement that also included a proprietary scheme for single manufacturer conferencing. In 1993, the number of products that shipped with the standard increased to more than 90%. By year-end 1996, the number was nearly 100%.

Although suppliers did not like to see standards transform highly differentiated products into commodities, customers quickly realized the benefits of interoperability. Interactive multimedia is not limited to exchanges within a single organization and, since not all organizations select the same platform, standards were essential to the broad video communications acceptance. Anyway, not all standards-based implementations deliver the same results. Many group systems manufacturers *enhanced* the H.320 standard with improvements that are not specifically called out in a recommendation, but that are transparent to it.

Although it was not the only factor in changing the video communication industry, H.320 standardization probably had the most impact. Prices plummeted by 50%, and the boardroom systems market underwent an evolution similar to that of the computer mainframe market. The group systems market segment is not likely to disappear completely, but personal conferencing has eclipsed it and does not seem likely to surrender dominance in the foreseeable future. Desktop, and now handheld, systems eliminate the requirement to compete for corporate facilities. Moreover, they build on the strength of video communication by reducing the link between *work* and *place.*

Personal conferencing systems must co-exist with the room-based systems but will not replace them. Although room systems may be derided for supporting *the meeting of two meetings*, they offer considerable value. Often, gathering people in a single room is the most effective and efficient way to get the job done. It only follows that linking to such gatherings could be quite productive. Personal conferencing products answer the need to work collaboratively, the need to share centrally-stored data, and the need to meet any time, anywhere. Group systems blend this with conventional workgroup dynamics.

GROUP-ORIENTED SYSTEMS TODAY

What are the components and features that one can expect to find in a group-oriented video communication system? Of course, all are built on standard codecs, and most include multiple silicon-based algorithms. One difference today is that most new video communications implementations are executed over IP networks (although not necessarily over the Internet). Whereas codec manufacturers used to charge a premium for their high-performance proprietary algorithms, the differentiators in the standards-based codec world are implementation and how fast vendors offer new codecs (e.g., H.264).

Group systems sell for anywhere between $3,000–50,000 (although one could conceivably use a $1,000 personal or even a $50 webcam for a group application). The additional for boardroom systems anymore primarily reflects aesthetics since most boardroom systems features are now available in set-top units. Still, boardroom systems have an edge in features. The price spread reflects differences in the following areas:

246

- Number and type of audio and video compression algorithms supported (H.261, H.263, H.264 G.711, G.722, G.728, G.729, G.729A)

- Picture resolutions supported (e.g., 4CIF, CIF, QCIF)

- Frame rates supported (60 fps high-end, 30 fps mid-range, 15 fps low-end)

- Network interfaces offered (e.g., Ethernet, USB, V.35/RS-530/RS-449 with RS-366 dialing, ISDN with NT-1 adapters, CSUs and inverse multiplexers)

- Transmission speeds supported (256 Kbps—2 Mbps)

- Audio-related features (e.g., microphones included, mixer, surround sound or stereo surround).

- Cameras (e.g., number and type included, lines of resolution, sensitivity to light, features such as voice-activated camera movement)

- Monitors (e.g., size, number, and resolution)

- User interface

- Security (e.g., secure password authentication, embedded AES encryption, H.233, H.234, H.235v3)

- Multipoint capabilities

- Peripherals (e.g., scheduling software, document camera, scan converter, inverse-multiplexer)

- Ongoing maintenance and support, and upgrade policy

The remainder of this chapter addresses the subtleties of these system components and features. Understanding the differences will help prospective buyers make accurate systems comparisons.

AUDIO AND VIDEO COMPRESSION ALGORITHMS

Group-oriented systems should support H.320 (including G.728 compression for low-bandwidth audio encoding), the ITU-T's H.231/H.243 protocols, and the T.120 data conferencing specification.

An H.320 system should be able to bridge through an H.323 gateway to conference with systems connected to packet-switched LANs and WANs (and vice versa). The H.323 standard is designed to facilitate point-to-point, multipoint, and broadcast calls among compliant LAN-, ISDN-, and room-based video communication systems. H.323 also supports T.120, which provides a standard means of communication for collaborative tools such as whiteboard applications.

Standards-compliance does not preclude systems from shipping with a high-quality proprietary compression algorithm for connections between homogeneous systems. Moreover, for connections between like systems, proprietary compression techniques usually deliver superior results. When the justification for a system is intra-organization use, standards may be less important than optimal quality. Nevertheless, standards capability is mandatory.

Group systems should offer automatic handshaking between codecs (to find the best compatible algorithm). It is not acceptable to ask an end user to fumble through the process of finding a coding technique that is common to both (all) systems. In H.320 systems, the H.221 protocol performs this task. Successful implementations of H.221 can be tricky. It is widely acknowledged as the standard within the H.320 family that is the most difficult to implement, because it is the most complex.

PICTURE RESOLUTIONS

Picture resolution is a major differentiation between group systems. Resolution is the number of horizontal and vertical pixels that comprise a frame of video. Because pixels are spread over a smaller space, a smaller monitor provides better resolution.

Although image clarity is, to a limited degree, determined by screen size, it is far more dependent on the processing techniques the codec uses to format the picture. The H.320 recommendation, specifically the H.261 codec specification, sets forth standards-based image resolutions, CIF and QCIF. The QCIF image format delivers resolutions of about 26,000 pixels-per-screen. CIF produces four times the QCIF resolution (a little over 100,000 pixels per screen). The H.320 standard guarantees compatibility; QCIF is mandatory, CIF is optional. Using either format, a vendor can claim standards-compliance.

In May 1996, the ITU-T approved the H.263 Recommendation, which it described five picture formats. Along with QCIF (mandatory) and CIF (optional) H.263 describes Sub-QCIF (SQCIF), 4CIF, and 16CIF (all optional and all covered in Chapter 7). SQCIF is not applicable to group systems because it produces very low-resolution images (it could be very useful in video communication from handhelds or in low-bandwidth applications). What is relevant to group systems is 4CIF, which delivers NTSC television quality, and 16CIF, which resolves an image with roughly the same clarity as HDTV. For obvious reasons, H.263 is replacing H.261.

Group-oriented systems may also offer proprietary formats. These will be described (should a buyer inquire) using vertical and horizontal pixel counts, relative to CIF (352 pixels by 288 lines) and QCIF (176 pixels by 144 lines).

FRAME RATES

The best compression techniques allow forwarding of more frames of information per second, therefore, better image, and sound quality and lower bandwidths. To achieve television's smooth motion handling, images must be refreshed at 30 frames per second (fps). At 30 fps, the eye cannot detect that the picture is being reassembled, so the brain perceives continuous motion. A refresh rate as low as 12 fps yields an acceptable picture. Below that, almost any motion appears jerky. When distinguishing between group systems, frame rate comparisons can be made, but without sophisticated test equipment, they will be subjective.

NTSC cameras capture video at 30 fps. What the codec does with those frames determines whether the motion on the screen will resemble a smooth waltz or the bunny hop. The H.320 standard does not specify a frame rate, and some companies slow the frame rate down to improve resolution. Nevertheless, all codecs (standards-based or proprietary) are optimized for a narrow range of network operating speeds. A discriminating evaluator will ask about a system's *sweet spot* (e.g., what operating bandwidth delivers the best picture resolutions and motion handling). It is unusual for a codec to operate at speeds less than 384 Kbps and consistently deliver 30 fps.

Regardless of transmission speed, a scene's *motion-component* will affect a codec's frame rate. Consequently, manufacturers discuss frame rates in terms of a codec's *average* and *maximum*. When there is little motion, a codec may achieve its maximum. When large portions of the picture change, or when intricacies in the image require rigorous computation, the codec must buffer the data while it performs intraframe compression. A codec may fall behind and receive a new frame before it finishes coding the previous one. If so, it will discard frames until it catches up.

A codec is sometimes called a compression engine. Like an automobile engine, it can deliver raw horsepower but can also be tuned to use its power in special ways. Motion compensation represents one type of tuning. Motion compensation is based on interframe coding. It is optional in the H.320 specification. An H.320 decoder must interpret a motion-compensated stream of pixels if they are sent, but an H.320 encoder does not have to be able to create such a stream. Motion compensation is computation-intensive, but it provides much better compression at comparable bandwidths (typically 30% better than intraframe-only encoding).

Even when two codecs support all the same optional sections of the ITU-T's H.320 Recommendation, *tuning* (signal pre-processing and post-processing) can improve performance. Powerful codecs take additional steps to enhance the H.261/263 process, and prepare video frames for compression in a way that is transparent to a standards-compliant decoder. This is known as *pre-processing. Post-processing* is also performed, to make the most of the work done during pre-processing. When a manufacturer contends that it enhances the standard, this is to what they are referring. Dissimilar codecs are oblivious to pre- and post-processing steps and, therefore, gain no advantage from supplementary processing as do homogeneous codecs. That is why in a one-manufacturer H.320 demonstration (in which codecs on both ends are identical) images are cleaner and motion handling is smoother.

NETWORK CONSIDERATIONS

Low-end group systems usually support transmission speeds of 56–384 Kbps. Large systems sold in the U.S. and Japan usually support speeds of 56 Kbps–1.536 Mbps. In the rest of the world, speeds range from 64–1.920 Mbps.

Of course, bandwidth is only one aspect of signal transmission. How bandwidth is packaged is another. Standard transmission services include ISDN (BRI and PRI), switched 384 Kbps (H0), switched 1.536 Mbps (H11), and DS0 multiples up to and including T-1/E-1. Any system should support IP.

An H.323 requires more bandwidth than an equivalent H.320 call, which uses out-of-band signaling. Consequently, one should plan for about 20% more bandwidth to allow for TCP/IP traffic overhead. For instance, for an H.320 256 Kbps equivalent call, one should allow 307 Kbps.

Bandwidth control is critical for an IP video communication session that might choke an entire data network. It is usual for a headquarters system to maximize its bandwidth and thereby choke branch networks. For that reason it may be useful to leverage advanced routing configurations or, better yet, to use sophisticated dedicated bandwidth management devices.

Network Interfaces

It is advisable to make network decisions before making system decisions. Information provided to a supplier should indicate network speed, and whether the network connection will be public or private, and whether it will be switched or dedicated. Varying image quality-requirements may require multiple speeds. A meeting between project managers and engineers may not demand the same quality requirements as a meeting between executives.

Those who use ISDN-based systems may want the option to select transmission rates (often running between 128 and 384 Kbps) depending upon the conference. The ability to deliver this flexibility will come from the codec itself (through H0 bonding) or from an external device (an inverse multiplexer). Some high-end group systems include a built-in inverse multiplexer, used to bond multiple B channels (or switched 56 circuits) together into higher bandwidth aggregates. Low-end roll-abouts do not bond beyond two channels. If speeds above 128 Kbps are required, bonding must be performed by a separate I-MUX. All else being equal, group systems that are data-rate agile (that function as a DTE device and can thereby follow a network clock) are preferable to those that are not. The ability to clock off the network allows the codec to detect changes and adjust its data rate accordingly.

During procurement, a customer should ask about the network (physical layer) interfaces that are offered. These might include Bell System interfaces (T-1) (e.g., RS-449, RS-422, RS-423, RS-366-A, RS-530) and ITU-T (V.35, X.21 and ISDN). Typically ITU-T V.35 connections are used in the United States for circuit-switched digital connections (excluding ISDN). The V.35 specification defines a high-speed serial communications interface (an electrical connection) between a network access device (e.g., a multiplexer or a codec) and network switching equipment. V.35 makes its connection using a large rectangular connector with large hold-down screws. Although the ITU-T has formally stated that the V.35 specification is out of date, it is still widely used and provides a solid connection that is almost impossible to dislodge accidentally. X.21 connections are used for circuit-switched digital connections in Europe and the rest of the world. Like V.35 X.21 is a digital signaling interface, defined by the CCITT/ITU-T, and used to connect DCE to DTE. X.21 describes the names and functions of eight wires. The physical connector has 15 pins. X.21 connectors are almost never provided in the U.S. ISDN connections rely on an RJ-11C plug on the network side of the NT-1 and an RJ-45 connection on the CPE side (the side that connects to the codec). It is also important to specify whether the customer or the supplier will supply the CSU/DSUs, NT-1s and cables.

Packet-Switched Interfaces

Group systems have historically relied upon circuit switching, but this is rapidly changing. Today, ITU-T standards address connections over IP. Although people associate them with personal conferencing systems, these standards apply to group systems as well. Group systems with Ethernet LAN interfaces facilitate multimedia exchanges (data collaboration).

AUDIO-RELATED FEATURES

Because audio quality is critical to the success of a videoconference, it is imperative to consider components such as microphones, speakers and echo cancellers that comprise the audio subsystem. Advances in audio conferencing technology have greatly improved sound quality; digital systems are clearer. Echo cancellation eliminates *clipping* (which causes sound to drop out when someone from one end interrupts the conversation at the other end). Modern systems are more tolerant of poor room acoustics, ambient noise, and degradation caused by the transmission line. Today's room-system video communication products deliver high-quality audio with digital signal processing (DSP), echo control, noise reduction, and full-duplex transmission.

Differences between various codec manufacturers' audio quality levels have substantially evened out. Nevertheless, there can be significant differences between high- and low-end rollabouts. Desktop systems present even more variation.

Audio Codecs

Digital audio is usually described by sampling rate, the number of bits per sample, the data rate (measured in bps), and the number of audio channels. The fundamentals of audio sampling, quantizing and encoding are covered in Chapter 6. The higher the audio bit rate, the less room will be left over for the video portion of the signal. Thus, it is important to look for efficient coding schemes.

Coding Standard	Coding Method	Bit Rate
G.711	PCM	64 Kbps
G.721	ADPCM	32 Kbps
G.722	ADPCM	64 Kbps
G.723	ADPCM	20, 40 Kbps
G.723.1	LD-CELP	5.3, 6.4 Kbps
G.726	ADPCM	16, 24, 32, 40 Kbps
G.727	ADPCM	16, 24, 32, 40 Kbps
G.728	LD-CELP	16 Kbps
G.729	CELP	8 Kbps

Figure 13-4. ITU digital voice standards for speech codecs

Almost all audio codecs included in video communication systems will comply with the ITU-T specifications described therein. The important ones for group-oriented systems are G.711, G.722, G.728, and G.729. As said earlier, G.728 audio compresses voice to 16 Kbps. G.711 and G.722 audio compress audio to 64 Kbps and, in an ISDN BRI connection, use half the network bandwidth for voice. G.711/G.722 compression makes sense at higher-end transmission speeds (712 Kbps). The difference between the two is that G.722 samples at twice the frequency of G.711 (7 kHz and 3.5 kHz, respectively) and thereby produces a more faithful representation of sounds other than simple speech.

There is also G.723 encoding, developed by the DSP Group of Santa Clara, California (standardized by the ITU-T). G.723 uses Multipurpose-Maximum Likelihood Quantization (MP-MLQ) to diminish voice to about 6 Kbps. CELP predicts speech through a process of vocal tract modeling. It sends only the *error* (difference between the model and real events). CELP is useful in low-bandwidth conferences, because it leaves more bandwidth for video transmission.

When inquiring about proprietary audio algorithms, it is important to ask the supplier to provide information on the range of frequencies encoded with each type and the respective data rates produced, as well as how bandwidth is assigned to audio. The more bits available for audio, the better the quality. Some systems allow the end-user to specify the audio bandwidth, and others offer flexible bit-rate audio that changes with the amount of video information transmitted. Some use rigid schemes that are (constant bandwidth is assigned no matter how much video is transmitted). Evaluating audio quality under different circumstances enables one to learn what works best. It is revealing to listen closely to audio quality during scene changes (e.g., someone moves around) and to compare that to the quality delivered when there is little or no motion.

Audio Channels

All group-oriented systems offer at least two channels (stereo sound). Additional channels are now common as well. Some provide a center channel; others provide up to five channels (known as surround-sound). When a center channel is added to a group system video communication unit it tends to anchor the sound field. As meeting participants move around the room, the audio portion of the conference appears to remain fixed. When only two channels are used (left and right speaker), it is possible to detect which speaker is producing which sounds by moving between them. A central loudspeaker also provides better frequency response matching across the stereo sound field and helps to match the action between the picture and the associated sound. When only two channels are used, the prominent an angular deviation between the sound and its picture can be detected by some listeners.

Microphones

A microphone is a *transducer* that converts sound into an electrical signal. Microphones vary in how they manage the ratio between signal and noise, how they prevent distortion, and how they use companding (noise-reduction circuitry) to improve signal-to-noise ratios. Of course, there are many different types and quality levels available. Cutting corners with inexpensive microphones is not wise. Microphones do not comprise a big part of the overall project cost, but good ones can make an enormous difference in the perceived quality of a system.

Suppliers and consultants can be helpful in identifying the ideal number and types of microphones for a given installation. Some microphones are built into the system control unit while others offer flexibility by being independent of it. Microphones are not always table-mounted; they can be suspended from the ceiling, built into (and coordinated with) a camera or attached to a lapel. They can be wired or wireless. Wireless lavaliere microphones are often preferred, but speakers may forget to remove them before leaving the room. They are also susceptible to RFI (radio frequency interference). A combination of microphones can be used in a single application, as long as they are arranged to avoid feedback.

The number of microphones supported by a system (e.g., the number of audio input ports) should be considered. To provide the best coverage, customers often order additional microphones and locate them strategically to maximize coverage. Moreover, when it comes to coverage, microphones should be evaluated in terms of pattern. They can be unidirectional or omnidirectional. Unidirectional microphones are designed to cancel out sound that does not emanate from a certain point. They are ideal for lecture applications but, because of their *dead spots*, speakers must be aware of where they are positioned in relation to them. Omnidirectional microphones detect sounds from all directions and are preferable for most group applications.

Audio Features

The audio-oriented features of video communication systems vary. The most basic include audio-only bridging, or the ability to include those without video communication equipment, and audio out-dialing. Out-dialing is an important feature that should be supplemented through a dialing keypad. An audio call can be used to dial someone on the other end when there are problems with establishing the video component of a conference. If audio dial uses a line from a PBX or switchboard, it is helpful to have a *switchhook flash* key that can be used to access PBX features such as three-way audio conferencing. A recorder jack is also a useful feature. All systems should include a mute button, which temporarily removes audio from the stream of bits sent to the distant end. When mute is activated, audio is removed from the information being transmitted to far-end sites (it is *not* generally removed from a taped recording of the proceedings).

Figure 13-5. Video Phone (courtesy of Tandberg).

Modern audio systems for video communication should include full-duplex digital audio, echo cancellation (with ability to turn this feature off), voice-activated automatic gain control, automatic noise suppression, an audio mixer, integrated tonal speaker test, real-time audio level meter for local and far-end microphones, microphone and VCR input audio mixing with VCR talk-over.

Echo Cancellation

Echoes have always posed a sound quality problem in conferencing applications. Echo elimination is critically important in true roll-about applications because one cannot depend upon room treatment alone to address the problem. It is important to understand how a given system eliminates echoes. There are two methods: echo suppression and echo cancellation. Echo suppression attenuates microphones and loudspeakers on an alternate basis to prevent feedback. However, it causes *clipping.* Echo cancellation provides for smooth, natural exchanges by modeling an echo and subtracting it from an incoming signal.

Echo cancellation systems must *train themselves* to room and network environments to operate effectively. Some units emit a burst of white noise to profile the room (sometimes called *whooshing* the room), which can be annoying to distant-end participants. Others use the voices of the meeting participants to train acoustic line-side echo cancellers. Some units automatically adapt to changing room and network conditions without losing stability. Still others become unstable when the unit is moved or when an object is sitting too close to the microphone. Increases or decreases in speaker volume may degrade performance, so it is good for prospective buyers to experiment.

When comparing echo cancellation equipment, it is important to compare convergence time, tail length, and acoustic echo return loss enhancement (AERLE) ratings. Convergence time is the measure of how long it takes a canceller to model the acoustic characteristics of the room in which it is installed. The time, measured in milliseconds, should be as short as possible (less than 70 ms) because echoes will sneak through until the canceller completes its model of reality. Tail length is measured in milliseconds and corresponds to how long an echo canceller will pause after an original signal's echo has been modeled to wait for the return echo. The longer the tail length, the less acoustic treatment a room will require. AERLE, measured in decibels (dB), describes the maximum echo cancellation produced by the canceller. The higher the dB rating, the better equipped the system is to remove echo. Typical values range between 6 and 18 dB. Evaluators should disable the variable attenuator, or *center clipper,* for accurate measurement.

CAMERAS

Any group systems should feature one or more color cameras. Cameras should provide about 65° field of view, about 25° tilt (up/down) range, about 100° pan (left/right) range, 265° field of view, 12x zoom (f-4.2 to 42mm), auto-focus, automatic white balance, presets, and far-end camera control. Nearly all are based on integrated circuits called charge-coupled devices (CCDs). A CCD camera focuses a scene sample not on an imaging tube (as in older cameras) but on a solid-state chip.

A CCD is much smaller than an imaging tube. A one-megapixel CCD camera module, including full camera functions and a mechanical shutter, measures about 7 mm, and uses pixels that measure less than 3 x 3 microns. The lens directs photons to strike the surface of the CCD chip, which is divided into individual elements (rather than the continuous photosensitive surface of an imaging tube). The surface of the CCD resembles a matrix of individual elements which, in turn, define the camera's spatial resolution capability. Each unique element samples the light that strikes it, converts that light to a charge, and then interprets that charge to camera circuitry.

Color video cameras sample the RGB components of a scene to reproduce color. Single chip CCD cameras place a color filter directly on adjacent rows of imaging elements, to create a series of *stripes*. With three stripes (R, G, and B), it is possible to sample the scene directly through the same lens at precisely the same time on the same chip. Because components are not registered precisely in line (e.g., the red element is a row above the green), the result is mild optical distortion.

In a three-CCD camera, the focused image is passed through a beam splitter, which allows each of the three chips to sample the full scene in each respective color for full resolution. The optical performance of the system is superior because each color sample uses the same precise optical alignment. If each chip were the same specification of the single chip camera, the spatial resolution in color could therefore be three times greater than that of the single chip.

Technical Considerations

Light sensitivity is important in cameras. Sensitivity is measured in lux; video cameras should offer, at least 2- 3-lux. This eliminates the need for special lighting and provides good depth of field, which keeps subjects in focus. Note that there is no standard method for measuring lux levels. However the camera's maximum aperture (f1.4, for instance) determines the lens' ability to capture light.

Typically, room system cameras focus best when objects are 5–20' in distance. Within this range, people can move around freely without concern that image capture will be compromised. Zoom lenses are capable of variable focal lengths that range from wide-angle to telephoto. A lens specified as 12:1 (12X), for example, offers a maximum focal length ten times that of the minimum focal length. A large room requires a higher zoom ratio (with longer available focal lengths) than in a small room. In a small room, telephoto capabilities do not matter; the lens must zoom out for a wide *angle of view*. Angle of view is measured in degrees. Because both the focal length of the lens *and* size of the camera's imaging area determine the angle of view, one should not assume that two cameras with lenses of the same focal length would offer the same angle of view.

A camera's arc of movement is its *range*. Range is measured in degrees, with 150∞ degrees being on the high-end. The time that it takes a camera to move from one point to another is described as its *repositioning time*. Moving the camera across its full range takes 1–2 second, and some cameras move more quietly than others do.

Some cameras offer *gain control* or *automatic gain control* (AGC). AGC boosts the video signal level during low light conditions. Most cameras calculate the required gain by averaging the levels in the overall image—this can cause problems in high contrast situations. AGC can also upset the levels and contrasts in an image when conference participants are wearing very bright (or white) clothing. These problems should not necessarily dissuade an organization from buying an AGC camera, but should be considered, particularly in boardroom applications.

White balance is another camera feature. Various lighting conditions produce different color components. Cool-temperature lighting casts a different tint on a room than does warm-temperature lighting. White balance adjusts the RGB components that, when combined, yield white in a video signal.

Auxiliary Cameras

Low-end video communication systems usually offer a single camera. High-end systems typically offer more than one. In addition to the camera built into the system cabinet, high-end installations include one or more auxiliary cameras.

Whereas system cameras work well to capture images of conferees seated at a table, auxiliary cameras perform better for other parts of the room. Some mount scissor-fashion on a wall while others freestand (usually on a tripod).

Mobile auxiliary cameras are helpful in a large room or in cases when a speaker uses a whiteboard or lectern. They offer the flexibility that a cabinet- or wall-mounted camera cannot provide.

Camera Control

A basic requirement of video communication systems is *camera control.* Two important camera control features are *far-end camera control,* and *camera presets.* Far-end camera control allows a person on one end of a videoconference to control a camera on the other end. It is useful when the participants on the other end are untrained in system operation or when they forget to move their camera to follow the action. Near-end participants can step in to help, and thus assume responsibility for cameras on both ends. Multipoint far-end camera control is generally not offered because controlling cameras in more than two sites can become confusing.

The camera-equivalent of telephone speed-dial, camera presets help speed camera movement. At the beginning of a meeting, a system operator frames key participants in the camera, makes any adjustments to zoom or centering that may be necessary, and then enters the position into memory. To recall the setting, the operator presses a button or touches an electronic pad to move the camera to its preset position.

When group discussion gets lively, presets allow an operator to capture most or all the action for the distant end. Most systems allow presets to be arranged for more than one camera (main and auxiliary, for instance). Some go farther to allow both near-end and far-end camera presets. The *number* of camera presets also varies between systems. While more presets can allow for more options, it can also make the system more confusing.

Many group-oriented video communication systems offer *follow me* camera features. In these systems, a software-controlled camera is synchronized with room or speaker microphones. Speech causes the camera to follow the action.

Graphics Support And Subsystems

Documents and graphics are exchanged in most in-person meetings. Video communication systems should provide some form of support for document transfer.

At a basic level, document stands (more often called document cameras) provide for standard resolution graphics exchange. Document cameras marry a small color video camera to an overhead projector. The camera points down at a baseboard with lighting to display physical objects, instructor's notes or transparencies, etc.

The capabilities of document cameras differ. Most systems offer single-chip CCD cameras. A three-chip CCD camera is required in cases where end-users want very high-quality graphics (remote inspection of parts, engineering applications, etc.). Some document cameras both focus automatically and offer manual adjustments. Others offer manual adjustments only. Controls are sometimes included as part of the video communication system's operator console. More often the user must make adjustments at the document camera itself.

Using a document camera or other image capture techniques, codecs may also deliver high-resolution display of *freeze-frame* graphics. Resolutions widely vary, from 2X. Picture formats generally include JPEG, TIFF (Tagged Image File Format) or TARGA (Truevision Advanced Raster Graphics Adapter). Graphics systems often comply with the ITU-T's H.320 Recommendation H.261 Annex D that doubles the CIF horizontal and vertical resolution for television-quality graphics.

Most group-oriented systems provide the ability to create and store images as slides. Program files can be retrieved during a videoconference and used to make presentations. Participants can view these visual materials and annotate them independently. Collaborative tools (which let participants share application control) vary in power. Almost all are now based on the ITU-T T.120 Recommendation.

Even when a video communication system does not integrate the ability to create, store, and retrieve documents and images, it can usually be equipped with a T.120 outboard system. This delivers similar functionality—at additional cost.

Figure 13-6. Workgroup Application (courtesy of Polycom).

Document and graphics subsystems vary in sophistication. Nearly all allow conference participants to annotate shared images in real time with electronic tablets and styli. Applications sharing systems are more sophisticated. These permit users at different sites to interactively share, manipulate and exchange or markup files and move images. Some systems offer high-resolution graphics, including CAD support for engineers to collaborate on product designs, and to view concurrently computer simulations. Most are PC-based, and connect through data ports on the codec that allow them to share the audio/video data stream. Images can be scanned in, retrieved from files, or captured with cameras.

When photorealism is required (X-rays, medical images and cases when an entire typed page or detailed diagram must be transmitted) enhanced-resolution systems are required. These systems are available in monochrome or color and go beyond the PC VGA resolutions of 640H x 480V pixels to produce images containing 3–13 times more pixels. Television cameras cannot scan at such high levels of detail. In very high-resolution applications devices called scanners are required.

Flatbed scanners resemble small photocopiers, and capture images that are placed facedown on the clear-glass surface, and scanned from underneath. The scanning area or size of the bed can affect the price of the scanner, particularly when scanners have a bed size of 11" x 17" or more. Within the scanner, one or more CCDs convert incoming light into voltages with amplitudes that correspond to the intensity of the light. Four-bit gray-scale scanners can capture 16 shades of gray; eight-bit gray-scale scanners can capture 256 shades of gray, and are often used in medical applications. Color scanners are usually required in multimedia work; they typically make three scanning passes, one each for red, green, and blue. A scanner should resolve an image at no less than 300 dots-per-inch. Images are stored on hard disk (requirements for saving images add up quickly), printed, or both.

MONITORS

Video monitors come in various sizes, configurations, and, depths. Obviously, monitor size is important. Beyond size, evaluating monitors is largely a subjective exercise. One should look for definition around the edges of the screen, crisp, clear images, true color, and straight horizontal and vertical lines.

A CRT (cathode ray tube, the big, old-fashioned kind) TV monitor can be evaluated in terms of its dot pitch. The dot pitch is the distance between same color (R, G or B) dots, in any direction. A smaller dot-pitch provides a higher resolution on a given monitor, but a smaller dot-pitch does not necessarily mean better picture quality. The goal is to match the dot-pitch with the pixels-per-inch delivered by the codec. Given that all monitors being compared are the same size (measured diagonally across the screen), a .21mm dot pitch is excellent; a .28mm dot pitch is acceptable (although text may look fuzzy).

Before buying a CRT monitor, measure the dimensions of its destination, and note the size of the room to ensure a comfortable viewing distance (you may need to split the difference between users at the near-end and far-end of the conference table). Note picture quality and the curvature of the picture tube; the flatter the tube, the less the shape distortion and light glare. If the monitor may be used for TV viewing, buy an HDTV-capable system to ensure usability after the FCC's expected digital conversion, in 2008. Check the remote control for ease of use, and note features (e.g., number of AV Inputs, picture-in-picture). Check the picture tube darkness when the TV is off; darker tubes display higher contrast images.

CRTs use three electron guns (one each for R, G and B) to excite the phosphors that comprise the pixels in an image. These guns must deliver signals of the same relative strength or *color balance* will be diminished. If the blue gun is adjusted to deliver a stronger signal the image will look bluish. Nearly all monitors can adjust the relative strength of the three electron guns to correct most problems.

Some group-oriented systems use PC monitors as opposed to television monitors. PC monitors are evaluated according to their Video and Electronics Standards Association (VESA) rating. There are four standard VESA resolutions. These are VGA 640H x 480V; Super VGA (or SVGA) 800H x 600V; and two XGA resolutions: 1,024H x 768V and 1,280H x 1,024V. Some monitors even go higher, to resolve an image at 1,800H x 1,440V. The ultrahigh resolutions primarily suit professional-level work in fields such as desktop publishing and computer-aided design, not video communication (for which SVGA delivers an excellent image).

Given the advantage of the smaller footprint, most people today will choose to buy a flat-screen monitor rather than a CRT. These come in two general types, LCD and plasma, each of which offers its own advantages and disadvantages.

The advantages of plasma over LCD include larger screen size, deeper blacks, and superior contrast and motion tracking. However, plasma is more susceptible to burn-in, generates more heat, offers a shorter life span, and is unreliable at higher altitudes. In contrast, LCD monitors require no burn-in, run cooler, last longer, and do not suffer from high altitude issues. Unfortunately, LCD monitors offer lower contrast ratio, grayer blacks, inferior motion tracking, and pixel burnout (which results in small black or white dots on the screen).

Picture-in-Picture

Although it is not really a monitor feature (it is a system feature), *picture-in-picture* (PIP) is an important in single monitor systems. Also known as quadrant display or windowing, PIP relies on a codec's receive circuitry to create and send two images to a monitor. One image occupies most of the screen except for a small corner where the second image, smaller but complete, can be inserted in a window. The window can be closed when it is not needed. PIP provides a mechanism for monitoring near-end image capture in single monitor systems.

Continuous Presence

Like picture-in-picture, *continuous presence* is not a function of the monitor but relies on it to display multiple images. It differs from PIP in that images describe far-end activity only. It differs from multipoint technology (in which each site can see only one other site at a time) by allowing a conference site to see several other sites simultaneously (usually two or four) and hear the audio from all sites.

Continuous presence is described in the ITU-T's H.324 specification. It relies on a codec to deliver a single composite video stream composed of the video signals from up to four sites (or four cameras in a single site). Most often, the receiving codec "stacks" these discrete images on a single monitor, where each occupies one quadrant (in Hollywood Squares™ fashion).

Figure 13-7. Workgroup Application (courtesy of Tandberg).

Continuous presence also works well in point-to-point applications in which a large group is seated around a table. In situations like this, a single camera cannot capture everyone in one image, so two are used. One camera is focused on people on one side of the table; a second is aimed at those on the other side. Again, the two images are stacked for transmission and offered on a split screen basis (where one-half of the table is displayed on the top half of the monitor and the other is displayed below) or in a *de-stack* mode where two monitors are used to display the images.

Continuous-presence requires double or quadruple the transmission bandwidth of single image applications. Hence, it works best at 384 Kbps and above (with 512 Kbps being the preferred minimum).

OPERATOR'S INTERFACE

The system operator's interface, also called the control unit, is critical because many people who would benefit most from video communication are not interested in learning how to use it. A group system control unit should be menu-driven and intuitive. Most allow the operator to dial calls using speed-call, switch between room and document cameras, activate the camera's pan, tilt and zoom features, control audio including volume and muting, and access peripherals such as slides projection systems, fax machines, and VCRs. Some systems allow the operator to move cameras at the remote site. This feature is called far-end camera control and is valuable when people on the distant end have not been trained to use the system.

Historically, a desktop-mounted device, the control unit now takes many forms. In the desktop category, there are control units that emulate a telephone, those that use an electronic template and stylus, and those that rely on a touch screen interface in which Mylar-covered electronics cause displays to change, depending on what part of the screen is pressed. This allows a great deal of control without a cluttered display. Wireless remotes are very common, particularly in small-group roll-about systems. These mimic a VCR or TV remote control device. Most provide an on-screen pointer and icon-driven controls. A user can move the pointer up and down by raising and lowering his hand, and select a feature by clicking a button.

It is important to consider the sophistication of the end users when choosing the interface. Because it can be so subjective, it is advisable to assemble an evaluation team and have participants rate different systems on ease-of-use. What may be intuitive to one may be mind-boggling to another. Finally, because documentation varies greatly, one should include the instruction manual in the evaluation process.

SECURITY

When conferences rely on the Internet or satellites, or when they include highly confidential content, encryption is advisable. AES (Advanced Encryption Standard), a NIST-standard (National Institute of Standards and Technology) secret key cryptography method that uses 128-, 192- and 256-bit keys, is ideal. AES is an U.S. government-sanctioned (NIST) private key cryptographic technique in which only authorized personnel know the secret key. In addition, the ITU-T has specified H.233/H.234 & H.235v3 for securing H.320 and H.323 compliant exchanges, respectively. Password protection for the operator's console and of the address book prevents unauthorized use of a system and can discourage theft.

MULTIPOINT CAPABILITIES

MCUs are something like a network depot where in which compressed video and audio signals terminate. Also known as bridges, they combine, exchange and manage inputs. Most MCUs can support multiple simultaneous bridged connections and are capable of supporting a variety of network interfaces and speeds. Not all systems support multipoint capabilities. For instance, low-end group systems often do not. MCUs eventually find their way into most organizations. They generally appear after the company has been renting bridge space from service bureaus and can thereby make a business case to purchase an MCU.

The ideal multipoint conference would provide every location with access to all available speech, video, and data from all locations. Unfortunately, continuous-presence multipoint conferencing is not historically been feasible because it requires connection-oriented network links between each and every site (as Internet bandwidth grows and quality-of-service improves, this will become a problem of the past). Fast switching provides an alternative to continuous presence. Consequently, video and data information have come to be provided selectively using criteria for determining which site should *have the floor.* In such cases, selection can be made manually (by, for instance, the conference facilitator) or it can be fully automatic.

Figure 13-8. Operator Interface (courtesy of Polycom).

MCUs have different operations modes. Most allow chairperson control in which a single conference manager determines who will be displayed on the screen. Others provide a rotating mode in which the MCU cycles through the conference sites, displaying each briefly on the screen. There is also a voice-activated mode in which the MCU fast-switches between sites based on who is speaking. In a voice-activated mode, some systems offer speech-checking algorithms provided for voice-activation to prevent loud noises (sneezes, coughs and doors shutting, for instance) from diverting the camera from the site with the speaker to the site with the loud noise.

MCUs must support all the video and audio algorithms present in the conference and, often graphics and still-image annotation. This includes the full H.320 suite. Within that suite the H.231 and H.243 protocols address multipoint conferencing. G.728 audio support is also desirable. Some MCUs limit connection speeds to only certain Px64 multiples and others do not support H0 multirate ISDN bonding.

MCUs are evaluated by size and capabilities, and described by the number of sites supported. Most low-end MCUs come equipped with four ports and can grow to eight ports. Larger systems can group up to 72 ports and be daisy-chained to link hundreds. Some smaller systems can be expanded by adding modules while others top-out at six or eight ports. Because all systems are limited by some maximum aggregate bandwidth, they may limit transmission speeds when called upon to connect too many sites. There is usually a perceptible end-to-end transmission delay in conferences in which three or more bridges are daisy-chained. There is usually a limit to the number of MCUs one can daisy-chain because doing so consumes a port on the MCU (used to make the daisy-chain connection to other devices).

MCUs generally support all common types of interfaces for public and private networks including Ethernet, T-1, F-T1, ISDN BRI, and ISDN PRI. Most allow *dial-in* (meet-me), *dial-out* or dedicated-access connections in any combination. Most also allow connection arrangements to be mixed in a hybrid conference. Today's MCUs uses tones to signal a conferee's entrance and exit. They also permit conferees to participate in the conference on an audio-only basis.

MCU management is an area in which large differences still exist between products. Most are managed in a Windows environment, but some are easier to use, offer richer diagnostics, and offer many more administrative features. Administrative features include scheduling utilities. Some systems allow conference scheduling in advance. Even more valuable is the ability for an MCU to configure itself automatically at a pre-determined time and, based on that configuration, to dial out to the sites involved in the conference. Also look for event logs and the ability to account for and bill back usage. In situations in which a single organization will have many MCUs, it is important to buy products that can be remotely managed.

EVALUATION CRITERIA

Service and Support

Video communication is, by nature, a dispersed application. It is critical that buyers arrange for comprehensive national and/or global product support. Support includes design, sales, installation, training and ongoing maintenance. Turnkey support is important when groups of people are waiting for a meeting. It is critical that the supplier will work with the carrier to deliver an operational product. It is also important to establish a national or global purchasing plan in which volume discounts apply to purchases made anywhere within the organization.

Structured Procurement

One way to procure equipment is to divide the evaluation process into two categories: objective and subjective. Objective evaluations are usually performed as part of a structured procurement process in which a Request for Proposal (RFP) or Request for Quote (RFQ) is issued to qualified suppliers. The RFP/RFQ describes exactly the features and capabilities that competing systems should include. When a buyer does not know what it needs, this becomes secondary to a Request for Information (RFI).

An RFI lays out the application and asks suppliers for their solutions. Once the responses come back, the good ideas are combined into a more formal procurement document. Appendix E lists some RFP questions.

Although video communication systems are reaching commodity status, differences exist in packaging and features, and especially between product categories. Most systems are similar within product categories but, when compared cross-category, reveal big differences. Small-group roll-abouts should be compared to each other, not to more fully-featured models unless comparisons are being made within a manufacturer's product line.

Equipment Demonstrations and Trade Shows

A fast way to compare all codecs at the same time and in an identical environment is to visit a carrier's demonstration facility. Some have worked cooperatively with manufacturers and/or distributors to provide a video communication system exhibit area. This is a useful way to investigate systems in quick succession. Of course, the carrier will promote its network services, but performing such a service grants them such license. It is critical to apply the same level of precision in choosing a carrier that one uses to select video communication equipment.

Trade shows provide good opportunities to view video systems in a similar environment. Systems vendors are excellent sources for learning about such events.

CHAPTER 13 REFERENCES

"Videoconferencing Audio Primer." *Network World Fusion*. Accessed January 2005. http://napps.nwfusion.com/primers/videoconf/videoconfscript.html

Silva, Robert. "History And Basics Of Surround Sound." *Accessed* January 2005.http://hometheater.about.com/od/beforeyoubuy/a/surroundsound_p.htm

Goad, Libe. "Camcorders & Portable Video: Smaller & Cheaper." *PC Magazine*. December 2004. http://www.pcmag.com/print_article2/0,2533,a=141848,00.asp

"Videoconferencing Endpoints." *Wainhouse Research*. May 2004.

Brown, Dave. "Bytes, Camera, Action!" *Network Computing* March 1, 1996 v7 n3 p. 46-59.

Frankel, Elana. "A Guide to Videoconferencing," *Teleconnect* September 1996 v14 n9 p. 66-82.

Judge, Paul C., Reinhardt, Andy. "PictureTel fights to stay in the picture," *Business Week* October 28, 1996 n3499 pp. 168-170.

Leonhard, Woody. "Say it face-to-face," *PC/Computing* November 1996 V9 n11 pp. 340-345.

Miller, Brian L. "Videoconferencing," *LAN Times* November 11, 1996 v13 n25.

Molta, Dave. "Videoconferencing: the better to see you with," *Network Computing* March 15, 1996 v7 n4 pp. 11-17.

14

IMPLEMENTING GROUP-ORIENTED SYSTEMS

*"What we anticipate seldom occurs; what we
least expected generally happens."*
Benjamin Disraeli

A good approach to managing the installation of a group-oriented video communication system is to set an installation date and work backwards to determine the project timeline. The next step is to expand the project plan through the inclusion of tasks (and task owners), deadlines, milestones, and critical success factors. Discussion of project management tools and methodologies transcends the scope of this book. Moreover, a Project Manager tends to use that with which he is familiar, and that which his organization endorses and provides.

In this chapter, we will consider the process of implementing a conventional group-oriented video communication facility that is suitable for routine organizational use. We will not address the special requirements of applications in areas such as telemedicine and distance learning, or those of unique facilities such as video command centers, mobile studios, or broadband television. Specialized applications warrant involvement of a consultant or video communication systems integrator. The same holds true for boardroom facilities, which generally require custom engineering. In some cases (particularly large installations) it may make sense to let the hardware or network vendor manage the entire project.

In review, the two categories of group-oriented videoconferencing systems are mobile stationary. We make these distinctions independent of the fact that one may equip a group-oriented system cabinet with wheels. It is generally not practical or advisable to move high-end roll-abouts. They can be heavy, and often do not feature very large wheels. Moreover, moving such systems may also require movement of peripherals such as DVD players, document and auxiliary cameras, and PC scan converters. Anyway, it is unlikely that random placement will result in the lighting and acoustical conditions that will allow for optimal audio and video quality.

Low-end cart-based and set-top systems are designed to be portable. Moving them still entails risk to delicate electronics, and jiggling may unseats boards and deteriorate wired connections. Nevertheless, some systems will be moved. When they are, it is important to schedule and track both the rooms and the equipment. The budget should include software that is designed for such management.

Appendix F includes a summary of tasks that one may encounter in implementing a group-oriented a video communication system. It is important to begin such a project with site selection and site preparation. Because they may require significant installation periods, network services should be ordered early in the project. Shortly before equipment installation, network services should be extended into the building and the equipment location, and should be tested and accepted. Next, equipment and peripherals should be installed, generally by a vendor or a system integrator. Before the production phase of the project, plans should be established for internal promotion and training. The project manager should develop scheduling and confirmation systems, and consider how users should allocate operational costs.

A haphazard implementation is likely to negate meticulously planning. Vendors can assume much of the burden, but overall success or failure falls to the customer. System promotion is most effective when an insider, someone who understands the inner workings of an organization, coordinates it. Senior management support is the key differentiation between long-term failure and success.

SITE SELECTION

Acceptance of video communication includes the willingness of users to change how they do things. To foster new habits, conference facilities should be convenient and accessible. In a large campus, the system should be located centrally or placed near those who may use it the most. Putting a fixed-location system in a room that is also used for non-video meetings may result in under-use of systems due to lack of room availability. The equipment may sit idle in high-demand space.

Executives play a lead role in developing a video communication culture, so astute systems integrators recommend putting group-oriented systems where executives can easily access them. Moreover, strategically-positioned signs can elevate awareness and help first-time users locate a facility.

To facilitate back-to-back meetings, a video conference facility should feature a lobby or adjacent anteroom for next-group gathering. It is a good idea to place a telephone and a directory in this area. Some facilities provide a smaller companion room for post-meeting discussions (particularly when executives use the facility). If there is a room coordinator, it is best to locate his or her workspace adjacent to the door. The coordinator will also require scheduling facilities.

Finally, the videoconferencing facility should include a fax machine and copier. The conference center should be situated such that food, beverages, and restrooms are within easy reach, since some meetings last all day.

Brokered Resale of Excess Capacity

The public can also share some corporate conferencing facilities. Public room network brokers seek sites with excess capacity so that they can resell blocks of time, take a brokerage fee, and pass the remainder of the revenues to the organization that owns the room. Organizations that contemplate resale should plan accordingly. Reception facilities, break areas, parking, and site security may require modification if the public will traverse corporate premises on a routine basis.

One should not take lightly the decision to go public. Once an organization begins to resell its excess capacity, it may encounter conflicts of interest between internal and external users. It is hard to *bump* a paying customer for a critical last-minute internal meeting. One can gain insight into what situations might arise and how they can be addressed by speaking to those who have already made such a decision. Finally, if resale is to account for significant room activity, the necessary additional technical support and training burden may offset derived revenue.

INTERIOR DESIGN

The goal is to make a video meeting comfortable for a broad range of users. Therefore, it is important to make the environment virtually indistinguishable from a conventional conference room. Unless the room is exclusively for the use of executives, one should avoid posh settings that may discourage others from using the facility. De-emphasize technology (cameras, monitors, microphones and cables) to avoid a *studio* feel. Cables should be concealed in conduit, ducts, or raceways where they cannot cause accidents. Conference tables should be modified to accommodate cables for microphones and wired control units.

Wall colors should be muted; grays and blues work best (system integrators generally recommend a light gray-blue). Many television backdrops are blue, because blue shades make skin tones appear warm and natural. It is best to avoid stark white, creamy yellows, and dark colors because these colors confuse the white balance functions a camera performs and causes skin tones to appear unnatural.

Wall, ceiling and floor treatments that do not greatly differ in reflectance minimize contrast ratios. Avoidance of highly reflective (chrome, polished metals or glass) or very light-color surfaces that cause glare minimizes *effulgence* values. Highly patterned wall coverings, carpets and upholstery fabrics, and surface textures or decorative details show up as distractions when viewed through a camera.

It is important to develop a schematic drawing of the conference room and to include all equipment and furniture in the plan. This helps to arrive at room dimensions that can comfortably support the maximum anticipated number of participants. One should not forget to place tables, chairs, lecterns and, of course, the video communication equipment itself in the floor plan. To accommodate the eye, one should maintain a distance between viewers and monitors of no less than four times the height of the picture. Most videoconferencing installation guides recommend separating the monitor from viewers by a distance of seven times the picture height.

Monitors should generally be at least 26 inches above the floor. This schematic can also show the coverage a camera affords with its lens set at the widest angle of view.

When large groups are expected to use the system, room designs are much more complex. Complete video coverage requires that auxiliary cameras be interspersed around a large room to capture different views. Monitors may be placed in corners or alcoves, mounted high on walls, or suspended from ceilings. Multiple microphones can be added to pick up sounds in various parts of the room. Sometimes they, too, can be suspended from the ceiling.

Videoconferencing seating arrangements vary widely. The size of the group is a critical factor in determining table placement and shape. There are V- and U-shaped designs, trapezoid arrangements, ovals and triangles. Trapezoids, four-sided figures with only two of the sides being parallel are popular for small groups. The widest end of the trapezoid is placed at the end facing the cameras. Participants sitting around the table can see each other and interact; all can see and be seen by viewers at the far-end. V-shaped tables also work well but require a larger room. Participants can face each other while swiveling slightly to see the screen.

One very difficult arrangement is to seat conferees in tiered rows, one behind the other. Conferees will turn their backs to the camera to converse with each other, and thereby exclude the distant-end from the conversation. No amount of training seems to overcome human nature.

Obviously, it is best select sites that require minimal demolition or reconstruction. One should also avoid dark and cavernous rooms, or those filled with windows that will cause lighting problems and echoes.

ACOUSTICS AND AUDIO

While planning for video, one should not overlook the importance of audio. It is more difficult to communicate without video than without audio. When audio is better on one of two identical videoconferencing systems, users also perceive picture quality as superior. The inverse is not as true.

Achieving good audio can be a challenge. Start the process by choosing a quiet location. Strive for minimal ambient (surrounding) noise. Ideally, room noise should be less than 50 dBa (where dBa refers to *decibels, adjusted*). Interior locations are best (street noise is not possible to control). Avoid high-traffic areas such as reception lobbies, cafeterias, bullpen environments, and heavily-used adjacent corridors. Soundproofing walls and double-glazing the windows can block ambient noise. This is worth the effort in the case of a boardroom installation but may not be in the case of roll-about or set-top systems. Acoustic treatment need not be expensive. Carpeting a room, hanging panels or drapes over windows, and placing screens in front of sound-generating equipment. Hang a partial curtain (so that it is well above people's heads when they are walking) from a high ceiling about mid-room from side to side to limit reflective noise.

Poorly designed HVAC (heating, venting and air conditioning) systems cause ambient noise. Installations should be equipped with proper-sized fans and should be physically isolated. Older fluorescent lights (which are also notorious for generating 60-cycle-hum in sound systems), computer fans, and vibrating objects can also generate unwanted noise.

Audio quality is primarily a product of the acoustical properties of the room, the amplification system, and the selection and placement of microphones and speakers. Loudspeakers should be placed at the level of the listeners' ears to create a direct path for the sound waves, and so that reflections are minimized. Regardless of design, to some degree, User B's microphone will capture sounds from User B's speakers, and will transmit User A's audio back to him. This is an undesirable effect called *direct acoustic coupling*.

Unidirectional microphones (that pick up sounds from one side, as opposed to *omnidirectional*, that that pick up sounds from all sides) are suited for conferencing because they minimize the sound picked up from the speakers and thereby reduce direct acoustic coupling. Unidirectional microphones respond differently depending on the angle of a sound's approach. If the sound comes from the front (relative to the unit's primary axis), the microphone receives it loud and clear. If it comes from behind, the microphone attenuates it (that is, it reduces the amplitude of the signal), and reduces transmitted reverberation. The drawback of unidirectional microphones is that they require conferees to speak within the microphone's capture range. An omnidirectional microphone can be superior, but only if the system counters direct acoustic coupling with good echo cancellation.

Walls and ceilings are launching pads for echoes; windows are even worse. In practice, speakers are often located too high or too low and, once sound waves bounce off of a couple of walls, the listener's microphone (which also is generally at about the same level as the participants' faces) captures those sound waves, transmits them to the sound system, and creates an undesirable effect called *indirect acoustic coupling*. Echo cancellation addresses problems caused by indirect acoustic coupling (also called *multipath echo*), in which a microphone captures sound waves that bounce off of objects. Because it involves sound bouncing off of surfaces and eventually approaching a microphone even from the front, unidirectional microphones can only do so much to control indirect acoustic coupling. The earliest approach to acoustic coupling was to avoid it altogether. *Echo suppression*, in which loudspeakers and microphones were attenuated on an alternating basis to prevent feedback, was the first real attempt to control indirect acoustic coupling. This caused words to be *clipped* as microphones switched back and forth. Interruptions, a natural part of any meeting, are not possible with echo suppression, which compromises the natural rhythm of speech.

Echo cancellation has largely replaced echo suppression. Speech is modeled to create a digitized replica of an echo. Using a digital signal processor (DSP) the synthesized echo of the signal is subtracted from the return echo. Modern echo cancellation systems are amazingly effective. With sophisticated echo cancellation systems, unidirectional microphones are no longer required. Continuous speech can be provided in both directions. Interruptions and side conversations (rude as they might be) are captured and transmitted. Audio is automatically adjusted to keep it from being too bright (too much residual echo) or too dead (no natural echo). Because they are processing-intensive, echo cancellation methods may introduce slight delays to interactive speech.

It is important to work closely with a supplier or multimedia systems integrator to develop a strategy for providing high-quality audio in a conferencing facility. Modeling can cut down the work, but there is no substitute for experimenting with microphone types, quantities, and placements. One should be particularly careful if the conference room is large or if multiple microphones will be used. Background noise and reverberation (the reflection of sound waves) are cumulative if all microphones are on at once. One solution (albeit cumbersome) is to use speak-listen control systems (sometimes called *push-to-talk* microphones). These do not allow interruption and thereby result in a condition called *capture*. However, they do provide good volume and are great in large, noisy rooms with big groups.

Systems are available that use automatic mixing techniques (for instance, *noise adaptive threshold*) to manage microphones. Speech detection circuitry provides a way to turn microphones on and off automatically, and to preclude the necessity for manual adjustment. Systems of this type give the floor to the loudest sound, even if it is something as disruptive as a sneeze. Auto adaptive technology seeks to counter such ill effects, but effectiveness varies with techniques. Some provide outstanding coverage and feedback rejection, while others deliver mediocre performance. One would be wise to arrange an installation as 30-day money-back-if-not-satisfied trial.

Many speakers prefer *lavaliere* microphones, which are worn on lapels or collars, or looped around the neck with an attached cord. The lavaliere can be wired or wireless. Both types uniformly capture a speaker's voice but wireless versions can be susceptible to radio frequency interference. If a lapel microphone is used, it is important to keep the speaker away from the loudspeakers to avert feedback problems. If a wireless microphone is used, it is important to remind speakers to remove it before leaving the facility. Many organizations start out with wireless microphones only to revert to wired varieties after replacement costs mount.

As an organization develops its audio conferencing strategy, it should ensure ample auxiliary inputs for video or audio capture, external audio bridges for multipoint audio, and voice-activated cameras. Finally, the audio component of any conferencing system should provide privacy and security.

LIGHTING

To render a natural scene, a videoconferencing facility must offer light of the right level, angle, and *color temperature*. There must be enough light to provide a *noise-free* picture with adequate depth of focus, and it must emit from the proper direction to avoid undesirable facial shadows. Lighting should also enhance image depth and contours by intentionally creating *desirable* shadows and highlights.

Light intensity is measured in foot-candles. One foot-candle offers illumination equal to the amount of direct light thrown by one candle on a square foot of surface. High light levels (125 foot-candles and above) support improved camera performance and depth of field, and thereby make it easier to focus. High light levels also reduce *noise* in the video signal, which appears as fine-grained static in the displayed image. Unfortunately, high light levels produce heat, which, in turn, can make the room harder to cool and can necessitate noisy HVAC equipment.

A minimum illumination level of 75 foot-candles is necessary for acceptable results. While low-light environments (under 80 foot-candles) generate less heat, they also wash out colors and accentuate shadows. Light levels between 75 and 125 foot-candles provide a good balance of comfort and camera performance.

Light angle is important factor. Most conference rooms have overhead fixtures that direct light down on the conference table. This causes undesirable facial shadows such as dark eye sockets, shadows under chins and noses, and excessive highlights (which is particularly unflattering on balding heads). Keep light sources in front of the participants and above the eyes. An angle of 45 degrees above the subject is low enough to avoid shadows. Multi-source lighting helps to maintain a constant level throughout the room (use a light meter to detect large differences). When installing lighting, be careful not to direct any lighting at the camera's lens.

When professionals discuss room lighting, they talk in terms of a light's temperature, which is measured in degrees Kelvin (K). The whiter the light, the higher the temperature; lower temperatures take on a reddish cast. If fluorescent fixtures are used, avoid the exclusive use of warm-yellow tubes; they make people appear jaundiced and may introduce 60 Hz flicker. Cool white or blue-white bulbs are more flattering when placed overhead. Some facilities combine bulbs of different temperature. If mixed properly, the result will be rich, pleasing colors (rendered by using cool low Kelvin temperature lights with a yellow or orange cast) and bright light (achieved with hotter, higher Kelvin blue or white bulbs).

VIDEOCONFERENCING INPUTS AND OUTPUTS

Video sources may include video cameras, DVDs, digital photography, VCR, PC-to-video, and PC-to-PC. Digital video cameras, or *room cameras*, are the most common type of system input. A room camera is generally included with a videoconferencing group system, but it limits placement options. Most have a limited field of view, and can only capture a portion of a conference room. Limitations in the size of an area covered are a function of the camera lens itself and the distance between the camera and the conference participants. *Zooming* allows a camera lens to be moved in and out to accommodate different group sizes.

Auxiliary cameras can be used when the maximum viewing angle (measured in degrees) of a system's camera limits room coverage. They can be wall- or tripod-mounted. Extensive whiteboard use may warrant training an auxiliary camera permanently on the presentation space. Electronic whiteboards (which are an invaluable option) may be another requirement. If one will be installed, be certain that there is an interface port on the videoconferencing system to support it.

Many implementations include a DVD or VCR. The ability to integrate computer graphics and documents into presentations is an important requirement. In the past, document cameras were used to capture printed images, blueprints and charts. Today, nearly all systems provide an interface to an audiographics sub-system, and many build the subsystem directly into the videoconferencing group system.

Nearly all room systems should include physical ports for audiographics equipment. Audiographics capabilities should conform to the ITU-T T.120 specification so that any end-point can share files. Document annotation capabilities are generally included. Consider providing space and wiring for the electronic tablets and/or workstations that provide the annotation interface.

If a document camera is needed, a placement decision must be made. Some are mounted in the ceiling to secure the field of view necessary for large documents. Some could be termed as graphics consoles. They provide *backlighting* to illuminate slides and transparencies. Additional features such as graphic image preview monitors or special control panels are convenient, but they add to a system's cost.

If applications are scientific, a facility plan should accommodate microscopes, high-resolution cameras, and other scientific equipment. If the goal is to accomplish engineering work over distance, the room may need CAD interfaces, as well as the ability to convey, and incorporate mechanical and electrical drawings, materials lists, specifications, and proposal preparation systems.

In addition, a phone interface for audio-only meeting participants is usually required. Audio add-on is usually provided as a feature of a videoconferencing system and requires a RJ-11 jack connection to a PBX or outside line.

Videoconferencing monitors must be selected according to the application. The greater the number of participants, the larger the monitor should be. Twenty-inch monitors will support up to four conference participants. A 27" monitor can support 6–8 participants. Where eight or more participants will be involved in a meeting, a 35" monitor is preferable. Today's flat-screen monitors significantly solve the space problems that legacy CRTs created.

Training environments, in which many viewers are involved and the room is large, often require projection systems (front or rear). In such a case, a RGB projection system need be procured. Systems in this category can use rear- or front-projection. Again, it is important to enlist the support of a professional for non-standard implementations of this nature.

SCAN CONVERTERS

Television monitors *paint* the screen using a technique known as interlace scanning. Interlace scanning divides a frame of video into two fields; one made up of odd lines and the other made up of even lines. To convey the frame, the field containing even lines is transmitted first, and the field containing odd lines is transmitted immediately thereafter. Sixty such fields are sent in a single second, which conforms to the 60 Hz electrical system used in the U.S. and Japan. This approach eliminates *flicker*, the shimmering effect that occurs when frames are delivered at a rate too slow to leverage human *persistence of vision.*

PCs and workstations use video graphics adapter (VGA) and Super-VGA (SVGA) formatted signals—better known as progressive scanning. Computer monitors do not interlace a signal to deal with flicker because their inherent frame rate of 60 fps and above is fast enough to fool the human eye. Unfortunately, this difference in scan rate creates an incompatibility between computer and TV monitors.

Computers and television systems differ in resolution, too. Again, resolution describes the number of horizontal and vertical pixels a monitor can display. The most commonly used formats in computing are 640-by-480 pixels (VGA) and 1,024-by-768 pixels (SVGA). SVGA can be extended to accommodate larger monitors, with 1,280-by-1,024 pixels (1.3 million pixels per screen).

In television systems, resolutions are either 525 vertical lines (NTSC) or 625 lines (PAL and SECAM). However, viewers do not see all the 525 NTSC scan lines on a television monitor because 42 of them are *blanked* during vertical retrace, which leaves 483 visible lines. The horizontal resolution of an NTSC video image depends upon the quality of the signal and monitor and generally ranges from about 300–450 dots per horizontal scan line. Consequently, the image quality of a picture displayed computer monitor will always be better than that displayed on a television monitor.

To display a computer-generated image on a television monitor, a scan converter is required. Scan converters also convert from *composite* to *component* video. With composite video, color is encoded using YIQ where the Y component describes luminance (a weighted average of the red, green and blue primaries of the image) and the I and Q components provide chrominance (color) information. VGA and SVGA monitors use a color encoding method called YUV in which luminance (Y) is stored *separately* from the chrominance (U and V).

CONNECTING TO THE NETWORK

Video communication facilities should be located close to the network interface (demarc). This keeps cabling costs down, simplifies troubleshooting, and eliminates the need for line drivers to regenerate a digital signal over extended distances. Plan and plan for power to avoid circuit overloads and limit problems caused by *dirty power*. Make sure that plugs and circuit locations are convenient.

Most videoconferencing group systems connect to an IP network or a private or public circuit-switched digital network. If, instead, the system will connect to satellite service the project manager will need to work closely with a supplier or room integrator. Such installations are potentially very complex and are outside the scope of this book.

The type of circuit that will support the video application determines how a videoconferencing system connects to the network. Physical interfaces include EIA RS-449/422 and its associated dialing standard RS-366-A, the ITU-T V.35 (also able to support RS-366-A dialing), and X.21 interfaces. Interfaces also include RJ-11 or RJ-45 (ISDN) and DS-1 (T-1). Of course, in the case of Ethernet, the connection will simply be RJ-45.

Despite the growth of IP, the most common transmission for group systems is still ISDN. That said, significant IP-based video is in use over existing frame relay, however native IP (in the form of MPLS) is likely to supplant frame relay as circuits come up for renewal. For ISDN, V.35, RS0-530, and RS-449 are the most widely used physical interfaces. For IP, RJ-45 is the standard interface.

The ITU-T X.21 interface, which is used primarily in Europe, operates at 56–384 Kbps and controls network dialing. One valuable feature of X.21 is its inherent dialing functions, which include the provision for reporting why a call did not complete. X.21 can be used to connect to both switched and dedicated networks.

ISDN BRI, the most complex network interface to arrange, requires a NT-1 interface. An RJ-11 plug from the network plugs into the NT-1. On the other side, connection to the equipment is made via an RJ-45 plug. The specific type NT-1 required depends upon which CO manufacturer's switch provides the connection. Several companies make an NT-1 that can interface to standard central office (CO) telephone switches. It is best to require the supplier to provide the NT-1 and take end-to-end responsibility. Provide the supplier with the LEC contact's telephone number and rely on them to resolve all issues related to ISDN connections.

Before buying ISDN-oriented interfaces, one should be certain that the LEC supports it. Allow a month or so to get ISDN service installed because some LECs are notoriously slow. Depending upon pricing and bandwidth, organizations may opt to use ISDN PRI. ISDN PRI terminates in an NT-2 interface, a PBX, or a specially-equipped network access device. Customers who buy PRI from an IXC in the U.S. can often use H0 and H11 bandwidth-on-demand service. This is true in some European

278

countries as well; overseas high-speed dial up is offered as H0 and H12 and is obtained from a PTT.

Some organizations use PBX stations to provide network support for videoconferencing systems. Generally, ISDN BRI stations are used in this case. On the *trunk* side of the PBX, connections are made to carriers or other corporate sites using T-1, fractional T-1, or ISDN PRI channels. Channels not used for video can carry inbound and outbound voice traffic. Putting videoconferencing behind a PBX allows an organization to move the system between conference rooms as desired (presuming the rooms are equipped with the proper station jacks).

Finally, a videoconferencing codec may interface to the network via T-1 or fractional T-1. Connections of this type are made through a T-1 multiplexer (T-1 MUX) or inverse multiplexer (I-MUX). Multiplexers connect to a codec using either a V.35 or RS-449/422 interface (again, V.35 is most common). Carriers bring T-1 service into the building demarc, where it is connected to a CSU. In review, the CSU is used for signal shaping and equalization, longitudinal balance, voltage isolation, and loopback testing. More than one installation has been halted because a customer forgot to order the CSU (and all the required cables) before starting the job.

Increasingly, organizations are deploying even room-based interactive video communications via IP (Internet protocol). While doing so over the Internet is risky business, doing so over leased broadband circuits can be viable. In such a case, a videoconferencing unit will connect via an Ethernet cable to a router, which either includes or is connected to a CSU/DSU.

The advantage in this case is that video traffic is packetized and can thereby share bandwidth with data traffic. This is not advisable without the assistance of an experienced network integrator that can ensure that all contenders for network bandwidth share nicely. A videoconference can hog so much bandwidth that the network cannot effectively support any other traffic. Similarly, a videoconference can fail because of extensive data traffic such as a large batch transfer.

DSL is generally not a practical means of supporting an interactive video session. ADSL, in particular, is impractical because, although download speeds may be ample, upload speeds may not. IDSL was largely created to provide ISDN BRI equivalent service where there was not ISDN service. It is adequate for a personal session, but not for a group session. Cable TV networks are becoming increasingly robust. Nevertheless, because the target is generally the consumer market, the network is not optimized for group video communications systems.

Leased frame relay circuits are commonly used for IP-based video communications. Although dicing bandwidth into frames and subsequently into IP packets is inefficient, it can work well if a knowledgeable integrator properly configures all ends of the network. Leased native IP circuits may be the most practical approach to IP-based video communications until the Internet develops with sufficient quality-of-service (QoS) mechanisms.

PROMOTING VIDEO COMMUNICATION

Before a video communication system is being installed, one should be formulating plans for fostering its use. The countless organizations that have failed in changing habits to include video communication demonstrate that a successful effort will be nothing short of an ambitious advertising campaign. Many companies put up signs and banners to announce the arrival of videoconferencing. It is helpful to get on the agenda at executive staff meetings, and to downplay technical aspects in favor of business *utility*. Wallet cards that provide instructions on scheduling a videoconference as well as troubleshooting can be helpful (send the same information in email announcements to enable cut-and-paste).

During the presentation, be sure to tell potential users how to find the corporate intranet Web page that describes how to arrange a videoconference. That site (or its paper-based equivalent) should cover the following subjects:

- A video communication fact sheet that describes systems and locations. This sheet should describe available services (e.g., room coordinator available to operate the system) and equipment (e.g., document camera and electronic whiteboard). It can be helpful to post such information in high-traffic locations to remind people that video communication is an option.

- A map that can be used to find the videoconferencing room (especially important in a campus environment).

- A simple set of instructions that lists how to make a reservation, and required lead-times for simple or complex conferences (e.g., a public room or conversion service might require additional time to schedule).

- An overview of video communication etiquette and tips for success. This can start with essentials such as being on time and adhering to a schedule. It can also include protocol tips for participants (e.g., speak naturally, look at the monitor, expect audio/video delay, avoid coughing or sneezing into the microphone, drumming fingers on the table, or carrying on side conversations). It is helpful to include tips regarding what to wear (wild plaids and complicated patterns can cause trouble for some systems). Finally, the *tips sheet* should address meeting materials such as how to prepare graphics (e.g., font sizes, landscape formatting, and centering text to leave white space around the edges), and should include a sample.

- A concise guide for multipoint conferencing etiquette. In advance, conferees should be briefed in the difference between point-to-point sessions and multipoint ones. This etiquette guide might suggest keeping the microphone muted when not in use. Other issues may include how to control the camera in a chairperson-controlled multipoint session, differences between point-to-point and multipoint systems and how they affect conference participants, how to dial into a bridge, and what to do if a site becomes disconnected during a multipoint conference. Before developing an etiquette guide, sit through a few multipoint sessions.

- It may be helpful to include a brief explanation of why multipoint conferences tend to offer lower quality images and sound. The decline in quality is related to a reduction in bandwidth—some is used for *overhead* so that multiple systems can coordinate their exchange.

- A description of the cost of typical videoconferences. Some identify the cost for point-to-point, multipoint and public room conferences between corporate locations. This is a useful selling tool; people are often surprised to learn how inexpensive a videoconference is compared to travel.

- A guide outlining questions to ask when attempting to arrange an inter-company videoconference. This document should provide a technical contact's name and number for a scheduler from another company to arrange a test. It should state the brand and type of codec installed in each room (including the software revision). It should identify which ITU-T protocols are supported (H.320, H.323, H.324, G.728, etc.), the long distance carrier, and the service. Since only switched access (e.g., ISDN BRI or switched 56) supports any-to-any connections, this is important information. It is important to note network speeds for each location and whether an I-MUX is available. If an I-MUX is used, describe the BOnDInG modes that are supported. If the organization is a member of a public room network or reservation service, provide relevant information. This guide should not be handed out to all users, but is invaluable to technical facilitators.

Organizations promote videoconferencing in various ways. One company temporarily installed its new videoconferencing equipment in cafeterias and lunchrooms. It encouraged employees to conduct impromptu conferences with whoever was on the other end as a means of getting comfortable with the medium. After a week of this arrangement, the company moved the equipment to its permanent home. Within a few months, equipment utilization reached 50%.

Web pages and newsletters that publicize video success stories are good promotion. Finally, to stimulate video communication, the organization may want to subsidize the cost of using the system for 6–12 months. Charge-back is easier to sell once the system is integrated into the culture. Rushing to make everyone pay his share may impede adoption and, consequently, return on investment.

SUPPORT SERVICES

System users will expect support at every conference site within an enterprise. Providing that support typically requires a team effort. System administrators are usually the focal point for customer service. They, in turn, direct the activities of the videoconferencing site coordinator assigned to each location.

The video communication system administrator's role varies between organizations. Some are responsible for all aspects of event coordination, including logistics, meal planning, scheduling participants, preparing handouts, and hosting the event. In other companies, the video communication system administrator provides scheduling support only. In most cases, s(he) simply schedules equipment, public rooms, carrier gateway and conversion services, and special equipment rental.

At larger facilities (e.g., headquarters and regional locations), there is generally a dedicated video communication system administrator. This individual trains system users, provides assistance with the equipment, schedules the rooms, coordinates with equipment and network services suppliers, and manages other related activities at the site. A video communication coordinator is often assigned to assist users at smaller sites (field or satellite locations) where a full-time person is not necessary. Some companies rely on participants to handle most of the technical operations on their own and provide training and support over the system. When this is the case, ease-of-use is crucial. Independent access to technical help must also be provided (e.g., a separate voice line for requesting assistance in case of system malfunction, and a serial line to allow a technician to remotely manage the system via modem. It is also important to install back-up systems in videoconferencing rooms such as speakerphones to support audio conferencing if the video fails.

A single individual at each site should assume responsibility for the system installed there. That person does not have to be technical, but must know the basics of system operation and troubleshooting. In locations where technical expertise is not strong, well-trained personnel and complete, accurate and easy-to-use system documentation is critical. Documentation should indicate what type of equipment is installed, the software revision, initial equipment settings, how to interpret on-screen error messages, and technical support information. It should also address the operation of peripherals (e.g., document cameras. The room should also include a checklist of common problems and how to resolve them.

Technical support is often more complex than suppliers imply. Vendors may not intentionally minimize the support burden associated with video communication, but it often takes on a life of its own after installation. Only set-top systems and good roll-abouts really deliver plug-and-play operation. Additional components such as inverse multiplexers, document cameras, VGA-to-NTSC scan converters, increase the support burden. Technical support becomes an issue only after the first important conference fails.

Technicians are generally responsible for site set-up, troubleshooting and diagnostics, ongoing maintenance and upgrades, technical user training, and system fine-tuning. If technicians are local, they usually perform these tasks themselves. If they are remote, they may travel to the site to perform installation and training. Subsequently, they are remotely available to assist moderately-knowledgeable users.

Job Duties	System Administrator	Technician	Field Site Coordinator
Business Case Development & Presentation	X		
Trial & Demonstration Management	X		
Project Management	X		
Installation Coordination	X	X	
Network Installation	X	X	
Equipment Scheduling	X		X
Public Rooms Scheduling	X		
Inter-Company Conference Coordination	X		
Usage Tracking & Management Reporting	X		X
Applications Development	X		X
Operational Training	X	X	X
Graphical Materials Preparation Assistance	X		X
Reservation Confirmation	X		X
Multipoints Set-up	X	X	X
Equipment Operation	X		X
Roll-About or Set-Top System Moving & Placement		X	
Cost Allocation	X		
Supplier Management	X	X	X
Trouble Shooting	X	X	X
Etiquette Training	X		X
Procurements (Upgrades)	X	X	
Newsletter Publishing	X		
Promotional - Other	X		X
Maintenance Contractor Tracking	X	X	
De-Installs & Re-Installs	X	X	
Migration Path Development	X		
Upgrade/Expansion Business Case Development	X		
Special Applications Support	X	X	X

Figure 14-1. Responsibilities Of Video Communication Support Personnel.

283

It is useful for a technician to be familiar with voice communications and network operations. Technicians can usually receive training from vendors, but a circuit-switched background is helpful. As video communication migrates to the desktop, the skills required will broaden considerably. With these systems a more in-depth understanding of LAN, particularly IP and Ethernet, technology becomes important.

The best time for a technician to develop a troubleshooting and system operation guide is immediately after receiving or providing system training, while details are fresh. The guide should list steps that anyone can follow to troubleshoot common problems. Perhaps the most helpful way to organize the list is by problem/cause. For example:

Problem: Can see but cannot hear participants

Solution: Ensure that microphone is plugged in

Another useful way to organize the document is to start with a "Top ten things to check" list with suggested corrective measures. For example:

1) *Power-cycle the equipment*

2) *Check that the monitor is turned on*

3) *Check that the right video source is selected*

Desktop-based video communication creates new support challenges. Desktop applications are harder to install and maintain because systems are under the control of various users. Rather than maintaining one system, administrators must maintain nearly as many systems as there are people in the organization. There are issues regarding compatibility between desktop systems and room systems (standards, networks, and applications). Once users become familiar with video, technical challenges will abound from people's experimentation and exploration.

SCHEDULING SYSTEMS AND SERVICES

Videoconferencing represents a large investment by most organizations' standards.

Even with declining costs, it is important to ensure that the system is often used. Therefore, scheduling processes and tools deserve serious consideration. Someone must assume responsibility for system scheduling which, as usage increases, becomes increasingly complex. It is a good practice to establish policies and procedures before system implementation rather than addressing scheduling *ad hoc.*

A primary issue is videoconferencing room exclusivity. When it is not possible to dedicate a room for videoconferencing, management must establish whether a videoconference can *bump* a scheduled non-videoconferencing meeting.

It is also important to address who will have access to the system and how usage will be prioritized. Organizations with international sites often reserve open hours when there is business-day overlap between time zones and reserve the remainder of the time for domestic connections, when hours of business overlap to a greater degree. For instance, early morning room bookings might be reserved for European connections, late afternoons held open for Asian conferences, and mid-day hours for Pacific, Mountain, Central, and Eastern time zones.

A business unit may buy its own equipment and prioritize its own use over that of others. Some groups make excess capacity available to others at a price. In other cases, policy states that the system is to be used exclusively by the department that purchased it. Others cannot use it even if it is sitting idle.

Excess corporate capacity is often *brokered* through a public room service. The decision to sell excess capacity introduces a host of priority-related issues. What if the room is scheduled for outside rental when an executive needs to conduct an urgent meeting? If the system is made available to outside groups, what is the cost, and how is the schedule arranged? Who is responsible for billing and collections?

What type of use is appropriate for the system? How does the importance of a meeting affect scheduling? Some companies allow important meetings to bump routine meetings. Policy should be established for asking someone to postpone a pre-scheduled session. Nevertheless, most companies authorize an employee to refuse such a request; doing so discourages executives from routinely *bumping* scheduled meetings. When they know they can get the room any time they want it, senior managers may be less likely to plan.

Many organizations levy cancellation charges on users who cancel a conference at the last moment. These are generally charged to the cost center as a thirty-minute minimum charge (and often waived if the facility can be otherwise re-scheduled).

Some companies that own numerous geographically separate videoconferencing rooms assign a coordinator to each local site. In such a case, the site coordinator should work through a central system administrator to coordinate use of his/her own room. An alternative is to use a multi-user-scheduling software package to manage multiple site reservations. Another issue that must be addressed is who schedules public rooms, codec conversion and IXC gateway services. This is typically done centrally but is sometimes delegated on a site-by-site basis. How conferences are scheduled is an issue that deserves consideration. Numerous companies provide service bureau scheduling for customers.

Sprint offers an on-line tool that allows an organization to schedule not only its own sites but also any other site also served by Sprint. Conferences can be arranged so that, at the appointed hour, all sites simply come up and are connected. Sprint also offers unique accounting and cost allocation services, and offers a wide range of management reports. AT&T and MCI WorldCom offer similar services.

If an organization chooses to provide its own scheduling, various scheduling packages exist to simplify the task. Nearly all are password protected and offer multiple levels of user privilege (e.g., view, schedule, and delete/change). Any package should schedule a conference based upon a request for a specific date and time and, if that time is not available, will offer alternate times. It should also perform automatic conference scheduling and schedule repeat conferences for a given time across a range of dates (e.g., every second Tuesday from 8:00 - 9:00). Typically scheduling packages show available time slots for all sites for a full day, and many will *window* to view multiple sites. All packages should produce usage and cost-related management reports, print a list of participants, and generate electronic messages to confirm participation.

Countless issues are involved in installing and supporting a videoconferencing system. It is best to develop the processes and policies to ensure success at the outset. The videoconferencing environment is involved, but manageable. An organization's best response to a distributed work force is to put in place capable information technology tools that will allow it to do more with less, and do it faster. If it does so, video communication can enable it to transcend such challenges.

CHAPTER 14 REFERENCES

Fritz, Jeffrey N. "From LAN to WAN with ISDN," *Byte* November 1996 v21 n11 p. 104NA3.

Hakes, Barbara T.; Sachs, Steven G.; Box, Cecelia; Cochenour, John. *Compressed Video: Operations and Applications* The Association for Educational Communications and Technology, a U.S. publication, 1993.

Noll, A. Michael. *Television Technology: Fundamentals and Future Prospects*, Artech House, Norwood, Massachusetts, 1988.

Portway, Patrick S. and Lane, Carla Ed.D. *Technical Guide to Teleconferencing and Distance Learning*. Applied Business teleCommunications, San Ramon, California. 1992.

Rossman, Michael and Brady, Kathleen. *MediaConferencing Classroom Designs for Education*, US West Communications, Business and Government Services, Portland, Oregon 1992.

15

Personal Conferencing

*If we are always arriving and departing, it is
also true that we are eternally anchored. One's
destination is never a place but rather a new
way of looking at things.*

Henry Miller

The worlds of computing and audio/visual communications have converged, and one of the most powerful outcomes of this union is *personal conferencing*. Personal conferencing leverages a variety of distance-collaboration tools that are easily used at a *personal* computer. Personal conferencing used to be synonymous with *desktop conferencing*, but wireless broadband and enhanced mobile devices are making personal conferencing possible virtually anywhere. They leverage components that are already part of the standard, multimedia desktop: a high-resolution monitor, a powerful microprocessor, speakers, a microphone, and a fast network connection. Personal conferencing simply requires the addition of software, a camera, a video capture board, and a network adapter.

The personal conferencing usage model is different from the group-oriented one. The culture and custom of boardroom video communication favored executives. Although this is changing, it is still common for a group-system videoconference to be comprised primarily of senior executives. Moreover, it is common for an unplanned executive conference to supersede a previously scheduled one.

Wherever they happen to occur, personal conferences do not require special rooms. Moreover, since desktop tools operate within the singular domain of IP networking, the range of applications is much broader. Individual contributors and members of a dispersed team are as likely (if not more) to use personal conferencing as are senior managers. It fits the needs of individual contributors such as project managers, engineers, graphic artists, consultants and any others who collaborate over distance. Personal conferencing is a flexible means of deploying any communication method that makes sense, when it makes sense. For instance, one can start a voice call first then spontaneously agree to a videoconference or share documents. The ability to hold impromptu meetings, as opposed to pre-scheduling a videoconferencing room, makes personal conferencing very powerful.

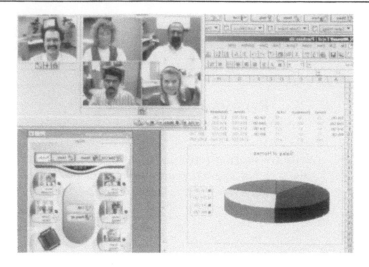

Figure 15-1. Application Conferencing. (courtesy of Intel).

Multipoint personal conferencing is an increasingly valuable tool for a distributed work groups. It allows for staff-meetings, problem-solving discussions, and strategy planning sessions without any two people being in the same room.

Personal conferencing products come in many shapes and sizes. Most offer a toolbox of real-time collaborative utilities that enhance group work. Apart from that similarity, the variations in implementation are broad. Products can be software-only, software and hardware, bundled or build-your-own, business quality or consumer-focused, audiovisual or audiographics. A veritable cornucopia of options exists. Data-only conferencing systems deliver capabilities that fall into five major categories:

- **Shared Whiteboard**

  Shared whiteboard tools are especially effective for geographically separated groups that need a shared drawing board. Products in this class work well for brainstorming sessions, project reviews, task force meetings, and product-design. They are ideal for a group that needs to explore ideas, and to document process. Most include an audio component, although some require participants to set up a separate telephone call. With a mouse, participants can cut, copy, paste, highlight, point to objects of interest, and type text-based messages. All participants can view changes in real-time. Products can be either object-oriented or pixel oriented (object oriented products are preferable). Electronic whiteboard enabled physical whiteboards may rely on a server, not on a desktop, to host computing. They use electronic markers and video feeds to create the illusion of a single site meeting across multiple sites. Electronic markers are much easier to draw with than a mouse.

Figure 15-2. Personal Conferencing (courtesy of Polycom).

- **Screen Sharing**

Screen sharing products are presentation-oriented, that is to say that they emulate *paper-like* text, graphic, and still images. Systems in a screen-sharing call permit an *initiator* to launch an application for the benefit of a remote viewer (or viewers). The initiator's machine is equipped with the conferencing software that allows the remote parties to view an application even though it is not running on their desktop. Remote viewers may or may not be able to make changes to the application, but can provide verbal input over an audio connection. Screen sharing applications are useful in distance learning and remote presentations, and are effective over the Internet or a LAN.

- **Application Sharing**

Application Sharing products enable distributed workers to turn any off-the-shelf application (word processing, slide presentation, and spreadsheet) into a shared one. Only one individual needs to run the application for everyone in the session to share it. After that one person launches the application (given that the application sharing session is already in progress), all conferees can both see a shared document and can make changes to it. Different products offer different ways of characterizing who is making what changes. Some identify users with colored fonts, others append names or initials, but nearly all allow changes to be made only after the initiator transfers control. Application Sharing differs from shared whiteboard in that the purpose of the former is to share an additional application to view or change a document, whereas the purpose of the latter is to capture rough notes and illustrations. Put differently, document creation and refinement is the objective in applications sharing; brainstorming and idea

291

capture is generally the focus of shared whiteboard applications. File transfer is also an applications sharing feature. Most products also offer chat tools for communicating via simple text messages, post comments to a conversation, or record meeting notes and action items as part of a collaborative process.

- **Interactive Multimedia/Video Servers**

Most interactive multimedia products support desktop-based just-in-time training. Some allow interaction with the compressed audiovisual stream. Others deliver a one-way-only stream of compressed digital video. Products in this class can take the form of audiovisual enhancements to Web-based applications. Interactive multimedia products typically compress the content using MPEG techniques. Stored material is available on-demand across a LAN or WAN. Interactive multimedia/video server products can be point-to-point or multipoint. They can be applied as integrated distance learning, performance support, and information management systems. Instructors often use such products in course preparation and information presentation.

- **Video Feeds**

Video feeds enable individuals to use PCs (and, increasingly, mobile devices) to receive live material as compressed digital broadcasts. One example is retransmission of television news programs. In-house examples might include special announcements by the CEO, quarterly results, and product announcements. Some video feed products allow viewers to send messages to which presenters can respond during the session. Others are simply receive-only systems that display motion video in a window on their viewer's device. Audio relies on the soundboard in the user's machine; headsets are helpful.

Of the applications listed above, the one with the most collaborative value is application sharing. Almost anything one can do with a whiteboard or a screen-sharing product, one can do with applications sharing. Not only can distributed participants view a slideshow, but they can also mark-up the slides via electronic whiteboarding. Participants can also conduct shared and private *conversations* via chat, and those chat threads can be saved as text documents.

The 80/20 rule applies to personal conferencing. Application sharing accounts for 80% of the communication and others participants' talking-heads account for the remainder. Conversely, the video component of the application demands about 80% of network bandwidth while the data components require the smaller remainder.

Nevertheless, if a camera, microphone, and speakers are part of the environment, a user can easily add the software necessary for desktop video communication. Because of the premium on desktop (or handheld) real estate, systems generally display a small image of the remote meeting participant's head and shoulders and leave the majority of the screen for applications sharing, text messaging, or other collaborative applications. While there is overlap, face-to-face interactive videoconferencing applications are either desktop-to-desktop or desktop-to-room-system.

- **Desktop-to-Desktop Video Communication Systems**

  Desktop-to-desktop video communication products offer features such as picture-in-picture, speed dialing, and mute. They either rely upon software-only codecs or incorporate additional hardware to assist in the compression process. Software-only compression designs warrant a *try it before you buy it* strategy to ensure that the quality of audio-video synchronization and the picture resolution and frame rate suit the applications. Moreover, one should never install a software-only compression product on a desktop with a slow processor. In business applications, it is best to adopt products that conform to one of the ITU-T's audiovisual standards (H.320, H.323, or SIP). H.320 systems use circuit-switched digital channels for transmission. H.323 and SIP systems use packet-switched LANs and WANs for signal transport.

- **Desktop-to-Group System Video Communication Products**

  When a desktop product is equipped with an H.320 compliant codec and has a network interface that allows it to establish circuit-switched videoconferencing connections, it can communicate with other H.320-enabled videoconferencing group systems. However, H.323-based products can also communicate with H.320 group-systems if they make the connection through an H.320-to-H.323 gateway or MCU. Such gateways should also accommodate SIP. The applications for group-to-individual conferencing continue to expand. It is a good idea to trial desktop-to-group systems and to view the implementation from both ends (e.g., from the desktop and from the boardroom system). The quality of the picture and the sound that a person perceives on one end largely reflects the capabilities of the system on the other end, where the compression takes place.

PERSONAL CONFERENCING MARKET DRIVERS

Virtually all technology and social trends are converging and thereby launching a renewed interest in personal conferencing. The Yankee Group anticipates that 17.5 million households will adopt VoIP calling by 2008.

Skype, a P2P voice-over-IP (VoIP) service, anticipates 950,000 subscriptions and 2.2 million downloads of its software in Taiwan by 2006. The SkypeOut service, which provides a paid service connecting calls to traditional landlines, includes call conferencing, video conferencing, short messaging service (SMS) and voice mail. In short, Skype provides subscribers with everything for the price of Internet access and a USB phone.

PC and Internet adoption is fueling the growth of IP communications, the most popular of which is *instant messaging*. Ironically, when Microsoft added IM to its videoconferencing product, the market was cold; when it added videoconferencing to its IM product, the market adopted the product with open arms. IM's symmetric, or real-time nature is allowing it quickly edge out email as the preferred communication mean. It is also an ideal means to initiate a video communication. Personal conferencing, after repeated false starts and years of over-optimistic predictions has taken hold. The huge installed base of Internet-connected PCs ensures its success. Other factors include:

- Abundance of multimedia-capable PCs and workstations

- Sophistication of even low end-users

- Proliferation and adoption of broadband Internet

- Dispersed enterprises, outsourcing, and acceptance of tele-work

- Advancements and deployment of wireless networking

- Acceptance and advancement of converged PDA/phones.

Modern PCs and mobile communications devices are giving workers more choices about how, when, and where they work. Success is attributable in part to ease of use. One of the most useful innovations in desktop computing was the introduction of the applications suite. Products such as Microsoft's Office offer a similar look and feel (and, arguably, similar logic) across applications. This sameness has led to an improved mastery of computing. Perhaps the greatest evidence of this collective expertise is the amazing growth of the Web. It took the GUI-based browser to make the Web accessible; now Web-based applications are supplanting PC-based ones.

As people gain familiarity with computers, they get a better return on the time they invest learning to use them. The homogeneity of the user experience has made the learning process more cumulative. The days in which the technical burden of personal computing was too great for the average user are gone. Mastering a new application is no longer difficult. Video communications are just one example of a family of applications that GUIs and browser-based interfaces are simplifying.

The other major piece of the puzzle is connectivity. Local telephone companies, cable television providers, carriers, and new hybrid companies are satiating demand with high-bandwidth transmission service based on coax, fiber, wireless WAN, and digital subscriber line (DSL) service. Most city-dwellers enjoy broadband local loop alternatives from all three of these options; pushing such choices to urban dwellers is a matter of time (and cost). Increased bandwidth leads to better multimedia transmission. From the point of view of a video conferee, this translates to improved resolution, motion handling, audio quality, and audiovisual synchronization. Broadband transmission makes personal conferencing communication realistic.

Coincidental with the upgrade of the U.S. telecommunications infrastructure, a similar undertaking is in progress within organizations. The LAN is the corporate equivalent of the local loop. Video communications over corporate LANs have been simplified by overwhelming standardization on IP over Ethernet. Like the Internet,

Ethernet was designed for character-oriented exchanges.

One popular trend is deployment of organizational intranets (networks within an organization) and extranets (networks that connect subsets of organizations). These enterprise-wide, hypertext-linked, multimedia-enabled networks connect PCs to server-based information caches, media servers, and communication gateways. LAN-based personal conferencing is a natural extension of corporate intranets.

Figure 15-3. Personal Conferencing (courtesy of Tandberg).

If the 1990s go down as *the decade of the desktop PC,* the following decade will go down as the decade of *the wired Personal Digital Assistant* (PDA). It is no wonder that today's enterprises are becoming increasingly distributed. With the tools to make one's office any place that is convenient at hand, why waste time traveling or going to *the office?*

That said, the synergy that can result from collective investigation and development of ideas over distance requires fine-tuning and requires special tools. Personal conferencing conveys audio and visual information, and adds to the richness of human interaction. It does not exactly duplicate in-person interaction, but it is much better than a simple phone call. Video communications can actually be more natural than in-person ones in that one can arrange a personal conference on the fly and from a comfortable setting. This results in a dynamic and fluid collaborative environment.

Besides a generally favorable climate for all desktop applications, certain events boost the popularity of personal conferencing *specifically*. These include:

- Maturity of personal conferencing standards
- Standards-compliant software and hardware
- Powerful compression algorithms
- Reduced cost of peripherals (e.g., cameras, speakers, microphones)
- IETF protocols and mechanisms for low-latency networking applications
- Collaboration between video communication suppliers and carriers.

The Picturephone foreshadowed the need for audiovisually-enhanced collaborative tools, but the installed base of Picturephones was so minute that none was of use. That could have been the situation personal conferencing faced were it not for interoperability standards. The ability to call or be called on-demand, without compatibility limitations has led to the acceptance of personal video communication.

Picturephone's other great misfortune was reliance upon a network that did not yet exist. Thirty years later, immense advances in video compression permit the use of much slower networks for signal transport while network advancements are making networks exponentially faster. At the same time, the cost and availability of personal conferencing peripherals have plummeted. It all adds up to fertile ground for the growth of anywhere-to-anywhere personal video communication.

PERSONAL CONFERENCING AND THE INTERNET

Streaming time-sensitive media (audio and video) over the Internet sends shivers, good and bad, down the spines of those most interested in such applications. On the plus side, IP-based streaming media tools hold tremendous promise. On the minus side, it presents a colossal challenge because video communication is perhaps the most bandwidth-intensive of applications. Video communication is challenging because there are many bits in a frame of video, because the delivery of this data must be smoothly paced, and because it involves multiple participants. In the worst-case scenario, everyone wants to interact with everyone else. In such a case, each participant-pair requires a stream of synchronized audio and video. The additional traffic crowds the network and creates traffic jams.

The problem of streaming media over a data network (e.g., a LAN or the Internet) has attracted a great deal of attention. User Data Protocol (UDP) offers no guarantee of packet delivery and allows packets to be dropped when the network is busy. On the other hand, Transmission Control Protocol (TCP) does guarantee delivery through the retransmission of dropped packets. However, TCP presents two essential problems. First, it introduces overhead delay that interrupts the smooth flow of media streams when congestion is present. Second, TCP retransmits dropped packets. This is valuable in text-based communications, but is a liability in streaming video. Streaming Media's pacing and sequence are sensitive; packets are only of use if they are available chronologically. Moreover, audiences hardly notice a single dropped video frame.

The Internet Engineering Task Force (IETF) and, specifically, the IETF's Audio Video Transport Working Group has developed protocols to deal with time-sensitive media streams. Three of those are the Real-Time Transport Protocol (RTP), the Real-Time Transport Control Protocol (RTCP), and RSVP (ReSerVation Protocol). The AVT group works to provide end-to-end network transport functions that suit applications that transmit real-time data (streaming media) via multicast (one-to-many) or unicast (one-to-one) services.

In the IETF Request for Comments (RFC) 1889, the RTP described *RTP: A Transport Protocol for Real-Time Applications*. RTP seeks to provide end-to-end network transport functions that are suitable for applications that transmit data with real-time properties, (e.g., audio and video) over multicast or unicast network services. Applications typically run RTP on top of UDP, although RTP can also be used with other underlying network or transport protocols.

RTP, a higher layer protocol, does not provide a mechanism to ensure that packets associated with a media stream arrive on time or in sequence. Lower layer services take care of those tasks. However, RTP does include a sequence number to allow the receiver to reconstruct packets in order. The sequence number can also be used to reconstruct a frame of video, for instance, an H.261 group of blocks (GOBs).

The User Location Service (ULS), developed in February 1996 by R. Williams of Microsoft Corporation, is a protocol for locating hosts and establishing real-time sessions between them. The ULS is a directory, a place in cyberspace where a client can publish connection information such as the transport address of an application or a person. The ULS enables communication between compatible applications by allowing other clients to retrieve these addresses.

The ULS is in not a part of the RTP, but the RTCP is. The RTCP, described along with the RTP in RFC 1889, is used to monitor the quality of service provided in a videoconference. It allows conferees to convey quality information to the meeting's chair or other person responsible for the conference's usability. RTCP provides the option for a sender to receive feedback from the recipients of a multicast stream on the success of RTP packet delivery. Since all participants get the same information, a single site can determine whether reception problems are unique to a site or are conference-wide. A sender can use the feedback from multiple recipients to adjust the broadcast, and scale back video quality or eliminate it entirely to reduce bandwidth requirements. Network administrators can use RTCP to see who is engaging in multicasts.

The RTP was adapted for use in videoconferences as RFC 1890: *RTP Profile for Audio and Video Conferences with Minimal Control*. RFC 1890 specifies a scheme to packetize the H.261 video stream for transport over a TCP/IP network. For instance,

RTP packets identify the type of payload contained in the packet. Depending upon the type of information the packet contains, the header may be appended with a timestamp. The header can also indicate encoding methods such as whether formal or de facto standards (G.728 or H.261, or MPEG) define the format of the media stream. The specification outlines the ports that are to be used for each payload encoding type and the port's clock rate. It also, on a per-port basis, defines whether the media stream consists of audio, video, or audio/video components and, when audio is present, how many channels are in use.

Another critical protocol for video communication over the Internet is RSVP, which enables a user to reserve bandwidth to provide a specified quality of service to a media stream or *flow.* Routers that offer RSVP support can accommodate such requests and reserve capacity for audio and video streams at a given time and day, and thereby lessen the chances that unpredictable delays will cause bothersome degradation. Once all routers have granted the reservation, the stream can move over the TCP/IP network free from frame-freezes and audio break-up. The result is usable and natural voice and video, to the user's desktop. Unfortunately, the operative phrase is, *"Once all routers have granted the reservation."* One router in the path that is not capable of or (more likely) is not programmed to recognize RSVP voids the work of the others.

In October 1996, T. Turletti and C. Huitema published RFC 2032, *RTP Payload Format for H.261 Video Streams*, in which they defined a way for H.261 information to be carried as payload data within the RTP protocol. The RTP payload format described in RFC 2032 specifies a header that contains the start and end bits used to package image frames. It also includes information on whether the packet contains an image that has been exclusively intraframe encoded, and an indication of whether motion vectors are used in the bit stream. The header also includes bits used for decoding a group of blocks and associated macroblocks, the value to be used during the quantization process, and horizontal and vertical motion vector data.

Internet standards track advances quickly. The Draft that defines an *RTP Payload Format for H.263 Video Streams* concisely states, "It is inappropriate to use Internet-Drafts as reference material or to cite them other than as work-in-progress."

ADVANCING THE PERSONAL CONFERENCING INDUSTRY

The personal conferencing industry was made not born. Many large and influential companies decided that it was time for the multimedia industry to take off and did something about it. Key in this group are Intel and Microsoft, both of which have a stake in seeing successive generations of increasingly powerful desktop applications. AT&T, MCI, Sprint, and the former Baby Bells all worked toward a time when average users would transmit multi-megabit traffic as routinely as they made telephone calls. In short, many different organizations are cooperating and embracing standards to provide the means for an interactive multimedia market. Driven by the need to simplify, they are bundling products into turnkey solutions.

Intel may have introduced the modern *one-stop shop* trend in personal conferencing. Its ProShare introduction barely preceded an announcement of an impressive partnership of five of the seven RBOCs, GTE, AT&T, and several networking companies. Intel offered steep price reductions when ProShare customers bought their systems through these partners. The discounts were especially attractive when ProShare was bundled with multi-carrier offerings (LEC and IXC combinations). One key aspect of Intel's partnership was its creation of the "ISDN Blue" standard configuration. It allowed a LEC to deliver a standardized ISDN BRI service that was ideally configured for use with ProShare.

Another noteworthy event was the July 1996 announcement that Microsoft and Intel had reached accord in their effort to cross-license personal video communication technology. Under the agreement, Intel provided its implementation of the H.323, RSVP, and RTP standards to Microsoft, and Microsoft, in turn, offered Intel its T.120 implementation, the ActiveX conferencing platform, and Microsoft's NetMeeting application. This created a standard that two industry giants endorsed.

Industry initiative and cooperation has enabled personal conferencing. However, as exciting as it might be, industry activity and technology events cannot eclipse the importance of applications. As always, the usage model is the chief determinant of personal conferencing purchasing decisions. In the next chapter, we will provide tips for evaluating and buying personal conferencing systems.

CHAPTER 15 REFERENCES

Rupley, Sebastian. "What's in Your Future?" *PC Magazine*. December 2004.http://www.pcmag.com/article2/0,1759,1744194,00.asp

Chuang, Bryan and Shen, Jessie. "Skype To Add Video-Conferencing Service In 2005? *DigiTimes.com*. December 2004.
http://www.digitimes.com/news/a20041228A6024.html

Braden, R., Zhang, L., Berson, S., Herzog, S., Jamin, S., "Resource ReSerVation Protocol (RSVP)—Version 1 Functional Specification," Internet Draft, November 1996.

Braden, R., Zhang, L., "Resource ReSerVation Protocol (RSVP)—Version 1 Message Processing Rules," Internet Draft, October 1996.

Czeck, Rita. "Desktop Videoconferencing: The Benefits and Disadvantages to Communication."

DataBeam Corporation. "A Primer on the T.120 Series Standards," Published on the World Wide Web.

DataBeam Corporation. "T.120 Primer—Ratification of the T.120 and Future T.130 Standards," Published on the World Wide Web.

Dunlap, Charlotte. "Conferencing vendors seal alliances," *Computer Reseller News* June 3, 1996 n686 p. 72.

Edmonds, Roger. "Desktop Videoconferencing at The Open Access College,"

Earls, Alan R. "PCs bring people face to face," *Computerworld* April 29, 1996 v30 n18 p. 95.

"Videoconferencing: AT&T WorldWorx service adds support for new Intel product; AT&T delivers standards-based multipoint voice/video/data desktop videoconferencing service for ProShare Video System 200," *EDGE, on & about AT&T* May 6, 1996 v11 n407 p. 34.

"New video conferencing options from Microsoft, Compaq, Intel," *Electronic News* June 3, 1996 v42 n2119 p. 24.

Estrin, Judy; Casner, Stephen. "Multimedia Over IP," *Data Communications on the Web* August 1996.

Gill, B. "Presentation Tools Take Multimedia Center Stage," *Gartner Group Multimedia Research Note--Markets* November 25, 1996

Hamblen, Matt. "Desktop video surge forecast," *Computerworld* October 14, 1996 v30 n42 p. 85.

Johnson, Edward G. "Video Guidelines," Published on the World Wide Web.

McCall, Tom. "Desktop Videoconferencing Drives, Doesn't Run Over, Demand for

Group Conferencing Systems Dataquest Study Indicates Strong Demand for Group and Rollabout Systems."

"MCI, Microsoft Link For Internet Conferencing," *Newsbytes* May 29, 1996 pNEW05290051.

O'Brien, Jim. "New Interoperability Standards Anticipate Universal Internet Conferencing," *Computer Shopper* September 1996 v16 n9 p. 630.

Pappalardo, Denise. "BBN to test RSVP," *Network World* December 9, 1996 V13, n50, p. 1.

Rupley, Sebastian, Levin, Carol. "Collaboration on call: videoconferencing takes a backseat to document sharing," *PC Magazine* September 10, 1996 v15 n15 p. 31.

Schroeder, Erica. "Fighting Problems of Interoperability; Groups Work on Conferencing Specs," *PC Week* February 5, 1996 v13 n5 p. 47.

Schulzrinne, H., et al. "RFC 1890: RTP Profile for Audio and Video Conferences with Minimal Control," January, 1996 Published on the World Wide Web.

Smith, Greg and Desmond, Michael. "What's Worse: Flying Coach or Setting Up An ISDN Videoconferencing System?" *PC World* November 1996 v14 n11 p. 630.

Turletti, T, Huitema, C. RFC 2032 "RTP payload format for H.261 video streams," October 30, 1996 Published on the World Wide Web.

Zhang, L., Deering, S., Estrin, D., Shenker, S., and Zappala, D., "RSVP: A New Resource ReSerVation Protocol," *IEEE Network* September 1993.

Zhu, C. "RTP Payload for H.263 Video Stream," Internet Engineering Task Force; Audio-Video Transport Work Group. INTERNET-DRAFT. World Wide Web.

16

Evaluating and Buying a Personal Conferencing System

There is science and applications of science,
bound together as the fruit of the tree which
bears it.

Louis Pasteur

The successful application of personal conferencing technology starts with the question, "What are we trying to accomplish" and continues with the question, "What does success look (and sound) like?" Chapter two and the product categories in the previous chapter can help define how one will use personal conferencing. If the purpose is to support communications between graphic artists or architects and their customers, products must focus on the presentation of visual materials (screen sharing). If visual cues such as facial expression and body language will enhance the exchange, one might consider videoconferencing. If the goal is to foster collaboration in a distributed workgroup, one might explore applications sharing and document collaboration tools.

Most personal conferencing offerings exist as part of a product family. At the low end, the package may include whiteboard or screen sharing capabilities. The next step up might add a document collaboration utility. Moreover, by adding a camera, a microphone, speakers, and one or more PC boards, one might obtain videoconferencing functionality. Other enhancements might include gateways that link LAN-video packages to circuit-switched digital systems or video servers for store-and-forward broadcast applications.

When developing an application, it is best to start with what is most necessary, and add features and capabilities as possible. People will probably have to change their habits to use these tools, so it is good to consider phasing-in new skills.

The savvy decision-maker, will thoroughly perform her homework, and will let her application be her guide. To assist in this process, we include the following guide and a Personal Conferencing RFP Checklist in the Appendices.

COMPARING DESKTOP CONFERENCING PRODUCTS

The remainder of this chapter provides a step-by-step process that one can use to compare personal conferencing products. The line of questioning is intentionally broad to include all product categories. The reader may want to ignore questions that do not pertain to her particular application. The checklist also omits questions such as the financial position of a supplier, its ability to provide support, warranty considerations, and the hardware any particular application requires. For ideas on these questions, refer to previous chapters.

Product And Manufacturer Background

Product (and supplier) background, industry position, and direction:

- Product name and the date it was introduced.

- Solution description (e.g., client-side, server-oriented, client/server).

- Is it offered as a cross-platform solution?

- What computing environments does it support?

- Assumptions that formed the foundation for product design and development (For a large installation, consider interviewing the product design & marketing groups, and signing non-disclosure agreements to obtain futures-information.).

- Product history, evolution to present position, and market status. Is there a history of backward compatibility? Have no- or low-cost upgrades been offered to stay current with standards?

- Size of the installed base in pertinent areas (e.g., Asia, Europe).

- Languages supported.

- Supplier's position in the industry. What other products does the supplier offer, and how do they relate to this one? Has the manufacturer created alliances and partnerships to enable ability to fully leverage personal conferencing? What is the nature of those agreements?

Product Family

Look at product families and how they inter-relate:

- How does the product fit within the desktop conferencing categories? Is it part of a family of products?

- Do members of the family interoperate across the entire offering? Can multiple conference types (e.g., shared whiteboard, applications sharing) be part of the same conference?

- What is the upward migration path between the lowest-cost product in the family and the most expensive?

- What are the plans to continually enhance the product family?
- Is the GUI consistent across all members of the product family?

Network Considerations

Network infrastructure is probably the biggest barrier to widespread adoption of personal conferencing, particularly videoconferencing. Ethernet, on the other hand, can handle compressed motion video rates with little problem, but does not offer isochronous service (guaranteed Real-Time delivery at a continuous bit rate). Various IETF protocols designed for IP networks (802.1Q, 802.1P, RTP/RTCP and RSVP) alleviate latency problems, but many networks incorporate routers and hubs that have not been configured to run these protocols. ISDN BRI to the desktop is fast enough for small screen applications, but it is only partially deployed, and is becoming eclipsed by Internet (or at least IP) deployment. Again, a few questions might help in comparing products:

- What type of transmission system does the product require? Will it work well over switched or dedicated digital networks and LANs?
- What switched digital network interfaces does the product support? In the case of ISDN, is the product N-ISDN-1 compliant? Does it ship with the network termination device (NT-1) and terminal adapter (TA)?
- Can videoconferencing, document conferencing or screen-sharing exchanges be conducted over the Internet? Provide detail.
- Does the product work over POTS lines?
- Do the products work over wireless networks? Which ones?
- Does the product offer a client for wireless enabled PDAs?

Data Conferencing Capabilities

Ask about multiple capabilities (e.g., whiteboarding, screen and application sharing, chat, videoconferencing):

- Does the product offer shared workspace or whiteboard tools? How many users can share the whiteboard simultaneously? Can users capture the information developed within the whiteboard and store it for later reference?
- Does the product provide for presentation-oriented screen sharing (the remote end sees a bit mapped image of the original)?
- Are there any tools provided so that the distant end can point to objects, highlight them, etc?
- Can the remote end save the document locally as an image file?
- Does the product permit file transfers? During performance testing, time file transfers and observe their effect on other applications that are running at the same time.

- Does the product integrate chat?

- Can individuals or teams use the product to interactively share and annotate files and documents (e.g., spreadsheets) or entire suites of applications (e.g., Microsoft Office). Must both (or all) devices be running an application, or can sharing take place when only one desktop has it installed? How is the application controlled, and by whom?

- Does the product provide store-and-forward multimedia messaging or networked video-on-demand capabilities? If yes, what implications does this have for the system's configuration? Have the supplier describe the system's capabilities in terms of how many video segments can be stored, maximum length/size, compression ratios, and algorithms used for compression, and other storage and playback-related attributes.

- Can the product be implemented as a multicast broadcast video arrangement? Have the supplier describe the environment. If the product captures and presents external content, what co-marketing agreements exist with content-providers (e.g., CNN)?

- Does the product maintain a directory to store IP addresses?

- Does the product support call notification that allows users to accept or deny a conference attempt based on who is calling?

- In what ways is the product unique in the marketplace? (look beyond the gimmicks to true business-oriented features and functionality).

- Is the product customized for vertical market environments? Which ones?

- Is the core technology licensable by third-party vendors such that they can create product enhancements?

Video communication Product Capabilities

If a given application will benefit from videoconferencing ask the following:

- Can the system support face-to-face personal videoconferencing?

- Is compression performed in software only or is it hardware-assisted? If the manufacturer includes hardware, can it be used to capture, compress and store video and audio for multimedia applications? In what file formats can video and audio be stored?

- If the product allows motion video images to be viewed in a PC window, what size is the window? Is the size adjustable? Can the user move the window between screen quadrants?

- How many video frames per second can be sent (maximum & average)?

- What is the image resolution range (in pixels)? How can users make resolution/motion handling trade-offs during a conference?

- How does a user initiate a video call? Does the product offer an on-line directory? Can video be added during a voice-only call?

- Does the product provide dynamic channel management and allocate available bandwidth depending on the type of communication being supported (data, audio, still-image, motion video)?

- Does the product depend on a separate telephone line for audio? If so, how is audio synchronized with video? Does the product offer an external speakerphone or a headset, to permit hands-free operation?

- What types of cameras does the product support (e.g., digital proprietary, analog NTSC, PAL, S-Video). How is the camera's focus adjusted (e.g., manually, a mouse, auto-focus)? How does the camera pan, tilt and zoom? Can the camera swivel to be a document camera?

- Does the product include simultaneous local and remote viewing capabilities? How are multiple windows managed on the PC screen?

- Identify the average and maximum frame rates (in fps) supported by the product and call them out separately for each network type and speed (e.g., 14.4 Kbps modem, 112 Kbps, 384 Kbps, TCP/IP) and network type (e.g., serial lines, ISDN/switched 56, or LAN).

- Can the system capture a high-resolution still image? Can these images be stored for later use? What is the maximum number of images that can be stored and recalled? Describe the resolutions supported including the color-depth (8-bit, 24-bit, etc.). Describe, including both picture resolution information and line speed, the range of transmission times required to transmit a still image. Can three-dimensional (3-D) objects be displayed, or does the system support video graphics only? Does the product leverage existing PC hardware to enhance quality?

- Can the system accept input from audio or digital video sources to include pre-recorded clips or materials in a conference? Is time-based correction supported?

- Can conferees see each other and converse while collaborating on documents?

- What type of camera is included? Does it provide wide angle and close-up view? Where is it mounted on the device? Can it pan, tilt and zoom?

- What kind of microphone does the system include? How is it muted (manual, software-controlled, icon for activation and deactivation)? Can a telephone or speakerphone be used with the system? Does the system support a headset or earpiece? Does the system require a separate telephone line for audio? How (and how well) is audio synchronized with lip movement? How are audio echoes controlled?

- Which video signals does it accept (NTSC, PAL, SECAM, S-Video)?

- Can the hardware be used for multimedia applications? Explain how video and audio input can be captured. Are additional boards required?

Standards Compliance And Openness

Look for signs that a supplier has complied, and will continue to comply, with standards. Since standards are not static, review the supplier's history of implementing the *original* standard and examine its ongoing efforts to stay current with enhancements. Has the supplier played a role in shaping a given standard? Here are a few sample questions that relate to standards:

- Are proprietary algorithms the *real* basis for the product or do standards provide the product's foundation? Is the product in transition from a proprietary implementation to a standards-based one? (Note: If a supplier is in the conversion process, ask for contractual assurances that you will be able to migrate gracefully and economically).

- Does the product support the ITU-T's T.120 Transmission Protocols for Multimedia Data (audiographics) suite?

- Is the product compliant with the entire H.320 standards family? What audio algorithms (e.g., G.711, G.722, and G.728) are supported? Does the product offer full CIF resolution or just QCIF/SQCIF? Does it implement error correction framing or motion prediction encoding?

- Does the product operate over a non-guaranteed quality-of-service packet-switched network (e.g., Ethernet, Internet)? If so, does it conform to the entire ITU-T H.323 Recommendation (H.225, H.245, H.261, H.263, G.711, G.722, G.723, G.728, and G.729)? Does it also support IETF protocols (RTP, RTCP, and RSVP)?

- If the product offers still-image compression, does it compress and decompress using JPEG? If not, what does it use?

- If the product supports playback video, is it stored and decompressed in MPEG format? If not, what compression techniques are used?

- Does the manufacturer belong to audiographics or video communication consortiums (e.g., IMTC) or ad hoc groups that develop standards? If so, to which group(s) does the manufacturer belong and what is the history and extent of its involvement?

- Do employees of the manufacturer sit on standards bodies (e.g., ECMA, IEEE, ISO, North American ISDN Users' Forum)? Has the supplier successfully contributed its technology to the ITU-T for adoption?

Technical Considerations

- What hardware does the product require? Specify the platform (e.g., Pentium

4). Describe the number of card slots required; specify the card size (half or full) and identify whether card slots must be adjacent. Define the type of monitor required (SVGA, etc.) and what products (e.g., graphics accelerators, video or audio boards) are also required.

- Does the system include all cabling required for complete installation?

Ease-Of-Use

- What is the procedure for installing and configuring the system?

- Does the product offer speed-dialing menus and activity logs?

- Does the product offer annotation tools (e.g., colored markers, highlighters, notepads and symbols)?

- What type of documentation is shipped with the product? An installation guide? Context-sensitive on-line help and a tutorial?

- Does the product offer a consistent user interface across all products?

- Does the supplier offer a Web page with installation and application tips, listing of known-bugs, information on product releases?

- Does the supplier sponsor a user's group?

Cost

- How does product pricing compare to that of others in the same group? Why?

- How many users are covered by any software licenses that must be purchased?

CONCLUSION

By carefully defining objectives, characterizing the best possible scenario, and researching technology, one can ensure satisfaction in implementing a personal conferencing system. Because of a thorough and well-conceived plan, screen sharing, videoconferencing, applications sharing, and document collaboration can facilitate the accomplishment of worthwhile work.

CHAPTER 16 REFERENCES

DiCarlo, Lisa. "Vendors zoom in on inexpensive add-in card, chip," *PC Week* March 11, 1996.

Hendricks, Charles E. and Steer, Jonathan P. "Videoconferencing FAQ."

Rettinger, Leigh Anne. "DT5: Desktop Videoconferencing Product Survey," *Cirriculum 21 Succeed.*

Rettinger, Leigh Anne. "Desktop Videoconferencing: Technology and Use for Remote Seminar Delivery," (Directed by Dr. Thomas K. Miller III).

Schroeder, Erica. "Vendors sharpen desktop image," *PC Week* May 13, 1996.

Sullivan, Kristina B. "Videoconferencing arrives on the Internet," *PC Week* August 19, 1996.

17

INFORMATION SECURITY
AND VIDEO COMMUNICATIONS

*"The only secure computer is one that is turned
off, locked in a safe, and buried 20 feet down in
a secret location."*

Bruce Schneier

CRASHING A VIDEOCONFERENCE?

Video communications have always been vulnerable to eavesdropping. However, when the only people on the call were a few executives, and the only people who could eavesdrop were telephone company central office administrators with nothing better to do, the subject attracted little attention.

The purpose of this chapter is not to provide step-by-step instructions for protecting a video communications system and network. The pace of technological change and the disparity between environments make that impractical. Regulation as diverse as California SB-1386, the Sarbanes-Oxley Act, and the Basel II Accord have underscored that securing communications is critical from personal, organizational, and global levels. The purpose of this chapter is to foster appreciation for security as video communication adoption grows, and especially as that communication is migrated to the Internet and other shared packet-switched networks.

Intercepting communications is not new. One can trace such noteworthy events as Hitler's defeat and Mary, Queen of Scots' execution to intercepted communications. Nor is *Hacking* new. During the 1960s, the *blue box* became popular for stealing long-distance minutes. One way it worked was thus: a telephone hacker or *phreaker* would place a toll-free call from a pay telephone. While the number was ringing, he would press a button on the blue box to generate a 2600 Hz tone, and thereby disconnect the ringing at the far end and leave the phreaker connected to the long-distance network. Then pressing touchtone pad keys in a certain sequence would allow the phreaker to place long-distance calls at no cost. Then there was the *black box*, a device that phreakers would use to attach to electromechanical gear in a central office (long before digital) to deceive the telephone company into thinking that a long-distance call had not been answered. Yet another device, the *red box* emulates the tones that coins dropping into a pay telephone generate.

Fraud has been around about as long as people have, but it is more sophisticated now, and more opportunities exist now for commiting it. Video communications pervade a plethora of applications and communications media. Packet-switched interactive multimedia intruduces new liabilities. While packet-switched networks offer a single channel over which data, voice, and video may travel, they also create more entries for compromise. The Internet is much like the *party-lines* of old in that one can never know who else is *on the line*, or what they intend. It is a challenge to to know who actually requests, sends, or receives data. Consequently, as we converge video onto data networks, we expose it to a host of new security threats.

On any given day, headlines about compromised military agencies, educational institutions, or large corporations fill newspapers. While some of this is still the pastime of bored teens or *hacker-wannabes* who can use any one of the countless hacking programs that are freely distributed on the Internet as easily as they use an Internet browser. Given what these folks have done to deface web pages, one can only imagine what they might like to do to a live broadcast or videoconference. Will is hard to stop.

It is likely, however, that recreational hacking has long-since been outpaced by corporate and governmental espionage or pirating. Disgruntled employees, professional spies, and pirates are increasingly responsible for information security breaches. In the 2004 CSI-FBI Computer Crime & Security Survey, respondents reported more than $141 million in related losses. These losses came despite firms with $10 million or less in annual sales spending an average of $500 per employee to prevent such losses.

Malicious code is a very serious threat. Although the Y2K bug dominated the headlines, the most common software bug of the 1990s was a form of malicious code known as a *buffer overflow*, that hackers use to control targeted computers. In a buffer overflow, a perpetrator floods an address bar or some other field with more characters than it can accommodate, then activates the excess characters as executable code through means that firewalls may not be able to prevent. Demand for new software often results in sloppy computer programming and elimination of quality control steps that might have prevented such weaknesses.

So what does this have to do with video communication? If it is video over IP networks, especially the Internet, a lot! In a November 1999 posting on SecurityPortal, Ronald L. Mendell reported on U.S. Army special agents who were showing their commanding officers how to remotely turn user-installed microphones and cameras on Microsoft Windows NT Workstations into remote spying devices. Doing so allowed them to anonymously and remotely watch and listen to people who had no idea that they were being watched. This particular attack is initiated in the form of a Trojan horse (malicious software that carries executable code) concealed in an e-mail attachment. Opening the attachment installed a remote administration tool and ordered the Trojan horse to turn on the video and/or audio of the targeted machine. Victims of this type of comprimise may be oblivious of eavesdropping because many user-installed cameras and microphones lack much if any indication that they are in use. In short, videoconferences that occur across IP networks are vulnerable to information security breaches.

In this section, we will ouline the basics of an information security posture. However, we will neither provide an exhaustive text nor a detailed technical explanation of how the related technologies work. We will provide some basic steps for developing information security strategies.

INFORMATION SECURITY STRATEGY

Firewalling and encrypting are critical, but are not in themselves an information security strategy. Such a strategy is built on a foundation of policies, guidelines, and procedures. An information security strategy can *mitigate* risk, but it cannot eliminate risk. Every organization must determine its own risk tolerances for each of its information assets. It needs to know to what extents it will go to protect those assets. An information security strategy should address three goals: confidentiality, integrity, and availability. In most cases, outside assistance is critical to establishing a strong security strategy, and to verifying the strategy is as effective in practice as it is in theory. As one works with a Certified Public Accountant (CPA) for financial accounting, one should work with a Certified Information Systems Security Professional (CISSP) to develop information security strategy.

To establish specific information security policies and guidelines, executive management must establish broad policies and standards regarding information assets, including network use. These constitute the foundation of an information security strategy. Nested under the security policies are procedures and policies. In this section, we will highlight areas in which the organization should focus.

Security Management Practices

Every organization should have a high-level security policy. We define an information security policy as an *explicit set of prescribed practices that fosters protection of the enterprise network and of information assets*. That policy should state that all information assets are the organization's property, and should define criteria upon which the organization classifies those assets.

A key aspect of security management practices is that of privileges and responsibilities. The basic policy regarding privileges is that no-one should have any more privileges than necessary to do their jobs. This minimum access policy fosters diligent division of responsibilities. Security policies should provide a methodology for identifying who is responsible for each information asset and the scope of each information asset owner's responsibilities. That information asset owner is responsible for establishing and reviewing the asset's classification, confirming that adequate security controls are in place, establishing and confirming appropriateness of access rights, determining the asset value, and mandating backup requirements.

Security Management Practices are the structure by which an organization determines its guidelines and operating procedures. The general security policy is the starting point for the various security policies that regulate file storage, data encryption, change management, backup and recovery, separation of responsibilities, access, auditing and testing, virus checking, and ongoing management. Top officials should demonstrate the importance of information security, and support efforts to formalize security management practices. Those practices are the foundation upon which the organization will build its information security strategy.

Law, Ethics, and Investigation

As important as any of an organization's assets is its ethical conduct. Ethics compel organizations to treat employees equally. If an organization accuses an employee of improper conduct, it must demonstrate that it is not singling out that employee. It must demonstrate that any monitoring of employees is a part of well-known and consistently-practiced policies and operating procedures.

If an organization believes anyone to be guilty of a crime, it must follow procedures regarding search and seizure, hearsay evidence (computer-generated evidence is often considered hearsay evidence), and chain of custody. Moreover, an organization should specify procedures for returning confiscated assets.

Risk Assessment

The basic business case for video communications is to support of the organization's business model. To be clear, video communications support the business model, not vice versa. The purpose of information security is, in turn, to support information technology applications such as video communication. Security cannot be so onerous that it impractically impedes business. Moreover, it cannot drain any more resources than necessary from investments that lead to revenue-generation. That said, video communications (especially Internet-oriented ones) introduce risks. Since the only alternative to taking risks is quitting business, one must formulate policy by classifying information assets and actions by value and by risk.

Risk is fundamentally a product of asset value, asset vulnerability (how easy it is to compromise it), and the threats that might exploit those vulnerabilities. Threats and vulnerabilities must be viewed in terms of confidentiality, integrity, and availability. In a videoconference in which doctors are discussing a patient's medical history, confidentiality may be the top priority. A buy or sell order from a stockbroker might score highest in integrity (note that the three are not mutually exclusive; they just change priority). A videoconference with a potential employee who presently works for a competitor may require more confidentiality than a weekly company-wide report from the company president. The Information Security Officer is responsible to work with data owners to establish values and to identify risks (including those that result from regulatory obligations).

With information asset classifications established, the ISO may prepare a high-level business impact analysis. This analysis should quantify the cost to the organization of the loss of each information asset. It should also compare and contrast how that loss relates to risks in confidentiality, integrity, and availability. However, determining the organization's plans for protecting its assets requires a guidelines and corporate information security policy that should include specific information security policies.

Physical Security

Physical security may be the least glamorous aspect of information security. However, it is too common to discover that a perpetrator could bypass an expensive firewall simply by walking in and out of the facility disguised as a building services attendant or by jacking into an unprotected Ethernet socket in a lobby. It is often easier to steal a computer than it is to hack into the system from the Internet. Polls blame insider threats for 70–90% of information security breaches.

A simple physical security survey begins by observing whether office doors, desk drawers and file cabinets are locked, workstations are shut down or locked, and that sensitive papers and media are secured. Stories abound of competitors viewing an organization's strategy through a conference room window. Moreover, it is simpler to steal a computer from a telecommuter's home than to break into corporate headquarters. Worse yet, a trusted friend may simply log on to the corporate network to obtain confidential information such as an initial public stock offering.

The primary goal of physical security is the safety of people. A physical security strategy begins with a business impact analysis that lists threats such as flood, loss of power, or earthquake, and rates each in terms of human impact, property impact, business impact, probability, and resources available in the case of such an event.

Historically, locked doors and managed keys have regulated physical access. Today, organizations require picture ID and electronic badges. Access to information assets often requires a personal ID number and possession of a one-time password-generator. Biometrics devices such as fingerprint scanners, retinal scanners, and hand geometry scanners are also becoming increasingly common.

Organizations should examine policies regarding management of mobile devices, such as wireless-enabled PDAs, which are now full-featured computers and video communications devices. Because it is better to design quality into a system than to retrofit it, organizations should think ahead. It is also important to review processes to ensure that terminated employees cannot exploit newly-unauthorized access.

Business Continuity and Disaster Recovery

As organizations increase reliance upon interactive multimedia, business continuity and disaster recovery become crucial components of an information security strategy. Without remote sites, there is little need for video communication. The information security officer will always be challenged to keep up with both physical and information asset growth at numerous premises. For this reason, the organization should establish standards in acquisition, configuration, and management. If the organization determines what hardware components will comprise its various offices, how those assets will be configured, and how those assets and configurations will be managed, it will simplify the process, ensure uptime, and save money.

The basis of the business continuity and disaster recovery plan is the risk assessment and the security policies. As a Project Manager finds critical paths by understanding task dependencies, the ISO can establish a business continuity and disaster recovery plan by establishing intradepartmental and interdepartmental dependencies. Steps to protect people include ensuring adequate emergency exits, emergency lighting, alarm systems, and fire suppression mechanisms. People should know where to meet to receive further instructions in an emergency evacuation. Bolting hardware to walls and floors may protect people from falling machinery in an event such as an earthquake or a hurricane.

Hardware protection steps include implementation of redundant power (ensure that the diverse-routed power does not follow the same path as the primary power), maintenance of spares, and installation of uninterruptible power supplies (UPS). UPS manufacturers such as American Power Conversion make it easy to determine the proper units for the hardware that one needs to protect. A protection strategy should include both on-site and off-site data archives. Network connectivity is generally achieved through the incorporation of redundant network connections (again, these should follow diverse paths), and hardware that is programmed to failover to a secondary path in the event of a circuit failure. For Internet connectivity, it may even be a good idea to incorporate a secondary service provider.

Again, the highest priority in a disaster is the protection of people. A good business continuity and disaster recovery strategy will ensure the protection of people as well as information assets and network assets.

User Education

Humans are the weakest link in security, but they are also the reason a communications infrastructure exists. Many of the aforementioned 70–90% of breaches that occur from the inside result from unintended consequences of poor operator practices. Users need to know the importance of such basics as regularly changing passwords, of choosing strong passwords, and of protecting passwords.

Organizations should inform users of policies that prevent loading of unauthorized software, whether unlicensed software or *freeware*. Organizations should make users aware of basic forms of attack so that they can recognize them and act accordingly (the acuity to unplug a network jack can save a network). Moreover, organizations should inform users of how to respond when they witness suspicious activity (e.g., a suspicious Email attachment or a program that acts different from usual). Code Red and Melissa (both of which are classified as *worms*) are two example of *malicious code* that began ravaging computer systems in 1999. Melissa affected more than 100,000 computers in a matter of a few days; Code Red took minutes. Organizations could have limited or prevented damage from either through good user education.

Organizations should regularly remind users of policies and guidelines through announcements from top executives because user awareness can be the best defense.

Cryptography and Network Security

Cryptography is getting easier to implement in interactive multimedia applications even though both encryption and decryption introduce latency. The H.323 standards family specifies security and encryption with the H.235 standard.

IPSec (IP Security) is a framework of open standards that eases interoperable establishment of secure private communications over networks. IPSec does this by establishing a virtual private network (VPN). A VPN session is one in which packets are encrypted before leaving the sender's system in such a way as to make them illegible until the desired recipient unencrypts them.

IPSec is a security standard that operates at the network or packet-processing layer, rather than at the application layer as does approaches such as SSL (Secure Sockets Layer). IPSec is a standard for implementing virtual private networks between networks or between users and networks. IPSec requires installation of a client, but no other significant changes to an individual user's computers.

IPSec defines Transport Mode and Tunnel Mode. Transport Mode protects only the (TCP, UDP, or ICMP) payload. Tunnel Mode Protects both the payload and the IP header. The elements of IPSec include the Authentication Header (AH), for source authentication, and Encapsulating Security Payload (ESP), for both source authentication and data encryption. Associated information is placed into the packet in a header that immediately follows the IP packet header. IPSec, in itself, does not provide security association, key distribution, or key management. It orchestrates multiple other protocols such as Internet Key Exchange (IKE, RFC 2409) which, in turn, relies upon ISAKMP for security association and key management and Oakley for key exchange (Oakley is based upon Diffie-Hellman key agreement protocol). IPSec can also be used in conjunction with a variety of encryption algorithms, including AES. Intent to standardize upon a cipher that will remain secure for the foreseeable future, NIST selected the Advanced Encryption Standard (based on the Rijndael algorithm, invented by Joan Daemen and Vincent Rijmen) as the successor to DES (Digital Encryption Standard) in 2001. A significant improvement upon DES' 56-bit key support, AES supports key sizes of 128 bits, 192 bits, and 256 bits.

IPSec enables secure packet transmission over shared networks. Although broadband access is making the Internet more viable for video communication, VPN encryption and decryption latency can be an impediment. However, it is clear that, as processors become faster, codecs become more sophisticated, and QoS techniques become more reliable, VPN based video communication will become common.

SECURITY: WHERE TO START

In its 2004 Global Information Security Survey, CIO Magazine, CSO Magazine, and PriceWaterhouse Coopers found that organizations that followed a best-practices approach to information security suffered more security incidents, but experienced less downtime and fewer financial losses. The study found that group to spend more on information security as a percentage of its IT budget, and to measure or review its own effectiveness more. It also found it to be more likely to integrate information security with physical security, to have more employees who are likely to comply with security policies, and to provide more reinforcement for information security efforts from top-level executives.

CIO Magazine interpreted the survey with a short-list of six practices that make the best-practices group more successful than the rest. First, it found the group to spend 14% of IT budgets on information security, in contrast with the overall average of 11%. Second, it found the group to have group information security with physical security rather than with IT. Third, it found the group to be 60% more likely to be current in software patching and more likely to have conducted a thorough threat analysis. Fourth, it found the group to be 50% more likely to have classified threats and vulnerabilities in the context of a risk assessment. Fifth, it found that more than 65% of the group had used the three previous steps to define overall security architecture. Finally, it found the group to have established processes to conduct quarterly reviews.

Security should be engineered in, not bolted on as an afterthought. Therefore, when building a video communications infrastructure, it is important to know what security measures the manufacturer has built in, how difficult it may be to deploy them, and what conflicts and trade-offs may result. Do not accept the marketing shtick, talk to competitors and integrators and do your reading to learn the whole story... or, as we may say, see *the whole picture*.

CONCLUSION

There is no practical way to prevent security incidents, only to mitigate the risk. Moreover, it is possible (although increasingly unlikely) to take no precautions and be lucky. However, regulations and litigation can make getting compromised and being found to have not taken diligent precautions expensive.

In its whitepaper, *Security for Videoconferencing*, Wainhouse Research warns, "PC-based videoconferencing systems are not considered ideal for secure videoconferencing applications." It continues, "Appliance based videoconferencing systems are better suited than PC-based systems for secure videoconferencing requirements." While such an assertion may be a helpful consideration, it is a typical example of a simplistic security recommendation. Security, like quality, is in the eye of the beholder. Security must be defined in terms of systems, requirements, risk, and risk tolerance.

An information security posture is critical in protecting the confidentiality, integrity, and availability of an organization's information assets. With the migration of video communications to data networks, comes vulnerability to unauthorized data collection, storage, access, and processing. While eavesdropping is nothing new, the Internet has fostered more sophisticated and widespread means. An information security strategy should begin with a broad security policy and a risk assessment. All users should understand the importance of security, and executive managers should emphasize the message. Once security policies, procedures, and guidelines are in place, however, the job is not done. Enforcement of the information security strategy requires periodic auditing and on-going auditing and monitoring.

CHAPTER 17 REFERENCES

Gordon, Lawrence. "2004 CSI-FBI Comp Crime & Sec Survey. *Computer Security Institute*. 2004. www.gocsi.com

"Security for Videoconferencing." *Wainhouse Research*. 2004.

"What is the AES?" *RSA Security*. January 2003.
http://www.rsasecurity.com/rsalabs/node.asp?id=2235

Krause, Micki and Tipton, Harold F. *Handbook of Information Security Management*, CRC Press 1999.

Bernstein, Terry, Bhimani, Anish B., Schultz, Eugene, Siegel, Carol A. *Internet Security for Business*, John Wiley 1996.

Pfleeger, Charles P. *Security in Computing*, Prentice Hall 1997.

Maitra, Amit K. *Building a Corporate Internet Strategy*, Van Nostrand Reinhold 1996.

Sullivan, Jennifer. "MP3 Pirate Gets Probation," *Wired Magazine*, November 24, 1999.
http://www.wired.com/news/

Paller, Alan. "Security's Vicious Cycle." *Computerworld*, November 1999.
http://www.computerworld.com

Sing, Simon. *The Code Book*, First Anchor, 2000.

Berinato, Scott. "The Six Secrets of Highly Secure Organizations." *CIO Magazine*, September, 15 2004. http://www.cio.com/

PART FIVE

THE FUTURE OF VIDEO COMMUNICATION

18

THE FUTURE OF VIDEO COMMUNICATIONS

*"We are looking at the dawning of the age of
visual communications."*

Dr. Norm Gaut

IT'S ABOUT TIME

Interactive video communication has overcome formidable barriers to become the practical and convenient solution it is today. Unfortunately, even though the technology is readily available, affordable and reliable, many of us have no concept of how to integarate it into our daily lives. This is consistent in the history of modern communications. Telephones did not become ubiquitous overnight; early carriers assumed substantial risk when they stretched telephone lines between cities with no guarantee that a person would pay a monthly fee for a telephone. Why should they? One could call from the post office, the corner store, or the local gasoline station any time he wanted... and run across other people at the same time. Fortunately for the carriers, people increasingly became accustomed to the telephone's ease and convenience, which meant more people became interested in using telephones more often and were willing to pay for the convenience of doing so in their own homes. Similarly, after significant resistance in the early stages, mobile phones have grown such that a home is likely to host at least two mobile phones for each wired phone. Email and chat have followed similar patterns.

Gordon E. Moore, a co-founder of Intel, made several observations in 1965 that formed the basis of his now famous *law*. That law effectively states that, at the semiconductor industry's rate of technological development and advancement, the complexity of integrated circuits doubled every eighteen months. A decade later, Moore revised his estimate and proclaimed that it was slowing to about two years. Decades later, it would appear that, from a cutting-edge, *high-tech* perspective, Mr. Moore is on-target. Today's PCs, Tablet Computers, PDAs and newly introduced Smartphones are exponentially more powerful, and packed with more features than their antiquated predecessors of only two years. Computing platforms become smaller, faster, and less expensive with each passing year. It is insightful to consider how Moore's Law might apply to software and application development.

Software development tools and techniques yield vastly more powerful products with each passing year. As a result, videoconferencing codecs (e.g., H.264) have improved significantly, and network architectures (e.g., SIP) have matured and are being widely deployed to support a host of premium network applications such as IM, VoIP, Push-To-Talk, and IP Video. Software enhancements alone offer the same video conference quality at 256 Kbps today that previoiusly required 384 Kbps. Increased understanding of security structures and encryption algorithms (e.g., AES) make Internet communications more secure and less bandwidth intensive. Additionally, whatever role Open Source software (e.g., BSD and Linux) takes in the long-run (we think it will be a significant one), it is providing needed competition and, therefore, incentive for incumbent OS and software vendors to provide better solutions to a more savvy and more demanding public.

It is also insightful to consider how Moore's Law might apply to connectivity advancements. In all but the most rural of areas, the Internet is no longer a novelty, but something nearly as common in homes as water and power. The question most homeowners in America face is not *if* one can get broadband, but *which* connectivity choice one prefers. Broadband connectivity and underlying infrastructures have amplified network consumer bandwidth capacity 25–75 times (an amazing 2500%–7500%) from forty Kbps to three Mbps in only five years! This explosion in capacity has fueled the development and adoption of bandwidth-consuming applications (e.g., Streaming Video, Video Mail, IM, MMS) and is fueling consumers' and business' demand for access to data and services anywhere, at any time, and in any way they choose. This demand is driving the need to offer always-on connections and mobile Internet solutions while at the same time fostering a preference for information and applications to be NOT installed in a single computer or server, but *on the network* as a service that can be globally—and securely—accessed from any device, at any time, from anywhere on the Internet.

Unlike the 1964 AT&T Picturephone, which demanded more than the networks and infrastructures of the time could deliver, today's communications infrastructure and computing have evolved with acceptance of their use. More importantly, we as a society have evolved to embrace new technology as never before. The world into which AT&T introduced PicturePhone marvelled at basic telephones and televisions; it's population could not comprehend how computers might fit into the world, other than to steal jobs and monitor behavior in an Orwellian society. Now, only forty years later, PDA-phones and personal computers are essential tools for executives, couriers, mechanics, cashiers, and countless others (and, yes, they are streamlining jobs and monitoring people as well!).

Worldwide, people leverage the Internet for messaging, research, entertainment, and commerce, and allow for *human* interaction between people in practically any city in the world. Personal computers and handheld communications devices are increasingly the physical links to this dynamic, evolving global communications system. Video communications tools are on our desks and in our bags, belts, and pockets. We simply await the right environment, the right networks, and the right carrier data service plan to make it practical to exploit these under-used resources.

Today we would view our great-grandparents' predisposition to send a telegram with a condescending eye; so will the next generation view its parents' pre-disposition with voice-only communication over *wired telephones*. Video communication has lived up to its promise to save time and money, but has not yet lived up to its potential to change behavior. However, the change *is* happening. Carriers and mobile phone makers have already made the camera phone a staple for a younger generation in the US and throughout the world. In Japan, where Mobile Video has been available to FOMA network users since 2004, millions of subscribers use video communications phones or simply *mobile videophones* every day. Moreover, 3G network upgrades that began in the US in 2004 will drive an explosion of personal and business mobile video use across the nation.

As exciting as it is, mobile video is not the only imminent application. In another development, the MPEG-7 Working Group intends to develop infrastructure and applications to leverage the same voice-recognition systems we use now to make calls such that it can perform data searches on virtually any imaginable criteria. Simply humming a tune to identify a recording that you would like to download, whether as lyrics, sheet music, or movies will yield results. Should we desire an image from about halfway through a given movie, we might simply describe it so our videophone can project it on our living room wall. Is this less likely today than AT&T's Picturephone was in 1964?

Although video communication is not today a high-dollar, high-end hardware and custom software enterprise solution, countless product vendors and service providers are investing in its enormous opportunity. The potential to sell consumer hardware, software, and services by enabling every PC, PDA, and phone for video communication is now driving the market forward at an accelerated pace. On-demand video, music, software and interactive video communication is causing the displacement of dial-up Internet connections with high-speed always-on services, and the enormous disruptive power of the Internet has undermined brick and mortar companies that physically distribute goods such as videos, music, and software. It is in response to growing competition from NetFlix™ that Blockbuster Video™ keeps changing its pricing model and introduced its own on-line service. With the firm entrenchment of the Internet, and the growth of broadband and PDAs, carriers must provide robust networks that support applications and offer unfettered and abundant access to services. Through the balance of this chapter, we will consider some technologies and the conditions that will transform video and interactive multimedia communications from being a *viable* tool today into being an *essential* tool tomorrow. We will begin with a look at the changes going on in the area of computer operating systems.

THE OPERATING SYSTEM

The Web has become the business platform-of-choice and Web-based architectures continue to make operating systems less significant. The growing popularity of Java, Linux, and even Symbian OS and Palm OS is evidence that the tides have changed and OS decisions are now much more likely to be considered on their merits—including features, reliability, supportability, security, and scalability—not market share.

Carriers as a whole have shown little more than frigid indifference to Microsoft's influence and have been among the first and loudest to decry Windows OS' well-documented lack of stability, flexibility, and security. Carriers are increasingly meeting their needs with Open Source solutions_ free and customizable OS's with flexible solutions for the demands of highly-available core networks. Intel, Novell, and IBM have all invested heavily in Linux and Linux-powered core solutions and in Java-based client solutions, leveraging the strength and cost-effectiveness of Linux in the core and Java's ubiquity and flexibility on diverse and rapidly evolving clients.

In many ways, Java is the programming language for distributed network applications that leverage Internet connectivity.

THE NETWORK

The Internet

The Internet is the preferred application network for allowing mobile devices to access rich multimedia content and interactive services such as video conferencing, IM and other presence-based services. In 1999, VON (Voice On the interNet) advocate Jeff Pulver accurately asserted, "It will be the technologies surrounding *instant messaging* and *presence management* that will be the true *killer app* for service providers who have invested in IP based communication networks." IP based technologies, applications and services have fostered new generations of mobile, handheld devices because they provide unique solutions that the public demands. One can already drag-and-drop a voice or video message into a task list on a PDA, and voice-to-text conversion is simply a matter of accessing a server-based unified messaging (UM) application. Instant messaging, unified messaging, and presence management will increasingly justify the investments that service providers make in IP networks. Carrier trunking switches are already relying upon SIP (session initiation protocol), which is quickly displacing H.323. IP networks offer integrated SS7 functionality, and allow carriers to supplant circuit-switch networks with MPLS-based, QoS rich, application-integrated IP networks.

Wireless Carrier Networks Evolve

In the past, the term *Wireless Carrier Network* might have simply represented a provider of cellular-based network telephony services (and limited data options). Today, data is the market driver, with most carriers now offer a host of *high-speed* data and even

326

data-only options with their ever-less expensive voice-only options. To maintain users and drive higher revenues, many traditional cellular carriers now offer connectivity and services over WiFi networks, and new alliances between conventional wireless carriers and wired broadband service providers will create a host of new and interesting *carrier-based* services.

A recent development in the evolution of carrier networks is the emergence of the Managed Virtual Network Operator (MVNO), a company that does not own licensed spectrum but resells wireless services under its own brand name using another company's network. MVNOs began in Europe, but are beginning to emerge globally. Several operate in the US and several more (notably AT&T Wireless, which, when it sold its own assets to Cingular, began reselling Sprint's network) have been announced. An example of an MVNO is Virgin Mobile, a *carrier* that purchases spectrum from Sprint in the US and other carriers across the globe to offer wireless e-commerce and communication services. MVNO *carriers* are in a unique position to offer demand-based services that emerging wireless standards such as WiMAX and VoWLAN enable, while conventional carriers prepare consumers for new cellular-based services with such technologies as WCMA, UMTS, and HSDPA.

Global sales of 3G services worldwide are projected to exceed $9 billion in 2005, and some estimates indicate that by 2008 there will two billion mobile users worldwide with 3G networks carrying traffic from half of them. Suffice it to say, 3G services are coming to America in a variety of forms and are already here in limited locales. However slowly, the networks we use will become increasingly more flexible, increasingly more available, more data–oriented and more wireless with each passing year.

CONSUMER ELECTRONICS

There is abundant evidence that the unparalleled rise and phenomenal success of cellular networks has been consumer appeal. Although cell networks are technology marvels and represent tens of billions of dollars of infrastructure investment, it is ultimately the appeal of a *cool* consumer gadget phone that motivates consumers to fuel the market by replacing old phones and by replacing old service plans with more suitable (and more expensive) ones.

Learning from the example of cellular networks, other technologies with obvious business applications are finding their warmest reception in the hands of gadget-savvy consumers. For example, Bluetooth is a Personal Area Network (PAN) technology with numerous business applications for data transfer and device synchronization, but the engine driving Bluetooth's development is the wireless earpiece one can purchase from any local electronics store. In another example, the various developments on the 802.x wireless Ethernet (what we will generically refer to as *WiFi*) realized its greatest success when Starbucks adopted it in 2001 to provide patrons with another reason to pay top dollar for coffee. In doing so, Starbucks became one of the largest national wireless broadband providers (might that make Starbucks an MVNO?). Similarly, other technologies are poised to take a significant role in the future of multimedia communications and, not surprisingly, most of these target the consumer space.

Wearable Networks

It has begun. Motorola announced, early in 2005, that it was designing embedded communications solutions into high-end sports gear—specifically snowboarder gear. Moreover, the whole solution is networked using Bluetooth radio technology. The idea of wearable computers is not new; it is just timely given the state of technology development. Although wearable computers were conceptualized years ago, the idea of networked components is an intelligent and workable solution that will yield benefits to countless consumers and businesses through the introduction of wearable communications products in 2005 and beyond.

Handheld Communications Devices Get "Pumped Up"

While the mobile telephone replaces wired telephones for many, smartphones and wireless PDA / phones (converged PDA Phones, or CPP) are displacing stationary PCs for others. Since the 1999 appearance of the PalmVII with mobile Internet access, the internet-connected CPP market has exploded to include Nokia, Motorola, Palm, Samsung, Research In Motion (RIM, a.k.a. the *Blackberry*) and too many more to list. Many manufacturers offer unique and innovative concept phones with a variety of form factors; some slide, some swivel, others flip... some flip and swivel! Since manufacturers initially launched camera integration into phones for still photos, early phone cameras pointed away from the user. This poses a problem for videoconferencing, in which the user may want to transmit his likeness while viewing the likenesses of others on the conference. Today there are phones with single or dual cameras, cameras on thumbwheels, video cameras and phones that are mostly high-end cameras with phone functionality added. Additional handheld device variety comes from the various innovative options for data input. Common offerings include full QWERTY keyboards (ironically, the enduring QWERTY keyboard was designed to not jam the hammers of a typewriter) and thumb-boards for a growing number of phones. The variety of these form factors is due to the unique ways in which different people work, play, and process information. It is also a testament to the fact that there is no single *best* form factor.

This innovation and variety will continue and even accelerate with the introduction of stronger and more energy efficient processors, better memory capacity and enhanced audio and video capabilities. Overall, screens will get larger and clearer, and camera resolution will increases to five Megapixels or better on camera-enabled smartphones. Audio capabilities will continue to improve and take advantage of PAN connections to *broadcast* to wireless stereo headsets and nearby wireless speakers.

Storage capacity will explode with the inclusion of mini-hard disk drives (HDDs) in high-end media phones, and removable media options will offer inexpensive multi-gigabit options for movies, pictures, MPEGs and just about everything else.

Software improvements will leverage the improved hardware and offer on-board data manipulation that in time, will parallel that of PCs. By 2008, voice recognition software will improve to the point of obviating the need for manual input in most cases (it could happen even sooner). Moreover, handheld devices will offer more multi-modal (simultaneous multi-band cellular + WiFi + Bluetooth) support and the capability to dynamically select the optimum network based on the user's preferences. Finally, enhanced security and *near-field* radio technologies will enable *point-of-sale* (POS) transactions to take place via compatible phones. The same technology will also enable keyless entry access to automobiles, offices, and homes.

Figure 18-1. Direct Media Broadcast TV Phone. Samsung SCH-B100.

MOBILE VIDEO

Mobile industry experts suggest that mobile video will take off in 2006, and that a quarter-billion people will generate more than $5 billion USD annually on mobile video services by 2008. Mobile video usage can mean anything from brief camera-phone video clips shared among friends to professionally produced commercial products such as movie trailers, music videos, sports highlights, and streaming video. Some examples include:

- Chat and personal communications including dating and community activities

- Mobile Marketing and location- or presence-based content and commercials

- News, special alerts, weather—all customized per user profiles

- Entertainment content, including movies and premium pay-per-view

- Regular TV clips, soap opera excerpts

- Music videos, including previews, and "special" features

- Video karaoke—you can "sing with the stars"

- Education and training, including how-to's and live demo's.

329

MOBILE TV—A BIG WINNER

Of course, no discussion of mobile multimedia content would be complete without *The Big One*, Mobile TV.

Globally, innovative technologies have been delivering Mobile TV for some time. These include the Korean Cell Broadcasting Service (CBS), which broadcasts news, entertainment, movie clips, and music to users within the cell broadcasting range. A separate effort sponsored by the Japanese and Korean governments has resulted in an agreement to reserve radio spectrum for Digital Multimedia Broadcasting (DMB), which will broadcast satellite signals directly to mobile handsets and, not surprisingly, Korean manufacturer Samsung released a phone that can receive satellite TV signals. Another option that is gaining momentum as a global standard is Digital Video Broadcast Handheld (DVB-H) which combines conventional television broadcast standards with elements specific to handheld devices. The World's first Direct Media Broadcast TV phone, the Samsung SCH-B100 with CDMA2000 1x EV-DO picks up TV signals directly from a geostationary satellite shared by Korea and Japan. It plays and records DMB-TV and Video on Demand (VoD). It features an MP3 player, Windows Media player, FM radio and an enormous QVGA TFT screen. Its 2.2 megapixel CCD camera offers 3x optical zoom, a glass lens, autofocus and a mechanical shutter. It also offers Bluetooth networking.

Countless hardware-based alternatives include Samsung and Vodaphone cell phones with integrated TV tuners that can pick up local analog television channels over the broadcast airwaves. In the US, cellular carriers Sprint and Cingular offer "MobiTV," a service that is enabled when a user downloads a special Java application to a service-compatible phone. This mobile television service offers real-time programs including MSNBC, CNBC, Discovery Channel and The Learning Channel among others. Programming on most MobiTV channels is identical to that on corresponding cable channels, although with a slight lag for the time it takes to process and transmit the content for a wireless device. Verizon's offers its Mobile TV service using a solution known as VCAST that uses Microsoft's Windows Media platform to power its new high-bandwidth video and audio streaming service. The client side of the solutions is enabled thru PacketVideo's media player with integrated Windows Media playback.

From a global perspective, Nokia has announced unwavering support for *IP Datacasting*, which leverages DVB-H as the path to Mobile TV. IP Datacasting delivers all content as standard IP data packets, to offer mobile users an experience that includes a combination of broadcast and multimedia content delivered to their mobile handset. This end-to-end solution is demonstrated with IP Datacasting network elements and Nokia's initial TV-optimized multimedia device, the 7710, an innovative handset that is designed specifically for TV.

Whereas wireless device manufacturers are designing and selling smarter and more feature-enabled mobile devices with access to (non-cellular) satellite and analog TV, wireless carriers will continue creating and offering enhanced services such as Mobile TV to fuel greater usage of data services over their increasingly robust and expensive cellular networks. This *cellular-service* model seems to have solid support from early deployments in Asia, where more than one million subscribers added cellular-based mobile TV service in the first month it became available.

There are a few obstacles that could disrupt Mobile TV; heading the list is concern over Digital Rights Management (DRM). Additionally, questions exist about consumers' willingness to pay for Mobile TV as well as technical and standardization issues, in-building coverage, battery life, and power consumption issues. Despite these hindrances, the Mobile TV market will explode in the years ahead. Broadcasters are becoming more interested in Mobile TV as it becomes profitable for them and they will begin to offer programming (and marketing campaigns) specifically aimed at the mobile market. Advertising dollars and the interactive nature of mobile communications enabled by SIP's presence- and location-based technologies will rapidly drive demand and rate of consumer adoption to an un-rivaled level. Expect Mobile TV's widespread consumer acceptance to take place commensurate with the widespread availability of 3G networks in the US in the latter part 2005 and 2006 and only to grow from there.

CONCLUSION

We can no more predict what our networks will look like in the year 2040 than could the designers of AT&T's Picturephone have predicted the current generation of handheld communications appliances. However, we do know that our network appliances will become more powerful, more portable, and more reliable, and that our networks will become more robust. To quote William James, "Change is the only absolute."

In the final analysis, technology only matters to the extent that it improves the quality of people's lives and breaks down barriers to worthwhile communication. We believe that today's technology offers an unprecedented opportunity to do just that.

We commend those noblest of aspirations that have led the most unlikely collaborators to find the necessary common ground to accomplish extraordinary achievements in improving our world, and thereby to improve people's lives. We offer our small contribution, this text, with hope that people will put those achievements to good use.

CHAPTER 18 REFERENCES

Wikipedia, English Version. http://en.wikipedia.org/wiki/, January 2004.

Bottger, Chris . "The Collaboration Revolution; Successful Meetings." Oct 2004 v53 i11 p32(1).

Pulver, Jeff. "The Pulver Report" October 21, 1999. http://www.pulver.com.

Kerner, Sean Michael. "Converged Devices Lead Mobile Demand." November 1, 2004 www.clickz.com/stats/sectors/hardware/article.php/3429541.

"Vendors Vary Wildly on Real Costs for VOIP Rollouts; Meanwhile, Customers Grow Weary of Carriers' Convergence Inertia" *PR Newswire Association*, Nov. 16, 2004.

Gold, Elliot. "Teleconferencing industry to hit $3.7 billion in revenues for 2004, growing to$4.0 billion in 2005: Market continues to be dominated by conference calls, with Webconferencing experiencing strongest annual growth" *PR Newswire Association*, September 9, 2004.

"Video Conferencing Finds New Popularity After A Few Years In The Doldrums" Rethink Research Associates, April 2004 p14(2).

Garfield, Larry. "Report expects big things for mobile video" *Infosync World*, www.infosyncworld.com/news/n/4978.html, 27 May 2004.

APPENDICES

A

INTERNATIONAL STANDARDS BODIES

Standards are published documents that describe, in useful detail, a technical specification. Competition, customer demand, and the convenience of developing interfaces that can be extended across multiple environments, are the driving forces that propel the adoption of standards.

Standards have value because they give governments and organizations the ability to reach beyond local implementations and conventions to construct unifying systems. The more widely-published a standard, the better its chances to make an impact on a technology or market. Wide publication is not enough, however. The organization that sponsors the publication of the standard, and its stature within a given 'community', is also of critical importance.

Standards can be established by law, formal recommendation, convention, or custom and are of two types; de jure and de facto. As the name denotes, a de jure standard is legitimized by a "jury" that assembles to develop it. A nationally- or internationally-recognized assembly, sanctioned by a government or governments, gathers in a series of formal meetings with the intention of developing a specification. These specifications become treaties between national governments that operate telecommunications networks.

De jure standards are not actually *laws* but, rather, technical *covenants*. Issued as *recommendations*, they offer a prescribed approach to connecting disparate systems and management domains. The word recommendation was chosen in order to stress the sovereign right of countries to deviate from a standard. It is usually expected that countries that do will make their reasons known in the form of a publication—but compliance is strictly voluntary.

De facto standards are more accurately described as conventions that are well established in practice, and made legitimate by virtue of their wide acceptance. De facto standards are 'created' by industry associations or consortia, representing key suppliers and, possibly, customers. These interested parties convene to resolve compatibility issues and later publish their results.

THE INTERNATIONAL TELECOMMUNICATIONS UNION

The ITU operates under the auspices of the United Nations. Headquartered in Geneva, Switzerland it is a de jure treaty organization. Its membership consists of "administrations" (where each is a country). Nearly every nation in the world is represented in the ITU. National delegates usually come from the largest telecommunication service provider in a member country. Each ITU administration has one vote.

The State Department represents the U.S. in the ITU. Elsewhere in the world, where telecommunications networks are not yet privately owned, government-owned Post Telephone and Telegraph entities (PTTs) provide members. Other organizations (non-government-owned carriers, manufacturers, scientific organizations, etc.) can participate, but only the government administration can cast a vote in an ITU plenary session.

The "one country-one vote" approach is meant to achieve harmony. The ITU's goal is not to create telecommunications standards but to encourage international cooperation in order to arrive at an efficient international telecommunications infrastructure. It operates in six languages and must have consensus before it can ratify a recommendation. This used to take years (four to be precise) but the ITU has accelerated its standardization work. It can now approve a standard in 15 months. This quickening of pace resulted in the approval of over 900 specifications during the period between 1993 and 1996. These are published in at least three languages (French, German, and English).

Within the ITU, two organizations promote recommendations that relate to multimedia and videoconferencing. These are the ITU's Telecommunications Standardization Sector and its Radiocommunications Sector. Until 1993, the ITU-T was known as the CCITT (Comité Consultatif International Téléphonique et Télégraphique) and the ITU-R as the CCIR (Comité Consultatif International Téléphonique des Radiocommunications).

The ITU-R generates recommendations that deal with the preparation, transmission, and reception of radio-based signals, with the term 'radio' being taken in the broadest sense, to include television and telephony. The ITU-R's contribution to multimedia primarily relates to television formats, specifically the CCIR Rec. 601-2 specification for digital video and the CCIR Rec. 709 –1 specification for HDTV.

The ITU-T is the largest body of the ITU. It generates functional and electrical specifications for telecommunications and data communications networks. Because standardization today is a global activity the ITU-T cooperates with other standards bodies. For instance, the Internet Engineering Task Force (IETF) is a member of the ITU-T and the European Telecommunications Standards Institute (ETSI) and International Standards Organization (ISO) also incorporate their protocols into the ITU standards.

The ITU-T has ratified two important groups of interactive multimedia standards: the H.32X and T.12X specifications. It also developed the G.7XX audio specifications and the V.80 modem specification. Along with its work on ISDN and B-ISDN, the

ITU-T has laid the foundation international audiovisual exchanges.

In October of 1996 the ITU-T convened its second World Telecommunication Standardization Conference (WTSC 96) in Geneva. At the Conference, the Questions for the next ITU study period (1997-2000) were drafted and Study Group 16 was created. Known as "Multimedia Services and Systems," Study Group 16 is intended to formally coalesce the informally linked spheres of computing, telecommunications and audiovisual entertainment. This structured convergence, and the standards that result, cannot help but have a profound effect on organizations, and society as a whole.

INTERNATIONAL STANDARDS ORGANIZATION (ISO)

The International Standards Organization was founded in 1947 to promote the development of international standards and activities that relate to the exchange of goods and services worldwide. Its membership consists of national bodies that are populated by representatives of relevant user and vendor communities. Members in a national body consist of corporations, companies, partnerships, and other firms that engage in industrial or commercial enterprises (or in education, research and testing). In the U.S. American National Standards Institute (ANSI) in the U.S.

The ISO is affiliated with the International Electrotechnical Commission (IEC). Like the ITU-T, the ISO/IEC is headquartered in Geneva, Switzerland. The ISO/IEC has developed many important standards including the Motion Pictures Experts Group (MPEG) and Joint Photographic Experts Group (JPEG) standards for high-resolution image compression.

JPEG is *the* international standard for still image compression. It begins with a digital image in the format YCbCr (such as is defined in CCIR-601-2) and provides several levels of compression. JPEG-1 is defined in ISO/IEC DIS 10918-1. MPEG (ISO/IEC 11172) outlines four different standards for handling tightly-compressed motion sequences: MPEG-1, MPEG-2, MPEG-3 and MPEG-4. MPEG is described in detail in Appendix B of this book.

According to ISO Internet information the term "ISO" is not an acronym. Rather, it is derived from the Greek word *isos*, which means equal. For example, isochronous transmission means, "equal in length of time," and an isosceles triangle has two equal sides. Standards are great equalizers and the ISO intends that they remain so. The name ISO offers the further advantage of being valid in all the official languages of the organization (English, French & Russian), whereas if it were to be an acronym it would not work for French and Russian.

INTERNET ENGINEERING TASK FORCE (IETF)

In addition to the ITU-T and the ISO there is another important standards development "community" involved in crafting recommendations for internetworked multimedia handling systems: the Internet Activities Board (IAB) and, through that body, the IETF. Standards for communications across the Internet evolve through the IETF's rather informal (in comparison to the ITU-T and ISO) standards-setting process that relies on circulars called Request for Comments (RFCs). RFC contributors are mostly scientists and academicians having an interest in some aspects of internetworking including, but not restricted to, multimedia communications. Often these individuals are involved in research funded by various grants or contracts with private corporations. RFCs are also developed in cooperation with other organizations and consortia, by employees of private organizations and by academicians.

THE IMTC

The IMTC is a non-profit corporation founded to promote the creation and adoption of international standards for multipoint document exchange and video conferencing. Its members promote a Standards First initiative to guarantee internetworking for all aspects of multimedia teleconferencing. The organization is open to all that support its goals.

The IMTC maintains an informative Web site that contains a wealth of information on the H.32X and T.12X Recommendations. See Chapter 8 for the URL.

The long-standing president of the IMTC is Neil Starkey of DataBeam, Inc.

B

MPEG Family of Specifications

The Moving Picture Experts Group (MPEG) is a committee of the ISO/IEC, or more specifically, the Joint Technical Committee 1, Sub-committee 29, Work Group 11. It has produced a series of four standards for lossy data compression and storage of full-motion video. These four standards are MPEG-1, MPEG-2, MPEG-3, and MPEG-4. MPEG-3, having been found to offer nothing to video, was merged into MPEG-2. Ironically, it has emerged as a popular music file format. Because that is the only common application for MPEG-3, we will not discuss it further in this text.

MPEG defines only the operations that a decoder must perform to interpret a compressed bit stream. It does not address what constitutes a valid bit stream (although it is implicit in the decoding process). Encoding is left up to the individual manufacturers. This allows a proprietary advantage to be obtained in spite of the fact that MPEG is a publicly available international specification.

The MPEG committee meets four times a year. During its meetings, it reviews the work that has been accomplished during the quarter and plans its next steps. To ensure that harmonized solutions to the widest range of applications are achieved the committee works jointly with the ITU-T Study Group 15 and its successor, Study Group 16. MPEG also collaborates with representatives from other organizations including the ITU-R, the SMPTE, and the ATSC.

Work Group 11 began meeting in 1988. Its original goal was to define a format for compressing audiovisual sequences for CD-ROM storage and playback. As it turned out, its work engendered a great deal of international interest. Technical experts from all over the world (more than 200 participants by 1992) collaborated on the development of a *syntax* for MPEG-1.

To represent an object using compact and coded data, it is necessary to develop a language that is capable of dealing with abstraction. The ISO refers to a unique language (the specific bit codes used to represent objects—and their meaning) as syntax. For example, a few tokens can represent a large block of data. Syntax is also used to decode a complex string of bits and to map those bits from their compact representation back into their original form (or a semblance thereof). In video coding, syntax exploits common characteristics (block-oriented motion, spatial and temporal redundancy, color-to-brightness ratios, etc.).

By the end of 1990, the MPEG-1 syntax had emerged. It defines a standard input format (SIF) that can be used to produce audio and video samples that are coded and compressed at a bit rate that is not to exceed 1.86 Mbps. Image sizes can be no larger than 720 pixels by 576 lines (although it is syntactically possible to encode picture dimensions as high as 4095 lines by 4095 pixels). In fact, MPEG-1 images are usually 352 pixels by 240 lines. They are captured using progressive scanning (such as is used by computer monitors). MPEG-1 samples two channels of audio, combines it with the video and delivers it (either over a direct connection to a CD-ROM storage device or via a telecommunications network connection).

After demonstrations proved that the syntax was flexible enough to be applied to bit rates and sampling frequencies much higher than those required for CD-ROM playback applications, a second phase (MPEG-2) was initiated within the committee. Its objective, to define a syntax for the efficient encoding of interlaced (broadcast) video, was more difficult to achieve than that of encoding progressive scanned images. Audio encoding also presented a challenge. MPEG-2 would introduce a scheme to decorrelate multichannel discrete surround sound audio.

In 1991, it appeared that HDTV had special needs that would require a separate syntax. Thus, plans were laid to define MPEG-3. Tests in late 1992 revealed that the MPEG-2 syntax would, with a small amount of fine-tuning, scale with the bit rate, obviating this third MPEG phase. As a result, MPEG-3 was merged into MPEG-2 in 1992. However, most music sites on the Internet now use MPEG-3 because its file sizes are approximately 30% smaller than are those of MPEG-2.

MPEG-4 (ISO/IEC 14496) was launched in late 1992 to explore the requirements of a more diverse set of applications. Its goal was to find a syntax that could be used to code very low bit rate (4.8–64 Kbps) audio and video streams. MPEG-4 encoding is aimed at videophone and small-screen terminal (176 pixels-by-144 lines) applications. Formal plans to pursue MPEG-4 were approved by unanimous vote of the ISO/IEC JTC1 in Brussels, Belgium in September 1993. A draft specification was delivered in 1997, and was approved in December 1999. In addition to videophones, MPEG-4 enables interactive mobile multimedia, remote sensing and sign language captioning.

In 1995, MPEG became virtually assured of dominance as a standard when Microsoft announced plans to bundle software MPEG-1 playback technology with its operating system, and Compaq announced that it would ship a motherboard with a hardware-based implementation of MPEG. In 1996, both IBM and LSI Logic announced large price reductions because of reduced number MPEG encoding chipsets. MPEG-2 is used for real time high-quality applications (e.g., digital-cable distribution systems and satellite broadcasts).

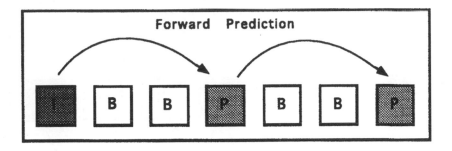

Figure Appendix B-1. Forward Prediction (Michel Bayard).

MPEG Picture Coding

MPEG manipulates motion video as image sequences. An image is divided into macroblocks. MPEG syntax identifies the three approaches that are used to code these macroblocks. MPEG manipulates motion video as image sequences. As such, a great deal of redundancy exists between frames. To achieve temporal compression, it is useful to compute some frames from other frames. Hence, MPEG defines three types of frames. The most basic type is the Intraframe or I-frame. I-frames are compressed using intraframe encoding (DCT). In addition there are Predicted frames (P-frames) and Bi-directional frames (B-frames).

I -frames are coded without reference to past or future frames and contain explicit information about every pixel in the image. I-frame coding generates significant amounts of data. A certain number of I-frames must be sent in an MPEG sequence (once every 400 ms is typical) to eliminate accumulated errors that result from interframe coding techniques. I-frames must also be sent for major scene changes. For highly efficient coding, the object is to send as few I-frames as possible.

P-frames are coded through a process of predicting the current frame using a past frame (an I-frame or another P-frame). The picture is broken into 16x16 pixel blocks. Each block is compared to a pixel block in the same horizontal and vertical position in a previous frame. The P-frame technique assumes all pixels in a block move together and does not describe each pixel individually. Only information about differences between pixels in the current frame and the previous frame is coded.

B-frame pictures use bi-directional interpolation coding. Interpolation is the process of determining that a given pixel is a certain color because a pixel in the same place in a prior frame was that color. B-frame coding uses hints to predict a current frame based on what was contained in a past and a future frame. B-frame coding includes background estimation—working with areas in the background that are uncovered as an object moves from side to side. B-frame coding also uses motion compensated prediction to guess at where a moving object will be, based on where it was in the prior frame.

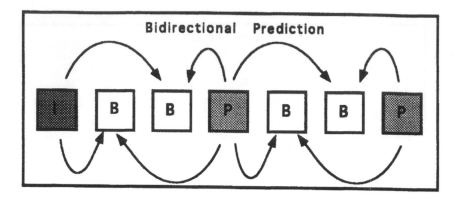

Figure Appendix B-2. Bi-Directional Interpolated Coding (Michel Bayard).

The process of MPEG encoding always starts with an I-frame. After a time interval that is defined by each particular implementation another frame is sent, typically, a P-frame. The elapsed time between the transmission of frames (of any type) is often too great to permit the illusion of smooth motion. Thus, B-frames are computed (interpolated) by comparing a past frame to a present frame. These interpolated frames are used to fill in the gaps between I-frames and P-frames (or some combination thereof).

B-frames tend to make motion sequences smoother on playback while conserving bandwidth. Unfortunately, their use forces a decoder to buffer the P-frames from which they are computed. This buffering increases the cost of a decoder and, consequently, presents a problem for some low-end, mass-market applications. I-frames anchor picture quality, because ultimately P and B-frames are derived from them. P-frames are the least demanding of the three, in terms of their impact on a decoder.

MPEG 1—ISO/IEC 11172

The official title of the ISO/Ice's MPEG-1 specification is, "Coding Of Moving Pictures And Associated Audio For Digital Storage Media At Up To About 1.5 Mbps." It was originally approved as a three-part specification in October of 1992. Later, two additional parts were added, one in 1994 and one in 1995.

MPEG-1 encoded video is optimized for playback on computers and set-top boxes. MPEG-1 uses progressive (non-interlaced) scanning to provide a typical picture resolution of 352H-by-240V pixels (although the resolutions can vary). Images are displayed at 30 fps (in North America and Japan only) and 25 fps. The MPEG-1 data rate is 150 Kbytes (not Kbps). This speed, which matches the data rate delivered by a single-speed CD-ROM player, is also compatible with T-1/E-1 network speeds.

Special MPEG-1 compliant chips provide near VHS-quality desktop video (with CD-

ROM quality stereo sound) at data rates of 150 Kbytes. This data rate conforms to T1/E1 speeds and matches the data rate from a standard, single-speed CD-ROM player).

The five MPEG-1 parts are as follows:

Part 1—SYSTEM (IS 11172-1). This part describes the syntax used to synchronize and multiplex audio and video signals. These signals can be transported over digital channels and/or stored as digital media. Part 1 also includes the syntax for synchronizing video and audio streams.

Part 2—VIDEO (IS 11172-2). This part describes compression of non-interlaced video signals. It specifies the header and bit stream elements for video and identifies the algorithms (semantics) used to code and compress video. Video handles the image by dividing it into nested layers. At the highest layer (the picture layer) frame rates and resolutions are defined. Below the picture layer comes slices, macroblocks, blocks, samples and coefficients. At each layer, coding and compression algorithms are defined. Methods are also described for synchronizing, accessing, buffering and error recovery.

Part 3—AUDIO (IS 11172-3). This part describes the compression of audio signals. It includes syntax and semantics for three compression methods (known as I, II, and III). These methods, or *classes*, define algorithms that vary in complexity and coding efficiency to produce streams of audio at different bit rates. Layer I (audio at 384 Kbps) is used for compact disc video. Layer II (audio at 224 Kbps) is used in satellite broadcasting and compact disc video. Layer III (128 Kbps audio) is used in ISDN, satellite and Internet audio applications.

Part 4—CONFORMANCE (IS 11172-4). This part describes the procedures for determining the characteristics of coded bit streams and how they are decoded. It also defines the meaning of MPEG-1 conformance for the first three parts of IS 11172 (Systems, Video, and Audio), and provides two sets of test guidelines for determining compliance in bit streams and decoders.

Part 5—SOFTWARE SIMULATION (IS 11172-5). This part contains a sample ANSI C language software encoder and compliant decoder for video and audio. It also provides an example of a system's codec, that can multiplex and demultiplex the separate video and audio streams that are used in computer files.

Brief Explanation of How MPEG-1 Works

MPEG-1 compresses relatively low-resolution source input format (SIF) video sequences (approximately 352-by-240 pixels) that are synchronized with relatively high-quality audio. The color images have been converted to the YUV space with the chrominance channels (U and V) further decimated to 176-by-120 pixels. MPEG-1 uses motion prediction (temporal compression) first and then follows it up with

DCT (spatial compression). DCT acts on 8-by-8 pixel blocks while motion prediction operates on 16-by-16 pixel blocks. After temporal and spatial compression are complete MPEG-1 uses fixed-table Huffman encoding and differential PCM. As is true with all MPEG types, MPEG-1 uses I frames, P frames and B frames for temporal compression. MPEG-1 files are usually smaller than QuickTime or Video for Windows files, however the quality is not as good.

MPEG 2—ISO/IEC 13818

The official title of the ISO/IEC's MPEG-2 specification is, "Generic Coding of Moving Pictures and Associated Audio." It was formally approved in November 1994. MPEG-2 was produced in collaboration with the ITU-T, which also published it as Recommendation H.262. The full title of Recommendation H.262 is "Information Technology—Generic Coding of Moving Pictures and Associated Audio."

MPEG-2 is intended to provide compression for studio-quality video applications that include digital broadcast television, digital storage media, digital VCRs and video on demand. MPEG-2 compressed media is suitable for distribution over coaxial and fiber optic cable and satellite.

MPEG-1 was forced to compromise on video quality so that data could play on a single-speed CD-ROM in real time. These compromises manifest as picture blockiness. By contrast, MPEG-2 generates higher quality images when compared to previous video CD players that utilize MPEG-1.

MPEG-2 goes beyond MPEG-1 to address the need for higher-resolution images and interlaced video formats. It does so by compressing a full CCIR-601 digital video image (480 lines by 720 pixels) at a rate of 30 frames per second. It also supports multiple programs in a single stream, interlaced and progressive scan (whereas MPEG-1 supports only progressive), interactive data transmission (for applications such as home shopping and distance learning) and surround-sound audio. MPEG-2 also differs from MPEG-1 in that it can use symmetric compression (MPEG-1 uses asymmetric compression techniques only). MPEG-2 is backward compatible with MPEG-1 and can incorporate MPEG-1 encoded streams.

The digital versatile disk (DVD) *standard* for storing large quantities of digital video and audio specifies MPEG-2 video compression (although it supports Dolby AC-3 audio and *not* MPEG-2 audio). DVDs use laser technology to store up to 17 gigabytes (GB) of data, enough for nine hours of studio-quality audio and video.

MPEG-2 Parts

The MPEG-2 ISO/IEC 13818 volume consists of a total of eight parts, 1-7 and 9 (part 8 was withdrawn).

Part 1—SYSTEMS (13818-1). Part 1 addresses how one or more elementary streams of video, audio and data are combined into single or multiple streams suitable for storage and transmission. These streams take one of two forms, the Program Stream and the Transport Stream. Each is optimized for a different set of applications. The Program Stream is similar to MPEG-1 Systems Multiplex. It results from combining one or more Packetized Elementary Streams (PES), which have a common time base, into a single stream that run at between 3-15 Mbps. The Program Stream is used for DVD encoding. The Transport Stream combines multiple PES with one or more independent time bases into a single stream. Elementary streams sharing a common time base form a program. The Transport Stream is designed for use in environments where errors are likely (broadcast and telecommunications). Transport stream packets are 188 bytes long.

Part 2—VIDEO (13818-2). Part 2 builds on the video compression capabilities of the MPEG-1 standard to offer coding tools. These have been grouped in seven profiles to offer different functions. The first five were approved with the standard in November of 1994. The sixth profile (the 4:2:2 Profile) and the seventh (Multiview Profile) were added in 1996.

Part 3—AUDIO (13818-3). Part 3 is a backward compatible multi-channel extension of the MPEG-1 Audio standard.

Part 4—COMPLIANCE TESTING (13818-4). Part 4 of MPEG-2 corresponds to Part 4 of MPEG-1. It was approved in March 1996.

Part 5—SOFTWARE SIMULATION (13818-5). Part 5 of MPEG-2 corresponds to Part 5 of MPEG-1 and was approved in March 1996.

Part 6—DIGITAL STORAGE MEDIUM COMMAND AND CONTROL (13818-6). This part defines a syntax for controlling VCR-style playback and random-access of bit streams encoded onto digital storage media (e.g., compact disc). Commands include fast forward, advance, still frame and go-to. Part 6 was approved in 1996.

Part 7—NON-BACKWARDS COMPATIBLE AUDIO (13838-7). This part extends the two-channel audio of MPEG-1 (11172-3) by adding a new syntax to efficiently decorrelate discrete multi-channel surround sound audio. Part 7 of MPEG-2 was approved in 1997.

Part 9— REAL-TIME INTERFACE (RTI) (13838-9). This part defines a syntax for video on demand control signals between set-top boxes and head-end servers. Part 9 was approved as an International Standard in July 1996.

MPEG-2 Levels

Each MPEG-2 defines "quality classifications" that are known as levels. Levels limit coding parameters (sample rates, frame dimensions, coded bit rates, etc.). The MPEG-2 levels are as follows:

Level	Resolution Sampled	Frame rate	Pixels per Second	Maximum bit rate	Purpose
Low	352 x 240	30 fps	3.05 M	4 Mbps	Consumer Tape, CIF
Main	740 x 480	30 fps	10.40 M	15 Mbps	CCIR-601 Studio TV
High 1440	1440 x 1440	30 fps	47.00 M	60 Mbps	Commercial HDTV
High	1920 x 1080	30 fps	62.70 M	80 Mbps	Production SMPTE

Figure Appendix B-3. MPEG-2 Levels

The two most popular MPEG-2 levels are the Low (or SIF) Level and the Main (or CCIR-601) Level.

MPEG-2 Profiles

MPEG-2 consists of six different coding tools that are known as *profiles*. Profiles are a defined sub-set of the MPEG-2 specification's syntax (algorithms). Different profiles conform to different MPEG-2 levels and are aimed at different applications (high- or regular-definition television broadcasts over different types of networks). Profiles also provide backward compatibility with other specifications such as H.261 or MPEG-1.

	Simple Profile	Main Profile	SNR Scalable Profile	Spatially Scalable Profile	High Profile	4:2:2 Profile
Low Level		X				
Main Level	X	X		X	X	X
High 1440		X	X		X	
High Level		X	X		X	

Figure Appendix B-4. MPEG-2 Profiles

The MPEG-2 Video Main Profile conforms to the CCIR-601 studio standard for digital TV, and is implemented widely in MPEG-2 decoder chips and in direct broadcast satellite. It will also be used for the delivery of video programming over cable television. It is targeted at a higher encoded bit rate of less than 15Mbps and specifies a resolution of 720x480 at 30 fps, allowing for much higher quality than is typical with MPEG-1. It supports the coding parameters as set forth in MPEG-2's high, high 1440, main and low levels. It uses the 4:2:0 chroma format. The Video Main Profile also supports I, P, and B frames. The U.S. ATSC Digital Television Standard (Document A/53—HDTV) specifies MPEG-2 Video Mail Profile compression.

The Simple Profile is nothing more than the Main profile without B frames. As we said earlier, B frames require a set-top or other MPEG decoder to have a certain amount of integrated circuit (IC) memory; that becomes too costly for some applications.

The SNR Scalable and Spatially Scalable Profiles are very complex and are useful primarily for academic research. Scalability allows a decoder to divide a continuous video signal into two or more coded bitstreams that represent the video at different resolutions (spatial scalability) or picture quality (SNR scalability). Video is also scalable across time (temporal scalability). Equipment manufacturers do not generally pursue scalable profiles. They require twice as many integrated circuits as non-scalable profiles, which approximately double their cost.

The High Profile is aimed at high-resolution video. It supports chroma formats of 4:2:2 and 4:2:0, resolutions that range between 576 lines by 720 pixels and 1152 lines by 1920 pixels, and data transfer rates of between 20 and 100 Mbps. It also supports I, P, and B frames.

The 4:2:2 Profile (that was added to MPEG-2 Video in January of 1996) possesses unique characteristics more desirable in the professional broadcast studio and post-production environment. The 4:2:2 Profile uses a chroma format of 4:2:2 (or 4:2:0), uses separate luminance and chrominance quantization tables, allows an unconstrained number of bits in a macroblock and operates at 608 lines/frame for 25 fps or 512 lines/frame for 30 fps.

The Multiview Profile, which was completed in October 1996, uses existing video coding tools for providing an efficient way to encode two slightly different pictures such as those obtained from two slightly separated cameras shooting the same scene. This allows multiple views of scenes, such as stereoscopic sequences, that are coded in a manner similar to scalable bit streams.

HOW MPEG-2 DIFFERS FROM MPEG-1

MPEG-2 is not intended to replace MPEG-1. Rather, it includes extensions to MPEG-1 to cover a wider range of applications. MPEG-1 deals only with progressive scanning techniques where video is handled in complete frames. The MPEG-2 standard supports both progressive scanning and interlaced displays such as those used for televisions. In interlaced video, each frame consists of two fields (half frames) that are sent at twice the frame rate.

The MPEG-1 specification was targeted nominally at single speed CD ROMs (1.5 Mbytes per second). MPEG-2 is targeted at variable frame rates, including those many times higher than MPEG-1.

Each of the two standards, MPEG-1 and MPEG-2, is divided into parts. MPEG-1 has five parts (listed in this document); MPEG-2 has eight, of which five are either identical to or extensions of MPEG-1 parts and three are additions.

The MPEG-1 stream is made up of two layers, the system layer and the compression layer. The MPEG-2 stream is made up of program and transport streams. MPEG-2 streams are subdivided into packets for transmission. The MPEG-2 program stream is designed for use in relatively error-free environments. As such, it is similar to the MPEG-1 system stream. The MPEG-2 transport stream is designed for error-prone environments (broadcast). MPEG-1 does not have an equivalent to the MPEG-2 transport stream.

Both the MPEG-1 and MPEG-2 standards define a hierarchy of data structures in the video stream. There is a group of pictures (a video sequence). The next level down in the hierarchy is the picture (the primary coding unit of a video sequence). MPEG-2 only defines a slice (used to handle errors in a bit stream). Both MPEG-1 and MPEG-2 define a macroblock (a 16 pixel by 16 line selection of luminance components and its corresponding 8 pixel by 8 line chrominance components), and a block (an 8-pixel by 8-line set of luminance and chrominance values).

MPEG-1 handles only two channels of audio. MPEG-2 handles up to five channels (surround sound). Because of this difference, MPEG-1 and MPEG-2 use different techniques to synchronize audio and video.

In summary, MPEG-2 extends MPEG-1 to handle audiovisual broadcasts (whereas MPEG-1 was aimed at playback applications). It can deliver audiovisual material at higher speeds, with greater resolution, with surround sound audio and in interlaced scanned environments. MPEG-2 may eventually replace MPEG-1 but it was not intended to do so.

MPEG 4—ISO/IEC 14496

The official title of the ISO/IEC's MPEG-4 specification is, "Very Low Bitrate Audio-Visual Coding." It was approved as a Working Draft in November 1994, finalized in October 1998, and became an International Standard in 1999.

MPEG-4 is targeted at low bit rate applications with frame sizes of 176x144 or less, frame rates of 10Hz or less, and encoded bit rates of 4.8–64 Kbps. The promise of

MPEG-4 is to provide fully backward compatible extensions under the title of MPEG-4 Version 2. MPEG-4 is designed to address digital television, interactive graphics applications that build on synthetic content, and interactive multimedia (e.g., distribution of, and access to, content). MPEG-4 provides standardized technological elements to enable the integration of the production, distribution, and content access paradigms of these three domains.

MPEG-4 enables extensive reusability of content across disparate technologies such as digital television, animated graphics, and World Wide Web (WWW) pages. It also offers manageability of content owner rights. Furthermore, MPEG-4 offers a generic QoS descriptor for dissimilar MPEG-4 media that can be interpreted and translated into the appropriate native signaling messages of each network. End-to-end signaling of the MPEG-4 media QoS descriptors enables transport optimization in heterogeneous networks. However, the exact translations from the QoS parameters set for each media to the network QoS are left to network providers. MPEG-4 offers end users higher levels of interaction with content, and allows the author to set those limits. It also extends multimedia to new networks, such as mobile networks and others that employ lower bit rates.

MPEG seeks to supersede proprietary, non-interoperable formats by providing standardized ways to represent units of aural, visual or audiovisual content, called *media objects*. These media objects could be recorded with a camera or microphone, or generated with a computer. MPEG describes the composition of these objects to create compound media objects that form audiovisual scenes. It multiplexes and synchronizes the data associated with media objects so that they can be transported over network channels and provide a QoS that is appropriate for the nature of the specific media objects, and it interacts with the audiovisual scene that is generated at the receiver's end.

Media objects may require streaming data that is conveyed in multiple elementary streams. Therefore, MPEG-4 defines an object descriptor that identifies all streams that are associated with one media object. This enables the handling of hierarchically encoded data, and the association of both object content and intellectual property rights information. Each stream is characterized by a set of descriptors for configuration information, (e.g., required decoder resources, or the required precision of encoded timing information). The descriptors may also describe the QoS it requests for transmission (e.g., maximum bit rate, bit error rate, or priority). MPEG-4 enables synchronization of elementary streams by time stamping individual access units within elementary streams. The synchronization layer manages the identification of access units and manages the time stamping. Independent of the media type, this layer allows identification of the type of access unit (e.g., video or audio frames, scene description commands) in elementary streams, and the recovery of the media object's or scene description's time base. It also enables synchronization among these.

The MPEG-4 ISO/IEC 14496 volume consists of four basic functions:

Delivery Multimedia Integration Framework (DMIF)

DMIF is a session protocol for the management of multimedia streaming over generic delivery technologies. It is similar to FTP except that, where FTP returns data, DMIF returns pointers toward (streamed) data. DMIF is both a framework and a protocol. The functionality provided by DMIF is expressed by an interface called DMIF-Application Interface (DAI), and translated into protocol messages that may differ based on any given network on which they operate.

MPEG-4 specifies an interface to the *TransMux* (Transport Multiplexing) models, the layer that offers transport services for matching the requested QoS. Any suitable existing transport protocol stack (e.g., RTP in UDP/IP, AAL5 in ATM, or Transport Stream in MPEG-2) over a suitable link layer may become a specific TransMux instance. The synchronized delivery of streaming data may require the use of different QoS schemes as it traverses multiple public and private networks. MPEG defines the delivery-layer FlexMux multiplexing tool to allow grouping of Elementary Streams (ES) with low multiplexing overhead. It may be used to group ESs with similar QoS requirements, to reduce the number of network connections, or to reduce end-to-end delay. The FlexMux layer may be empty if the underlying TransMux instance provides adequate functionality.

The DMIF framework enables control functions such as the ability to identify access units, to transport timestamps and clock reference information, and to recognize data loss. It also enables interleaving of data from different elementary streams into FlexMux streams. Moreover, it permits the conveyance of control information to indicate the required QoS for each elementary stream and FlexMux stream, translation of such QoS requirements into actual network resources, association of elementary streams to media objects, and conveyance of the mapping of elementary streams to FlexMux and TransMux channels.

Systems

MPEG-4 supports scene description for composition (spatio-temporal synchronization with time response behavior) of multiple media objects. The scene description provides a rich set of nodes for two-dimensional and three-dimensional composition operators and graphics primitives. It also supports text with international language support, font and font style selection, timing and synchronization. Moreover, MPEG-4 supports interactivity such as client and server-based interaction, an event model for triggering events or routing user actions, and event management and routing between objects in the scene, because of user or scene triggered events.

The FlexMux tool provides interleaving of multiple streams into a single stream, including timing information, and the TransMux provides transport layer independence through mappings to relevant transport protocol stacks. MPEG-4 also provides the initialization and continuous management of the receiving terminal's timing identification, synchronization and recovery mechanisms, as well as other receiving terminal buffers. It also recognizes data sets such as the identification of Intellectual Property Rights that relate to media objects.

Audio

MPEG-4 Audio permits a wide variety of applications that could range from simple speech to sophisticated multi-channel audio, and from natural sounds to synthesized sounds. In particular, it supports the highly efficient representation of audio objects consisting of:

Speech signals: Speech coding can be done at bitrates from 2 Kbps up to 24 Kbps using the speech coding tools. Lower bitrates, such as an average of 1.2 Kbps, are possible through variable rate coding. Low delay is possible for communications applications. When using the HVXC tools, speed and pitch can be modified under user control during playback. If the CELP tools are used, a change of the playback speed can be achieved by using an additional tool for effects processing.

Synthesized Speech: Scalable TTS coders bitrate range from 200 Bps to 1.2 Kbps that allows a text, or a text with prosodic parameters (e.g., pitch contour, phoneme duration,), as its inputs to generate intelligible synthetic speech.

General audio signals: Support for coding general audio ranging from very low bitrates to high quality is provided by transform coding techniques. With this functionality, a wide range of bitrates and bandwidths is covered. It starts at a bitrate of 6 Kbps and a bandwidth below 4 kHz, but also includes mono or multichannel broadcast quality audio.

Synthesized Audio: Synthetic Audio support is provided by a Structured Audio Decoder implementation that allows the application of score-based control information to musical instruments described in a special language.

Bounded-complexity Synthetic Audio: This is provided by a Structured Audio Decoder implementation that allows processing of a standardized Wavetable format.

Examples of additional functionality include speed control, or change of the time scale without altering the pitch while decoding, and pitch change, or change of the pitch without altering the time scale while encoding or decoding. Audio Effects provide the ability to process decoded audio signals with complete timing accuracy to enable such functions as mixing, reverberation, and spatialization.

Visual

The MPEG-4 Visual standard will allow the hybrid coding of natural (pixel based) images and video together with synthetic (computer generated) scenes. This will, for example, allow the virtual presence of video communication participants. To this end, the Visual standard will comprise tools and algorithms supporting the coding of natural (pixel based) still images and video sequences as well as tools to support the compression of synthetic 2-D and 3-D graphic geometry parameters (i.e. compression of wire grid parameters, synthetic text). The MPEG-4 visual standard will support bitrates typically between 5 Kbps and 10 Mbps, progressive and interlaced video, and resolutions ranging from SQCIF to DTV. Moreover, it supports compression efficiencies from *acceptable* to *near lossless*, and will allow for random access of video such as fast forward and reverse.

Spatial scalability allows decoders to decode a subset of the total bitstream generated by the encoder to reconstruct and display textures, images and video objects at reduced spatial resolution. For textures and still images, a maximum of 11 levels of spatial scalability is supported. For video sequences, a maximum of three levels is supported. Temporal scalability allows decoders to decode a subset of the total bitstream generated by the encoder to reconstruct and display video at reduced temporal resolution. A maximum of three levels is supported. Quality scalability allows a bitstream to be parsed into a number of bitstream layers of different bitrate such that the combination of only a subset of the layers can be decoded into a meaningful signal. The bitstream parsing can occur either during transmission or in the decoder. The reconstructed quality, in general, is related to the number of layers used for decoding and reconstruction.

MPEG-4 supports shape coding to assist the description and composition of conventional images and video as well as arbitrarily shaped video objects. Applications that benefit from binary shape maps with images are content-based image representations for image databases, interactive games, surveillance, and animation. Error resilience assists the access of image and video over a wide range of storage and transmission media such as the useful operation of image and video compression algorithms in error-prone environments at low bit-rates (i.e., less than 64 Kbps). Tools are provided that address both the band-limited nature and error resiliency aspects for access over wireless networks. The Face Animation part of the standard allows sending parameters that calibrate and animate synthetic faces.

Tools	AAC main	AAC LC	AAC SSR	AAC LTP	AAC Scalable	TwinVQ	CELP	HVXC	TTSI	Main syn- thetic	Wave- table synth- esis	General MIDI	Algo- rithmic Synthesis & Audio FX
MPEG-2 AAC Main	X												
MPEG-2AAC LC		X		X	X								
MPEG-2AAC SSR			X										

	1	2	3	4	5	6	7	8	9	10	11	12
Noise Shaping	X	X	X	X	X							
Long Term Prediction			X	X	X							
Tools for Large Step Scalability				X								
TwinVQ					X							
CELP						X						
HVXC							X					
TTSI								X				
Structured Audio tools									X		X	
SA Sample Bank Format									X	X		
MIDI									X	X	X	

Figure Appendix B-5. MPEG-4 Tools

MPEG-4 only standardizes the parameters of these models, not the models themselves.

MPEG-4 Parts

The MPEG-4 requirements have been addressed by the six parts of the MPEG-4 Version 1 standard:

Part 1: Systems - specifies scene description, multiplexing, synchronization, buffer management, and management and protection of intellectual property

Part 2: Visual - specifies coded representation of natural & synthetic visual objects

Part 3: Audio - specifies the coded representation of natural and synthetic audio objects

Part 4: Conformance Testing - defines conformance conditions for bitstreams and devices; this part is used to test MPEG-4 implementations

Part 5: Reference Software - includes software corresponding to most parts of MPEG-4 (normative and non-normative tools); it can be used for implementing compliant products as ISO waives the copyright of the code

Part 6: Delivery Multimedia Integration Framework (DMIF) - defines a session protocol for the management of multimedia streaming over generic delivery technologies.

Parts 1–3 and 6 specify the core MPEG-4 technology; Parts 4 and 5 are *supporting parts.* Parts 1, 2 and 3 are delivery independent, leaving to Part 6 (DMIF) the task of dealing with the idiosyncrasies of the delivery layer. The MPEG-4 parts are independent and can be used independently or in conjunction with proprietary technologies. However, they were developed to ensure that the maximum benefit results when they are used together.

Participants in MPEG-4 represent broadcasters, equipment and software manufacturers, digital content creators and managers, telecommunication service providers, publishers and intellectual property rights managers, and university researchers. Hundreds of experts, from organizations throughout the world, developed MPEG-4 for the next generation of multimedia products and services. MPEG-4 Version 1 became available at the end of 1998; MPEG-4 Version 2 extends the standard in a backward compatible way.

MPEG-4 Audio Object types

For coding natural sound, MPEG-4 includes the Advanced Audio Coding (AAC) and Twin Vector Quantization (Twin VQ) algorithms. The following object types exist:

1. The **Advanced Audio Coding (AAC) Main** object type is similar to—and compatible with—the AAC Main profile that the MPEG-2 (ISO/IEC 13818-7) defines. MPEG-4 AAC adds the Perceptual Noise Shaping tool and supports five channels plus a low frequency channel in one object.

2. The **MPEG-4 AAC Low Complexity** object type is a low complexity version of the AAC Main Object type.

3. The **MPEG-4 AAC Scalable Sampling Rate** object type is the counterpart to the MPEG-2 AAC Scalable Sampling Rate profile,

4. The **MPEG-4 AAC LTP** object type is similar to the AAC Main object type but replaces the MPEG-2 AAC predictor with the long-term predictor to provide the same efficiency at a lower implementation cost.

5. The **AAC Scalable** object type allows a large number of scalable combinations such as combinations with TwinVQ and CELP coder tools as the core coders. It supports only mono or 2-channel stereo sound.

6. The **TwinVQ** (Transform domain Weighted Interleave Vector Quantization) object type is based on fixed rate vector quantization instead of AAC's Huffman coding. It operates at lower bitrates than AAC and supports both mono and stereo sound.

7. MPEG-4 includes two different algorithms for coding speech; each operates at different bitrates, plus a Text-to-Speech Interface.

- The **CELP** (Code Excited Linear Prediction) object type supports 8 kHz and 16 kHz sampling rates at 4–24 Kbps. CELP bitstreams can be coded in a scalable way through bit rate scalability and bandwidth scalability.

- The **HVXC** (Harmonic Vector Excitation Coding) object type gives a parametric representation of 8 kHz, mono speech at fixed bitrates between 2–4 Kbps and less than 2 Kbps using a variable bitrate mode, and supports pitch and speed changes.

8. The **TTSI** (Text-to-Speech Interface) object type gives an extremely low-bitrate phonemic representation of speech. The actual text-to-speech synthesis is not specified; only the interface is defined. Bit rates range from 0.2 to 1.2 Kbps. The synthesized speech can be synchronized with a facial animation object (see below).

9. Lastly, a number of different object types exist for synthetic sound.

- The **Main Synthetic** object type collects all MPEG-4 Structured Audio tools. Structured Audio is a way to describe methods of synthesis. It supports flexible, high-quality algorithmic synthesis through the Structured Audio Orchestra Language (SAOL) music-synthesis language, efficient Wavetable synthesis with the Structured Audio Sample-Bank Format (SASBF), and enables the use of high-quality mixing and postproduction in the Systems Audio BIFS tool set. Sound can be described at '0 Kbps' (i.e., sound continues without input until it is stopped by an explicit command) to 3-4 Kbps for extremely expressive sounds in the MPEG-4 Structured Audio format.

- The **Wavetable Synthesis** object type is a subset of the Main Synthetic object type that makes use of the SASBF format and the widely used MIDI (Musical Instrument Digital Interface) Wavetable format tools to provide simple sampling synthesis.

- The **General MIDI** object type gives interoperability with existing content (see above). Unlike the *Main Synthetic* or *Wavetable Synthesis* object types, it does not provide completely predictable (i.e., normative) sound quality and decoder behavior.

10. The **Algorithmic Synthesis and AudioFX** object type provides SAOL-based synthesis capabilities for very low-bitrate terminals (*FX* stands for *effects*).

11. The **NULL** object type provides the possibility to feed raw PCM data directly to the MPEG-4 audio compositor to allow local sound enhancement at the decoder. Support for this object type is in the compositor, not in the decoder.

MPEG-4 Profiles

Although there are numerous object types in the audio area, there are only four distinct profiles (as defined below). Codec builders can claim conformance to profiles at a certain level, but cannot claim conformance to object types.

Two levels are defined, determining whether either one or a maximum of 20 objects can be present in the (audio) scene.

The *Scalable* profile was primarily defined to allow good quality, reasonable complexity, low bitrate audio on the Internet (an environment in which bitrate varies between users and over time). Scalability enables optimal use of limited and of varying bandwidth without encoding and storing the material multiple times. The scalable profile has four levels that restrict the amount of objects in the scene, the total amount of channels, and the sampling frequency. The highest level employs the concept of complexity units.

The *Synthetic* profile groups the synthetic object types. The target applications are defined by the need for good quality sound at very low data rates. There are three levels that define the amount of memory for data, the sampling rates, the amount of TTSI objects, and some further processing restrictions.

The *Main* profile includes all object types. It is useful in environments in which processing power is available to create rich, highest quality audio scenes that may combine organic sources with synthetic ones. Two applications are the DVD and multimedia broadcast. This profile has four levels that are defined in terms of complexity units. There are two different types of complexity units: processor complexity units (PCU), which are specified in millions of operations per second, and RAM complexity units (RCU), that are specified in terms of the number of kWords. The standard also specifies the complexity units required for each object type. This provides authors with maximum freedom in choosing the right object types and allocating resources among them. For example, a profile could contain main AAC and Wavetable synthesis object types, and a level could specify a maximum of two of each. This would prevent the resources reserved for the AAC objects to be used for a third and fourth Wavetable synthesis object even though it would not break the decoder. Complexity units enable the author freedom to use decoder resources for any combination of profile-supported object types.

How MPEG-4 Differs from MPEG-2

The difference between MPEG-4 and MPEG-2 is not improvement in video or audio reproduction for a single application (as MPEG-2 specifically supports DTV). MPEG-4 represents the establishment of an adaptable set of tools for combining various types of video, audio, and interaction to provide environments beyond what anyone may now imagine. The essence of MPEG-4 is an object-based audiovisual representation model that allows authors to build scenes using individual objects that have relationships in time and space. It addresses the fact that no single coded representation is ideal for all object types. For instance, animation parameters ideally address a synthetic self-playing piano; an efficient representation of pixel values best suits organic video such as a dancer. MPEG-4 facilitates integration of these different types of data into one scene. MPEG-4 also enables interactivity through hyperlinking with the objects in the scene (e.g., through the Internet). Moreover, MPEG-4 enables selective bit management (pre-determining which bits will be forfeited if bandwidth becomes less than desirable), and straightforward re-use of content without transcoding.

Object Type	Speech	Scalable	Main	Synthetic
AAC Main			X	
AAC SSR			X	
AAC LC		X	X	
AAC LTP		X	X	
AAC Scalable		X	X	
TwinVQ		X	X	
CELP	X	X	X	
HVXC	X	X	X	
TTSI	X	X	X	
Main Synthetic			X	X
Wavetable Synthesis			X	X
General MIDI			X	X
Algorithmic Synthesis			X	X
Number of levels	2	4	4	3

Figure Appendix B-6. MPEG-4 Object Types

MPEG-4 benefits myriad applications in various environments. Whereas MPEG-2 is constructed as a rigid standard, MPEG-4 represents a set of tools, or *profiles* that address numerous settings, and countless combinations. MPEG-4 is less an extensive standard than an extensive collection of standards that authors can select and combine as they choose to improve existing applications or to deploy entirely new ones. It facilitates creation, alteration, adaptation, and access, of audiovisual scenes. With the advent of MPEG-4, one can expect richer virtual environments that will benefit such applications as conventional terrestrial multimedia entertainment, remote multimedia, broadcasting, and even surveillance. MPEG-4 is designed to enable convergence through the coalescence of such different service models as communication, on-line interaction, and communication.

To foster competition in the non-normative areas, MPEG-4 specifies normative tools only as interoperability necessitates. For instance, decoding is specified in the standard because it must be normative; video segmentation and rate control are not strictly specified because they can be non-normative. The MPEG-4 groups "specified the minimum for maximum usability." This strategy facilitates creativity and competition, and ensures that authors can make optimal use of the continuous improvements in the relevant technical areas. Competitors will continue to establish differentiation through the development of non-normative tools.

MPEG-7

Formally called "Multimedia Content Description Interface," MPEG-7 will be a standardized description of various types of multimedia information (it is worth noting that the group apparently arbitrarily chose the number seven; at the time of this writing, MPEG-5 MPEG-6 have not been defined). It will complement MPEG-1, MPEG-2, and MPEG-4. This description will be associated with the content itself to enable fast and efficient searching for material that is of interest to the user. The MPEG-7 Group is comprised of broadcasters, equipment manufacturers, digital content creators and managers, transmission providers, publishers and intellectual property rights managers, and university researchers who are interested in defining the standard. Participants include Columbia University, GMD-IPSI, Instituto Superior Técnico, Kent Ridge Digital Labs, KPN Research, Philips Research, Riverland, and Sharp Labs.

The purpose of MPEG-7 is to establish a methodology for searching for video and audio context on the Internet as one can search for text now. The group recognized the immense amount of audio and video content that may be accessible on the Internet, the interest in locating that content, and the current lack of any way to locate that content. The problem is that no methodology exists for categorizing such information. One can search on the basis of the color, texture, and shape of an object in a picture. However, one cannot effectively search for *the moment in which the wicked witch melts* in The Wizard of Oz. Because the same limitations apply to audio, the group alluded to enabling a search based on humming a portion of a melody. It is not hard to understand how this applies to the ongoing cable television paradox of *500 channels but nothing worth watching* by allowing users to search on characteristics that suit their whim.

358

This newest member of the MPEG family, called "Multimedia Content Description Interface" (MPEG-7), may extend the limited capabilities of existing proprietary solutions in identifying content by including more data types. MPEG-7 will specify a standard set of descriptors to describe various types of multimedia information, ways to define other descriptors, and structures, or Description Schemes, for the descriptors and their relationships. The combination of descriptors and description schemes will be associated with the content itself to allow fast, efficient searching. MPEG-7 will also standardize a Description Definition Language (DDL) to specify description schemes. MPEG-7 will allow searches for AV content, such as still pictures, graphics, 3D models, audio, speech, video, and *scenarios* (combinations of characteristics). These general data types may include special cases such as personal characteristics or facial expressions.

The MPEG-7 standard builds on other representations such as analog, PCM, MPEG-1, -2 and -4. The intent is for the standard to provide references to suitable portions of such standard representations as a shape descriptor that is used in MPEG-4, or motion vector fields used in MPEG-1 and MPEG-2. However, MPEG-7 descriptors are not dependent upon the ways the described content is coded or stored. One can attach an MPEG-7 description to an analog movie or to a picture that is printed on paper. Even though the MPEG-7 description does not depend upon the representation of the material, the standard builds on MPEG-4, which provides the means to encode audio-visual material as objects that have certain relations in time and space (either two- or three-dimensional). Using MPEG-4 encoding, it will be possible to attach descriptions to audio and visual objects within a scene. MPEG-7 will offer different levels of discrimination by allowing variable degree of granularity in its descriptions.

To ensure that the descriptive features are meaningful in the context of any given application, MPEG-7 will allow description of the same material using different types of features. In visual material, for instance, a lower abstraction level may describe shape, size, texture, color, movement (trajectory) and position (where the object can be found in the scene). In audio, the lower abstraction may describe pitch, timbre, tempo, or changes and modulations. The highest abstraction level would provide semantic information such as, "This scene includes a child playing a Mozart sonata on a piano, a gray and yellow cockatiel squawking in a gold cage, and an adult trying to calm the bird." All these descriptions would be coded to allow for searching. The level of abstraction is related to the way the features can be extracted: many low-level features can be extracted in fully automatic ways, whereas high-level features may require human interaction. In addition to a description of the content, other types of information about the multimedia data may be required. Examples may include:

The form—Such as the coding scheme used (e.g. JPEG, MPEG-2), or the file size. This information helps determine whether the user can read the material.

Conditions for accessing the material—Such as copyright and price information.

Classification—Content sorting into pre-defined categories (e.g., parental rating).

Links to other relevant material - To assist the user in conducting a timely search.

Context - For recorded non-fiction content, it is important to know the occasion of the recording (e.g. Superbowl 2005, fourth quarter)

The Group seeks to make the descriptions as independent from the language area as possible, textual descriptions will be desirable in many cases — e.g., for titles, locations, or author's name. MPEG-7 data may be physically located with the associated AV material, in the same data stream, or on the same storage system. The descriptions may reside remotely. When the content and its descriptions are not co-located, mechanisms that link MPEG-7 will address applications that can be stored on-line or off-line, or streamed, and can operate in both real-time and non real-time environments. Digital libraries (e.g., image catalog or musical dictionary), multimedia directory services (e.g., yellow pages), broadcast media selection (e.g., radio or TV channel), and multimedia editing (e.g., personalized electronic news service or media authoring) will all benefit from MPEG-7.

MPEG-7 allows for any type of AV material to be retrieved by means of any type of query material. For example, video material may be queried using video, music, or speech. MPEG-7 does not dictate how the search engine may match the query data and the MPEG-7 AV description, but it may describe a standard programming interface to a search engine. A few query examples may include:

1. *Music*- Play a few notes on a keyboard to receive a list of musical pieces that contain the required tune or images that in some way match the notes.

2. *Graphics*- Draw a few lines on a screen and receive a set of images that contain similar graphics or logos.

3. *Image*- Define objects, including color patches or textures and receive examples from which to select.

4. *Movement*- On a given set of objects, describe movements and relations between objects and receive a list of animations that match the described temporal and spatial relations.

5. *Scenario*- For a given content, describe actions to receive a list of scenarios in which similar actions occur.

6. *Voice*- Using an excerpt of a given vocalist's voice, receive a list of that vocalist's recordings or video clips.

MPEG-7 is still at an early stage and is seeking the collaboration of new experts in relevant areas. The preliminary work plan for MPEG-7 projects a committee draft in October 2000, a final committee draft in February 2001, a draft international standard in July 2001, and an international standard in September 2001.

C

CCIR-601

CIR-601 is the ISO/IEC standard that defines the image format, acquisition semantic, and parts of the coding for digital "standard" television signals. Because many chips that support this standard are available, CCIR-601 is commonly used in digital video applications for computer systems and digital television. It is central to the MPEG, H.261, and H.263 compression specifications.

CCIR-601 is applicable to both NTSC and PAL/SECAM systems. In the U.S., CCIR-601 is 720x243 fields (not frames) of luminance information, sent at a rate of 60 per second. The fields are interlaced when displayed. The chrominance channels are 360x243 by 60 fields a second, again interlaced.

CCIR-601 represents the chroma signals (Cb, Cr) with half the horizontal frequency as the luminance signal, but with full vertical *resolution*. This particular ratio of sub-sampled components is known as 4:2:2. The sampling frequency of the luminance signal (Y) is 13.5 MHz. Cb and Cr are sampled at 6.75 MHz.

CCIR-601 describes the way in which analog signals are filtered to obtain the samples. Often RGB signals are converted to YCbCr. The formulas given for the CCIR-601 color conversion are for gamma corrected RGB signals. The gamma for the different television systems are specified in CCIR Report 624-4.

The encoding of the digital signal is described in detail in CCIR Rec. 656. The correspondence between the video signal levels and the quantization levels is also specified. The scale is between 0 and 255, the luminance signal provides for 220 quantization levels; for the color-difference signals, it provides for 225 quantization levels. The signals are only coded with 8-bits per signal.

D

WORLD TELEVISION AND COLOR SYSTEMS

Country	Television System	PTT Digital Service Network Interface
Abu Dhabi	PAL	No digital services
Afghanistan	PAL B, SECAM B	No digital services
Albania	PAL B/G	No digital services
Algeria	PAL B/G	No digital services
Andorra	PAL	No digital services
Angola	PAL I	No digital services
Antarctica	NTSC M	No digital services
Antigua and Barbuda	NTSC M	No digital services
Antilles	NTSC	No digital services
Argentina	PAL N	No digital services
Australia	PAL B/G	X.21
Austria	PAL B/G	X.21
Azerbaijan	SECAM D/K	No digital services
Azores	PAL B	No digital services
Bahamas	NTSC M	No digital services
Bahrain	PAL B/G	No digital services
Bangladesh	PAL B	No digital services
Barbados	NTSC M	No digital services
Belgium	PAL B/H	X.21
Belgium (Armed Forces Network)	NTSC M	X.21
Belize	NTSC M	No digital services
Benin	SECAM K	No digital services

Bermuda	NTSC M	No digital services
Bolivia	NTSC M	No digital services
Bosnia/Herzegovina	PAL B/H	No digital services
Botswana	PAL I, SECAM K	No digital services
Brazil	PAL M	No digital services
British Indian Ocean Territory	NTSC M	No digital services
Brunei Darrussalam	PAL B	No digital services
Bulgaria	PAL	No digital services
Burkina Faso	SECAM K	No digital services
Burma	NTSC	No digital services
Burundi	SECAM K	No digital services
Cambodia	PAL B/G, NTSC M	No digital services
Cameroon	PAL B/G	No digital services
Canada	NTSC M	V.35/V.25
Canary Islands	PAL B/G	No digital services
Central African Republic	SECAM K	No digital services
Chad	SECAM D	No digital services
Chile	NTSC M	V.35
China	PAL D	No digital services
CIS (formerly USSR)	SECAM (V)	No digital services
Columbia	NTSC M	V.35
Congo	SECAM K	No digital services
Cook Islands	PAL B	No digital services
Costa Rica	NTSC M	No digital services
Cote D'Ivoire (Ivory Coast)	SECAM K/D	No digital services
Croatia	PAL B/H	No digital services
Cuba	NTSC M	No digital services
Cyprus	PAL B/G	No digital services
Czech Republic	PAL B/G (cable) / PAL D/K (broadcast)	No digital services

Denmark	PAL B/G	X.21/V.35
Diego Garcia	NTSC M	No digital services
Djibouti	SECAM K	No digital services
Dominica	NTSC M	No digital services
Dominican Republic	NTSC M	No digital services
Dubai	PAL	No digital services
East Timor	PAL B	No digital services
Easter Island	PAL B	No digital services
Ecuador	NTSC M	No digital services
Egypt	PAL B/G, SECAM B/G	No digital services
El Salvador	NTSC M	No digital services
Equitorial Guinea	SECAM B	No digital services
Estonia	PAL B/G	No digital services
Ethiopia	PAL B	No digital services
Falkland Islands	PAL I	X.21
Fiji	NTSC M	No digital services
Finland	PAL B/G	X.21
France	SECAM L	X.21
France (French Forces TV)	SECAM G	X.21
Gabon	SECAM K	No digital services
Galapagos Islands	NTSC M	No digital services
Gambia	PAL B	No digital services
Georgia	SECAM D/K	
Germany	PAL B/G	X.21
Germany (Armed Forces TV)	NTSC M	X.21
Ghana	PAL B/G	No digital services
Gibraltar	PAL B/G	No digital services
Greece	PAL B/G	No digital services
Greenland	PAL G	No digital services
Grenada	NTSC M	X.21

Guadeloupe	SECAM K	No digital services
Guam	NTSC M	V.35
Guatemala	NTSC M	No digital services
Guinea	PAL K	X.21
Guyana (French)	SECAM M	No digital services
Haiti	SECAM	No digital services
Honduras	NTSC M	No digital services
Hong Kong	PAL I	V.35
Hungary	PAL K/K	No digital services
Iceland	PAL B/G	No digital services
India	PAL B	No digital services
Indonesia	PAL B	V.35
Iran	PAL B/G	No digital services
Iraq	PAL	No digital services
Ireland	PAL I	X.21
Isle of Man	PAL	X.21
Israel	PAL B/G	V.35
Italy	PAL B/G	X.21
Ivory Coast	SECAM	No digital services
Jamaica	NTSC M	No digital services
Japan	NTSC M	X.21
Johnston Island	NTSC M	No digital services
Jordan	PAL B/G	No digital services
Kazakhstan	SECAM D/K	No digital services
Kenya	PAL B/G	No digital services
Korea, North	SECAM D, PAL D/K	No digital services
Korea, South	NTSC M	V.35
Kuwait	PAL B/G	No digital services
Kyrgyz Republic	SECAM D/K	No digital services
Laos	PAL B	No digital services
Latvia	PAL B/G, SECAM D/K	No digital services
Lebanon	PAL B/G	No digital services

Lesotho	PAL K	No digital services
Liberia	PAL B/H	No digital services
Libya	PAL B/G	No digital services
Liechtenstein	PAL B/G	X.21
Lithuania	PAL B/G, SECAM D/K	X.21
Luxembourg	PAL B/G /SECAM L	X.21
Macau	PAL I	No digital services
Macedonia	PAL B/H	No digital services
Madagascar	SECAM K	No digital services
Madeira	PALB	No digital services
Malaysia	PAL B	V.35
Maldives	PAL B	No digital services
Mali	SECAM K	No digital services
Malta	PAL B	No digital services
Marshall Islands	NTSC M	X.21
Mauritania	SECAM B	X.21
Martinique	SECAM K	No digital services
Mauritius	SECAM B	No digital services
Mayotte	SECAM K	No digital services
Mexico	NTSC M	V.35
Micronesia	NTSC M	No digital services
Midway Island	NTSC M	No digital services
Moldova	SECAM D/K	No digital services
Monaco	SECAM L, PAL G	No digital services
Mongolia	SECAM D	No digital services
Montserrat	NTSC M	V.35
Morocco	SECAM B	No digital services
Mozambique	PAL B	No digital services
Myanmar (Burma)	NTSC M	No digital services
Namibia	PAL I	No digital services
Nepal	B	No digital services

Netherlands	PAL B/G	X.21
Netherlands (Armed Forces Network)	NTSC M	X.21
Netherlands Antilles	NTSC M	X.21
New Caledonia	SECAM K	No digital services
New Zealand	PAL B/G	X.21
Nicaragua	NTSC M	No digital services
Niger	SECAM K	No digital services
Nigeria	PAL B/G	No digital services
Norfolk Island	PAL B	No digital services
North Mariana Islands	NTSC M	No digital services
Norway	PAL B/G	X.21
Okinawa	NTSC	No digital services
Oman	PAL B/G	No digital services
Pakistan	PAL B	No digital services
Panama	NTSC M	No digital services
Papua New Guinea	PAL B/G	No digital services
Paraguay	PAL N	No digital services
Peru	NTSC M	No digital services
Philippines	NTSC M	V.35
Poland	PAL D/K	No digital services
Polynesia	SECAM K	No digital services
Portugal	PAL B/G	No digital services
Puerto Rico	NTSC M	V.35
Qatar	PAL B	No digital services
Reunion	SECAM K	No digital services
Rumania	PAL D/G	No digital services
Russia	SECAM D/K	X.21
Sabah and Sarawak	PAL	No digital services
Samoa	NTSC M	No digital services
Sao Tomé E Princepe	PAL B/G	No digital services
Saudi Arabia	SECAM B/G, PAL B	No digital services

Senegal	SECAM K	No digital services
Serbia	SECAM	No digital services
Sierra Leone	PAL B/G	No digital services
Singapore	PAL B/G	V.35
South Africa	PAL I	V.35
Spain	PAL B/G	V.35
Sri Lanka	PAL	No digital services
St. Kitts	NTSC M	No digital services
St. Lucia	NTSC M	No digital services
St. Pierre and Miquelon	SECAM K	No digital services
St. Vincent	NTSC M	No digital services
Sudan	PAL B	No digital services
Surinam	NTSC M	No digital services
Swaziland	PAL B/G	No digital services
Sweden	PAL B/G	X.21
Switzerland	PAL B/G	X.21
Syria	SECAM B, PAL G	No digital services
Tahiti	SECAM	No digital services
Taiwan	NTSC	V.35
Tajikistan	SECAM D/K	No digital services
Tanzania	PAL B	No digital services
Thailand	PAL B/M	V.35
Tibet	PAL	No digital services
Togo	SECAM K	No digital services
Trinidad and Tobago	NTSC M	No digital services
Tunisia	SECAM B/G	No digital services
Turkey	PAL B	No digital services
Turkmenistan	SECAM D/K	No digital services
Uganda	PAL	No digital services
Ukraine	SECAM D/K	X.21
United Arab Emirates	PAL B/G	No digital services

United Kingdom	PAL I	X.21
United States	NTSC M	V.35
Uruguay	PAL N	No digital services
Uzbekistan	SECAM D/K	No digital services
Venezuela	NTSC M	V.35
Vietnam	NTSC M, SECAM D	No digital services
Virgin Islands	NTSC M	No digital services
Wallis & Futuna	SECAM K	No digital services
Yemen	PAL B, NTSC M	No digital services
Yugoslavia	PAL B/G	X.21
Zaire	SECAM	No digital services
Zambia	PAL B/G	No digital services
Zanzibar	PAL	No digital services
Zimbabwe	PAL B/G	No digital services

Note (1). Digital services can be obtained via private satellite services in some cases. Check with the PTT to determine which countries will allow private networks and what the conditions of service are.

E

VIDEOCONFERENCING RFP CHECKLIST

Define: Application and locations involved

√ Category of system requested for each location

√ Features required for each site

Ask: Basic System Specifications

√ Cameras (e.g., room, auxiliary, document)

√ Monitors (e.g., single, dual or other)

√ Algorithms (e.g., H.264, G.729a)

√ Network supported (e.g., ISDN, T-1, IP)

√ Network interfaces (e.g., V.35, RS-449/422)

√ Network type (e.g., dedicated, switched or hybrid)

√ If behind PBX, who provides interface components?

√ Who provides CSU/DSUs, cables & incidental equipment?

√ Who provides NT-1 (NT-2) in ISDN applications?

√ If an MCU is required, how is it configured?

√ If an inverse multiplexer is required, how is it configured?

Ask: History of manufacturer

√ Request financial data on supplier

√ Sign non-disclosure to determine mfg. future plans

√ Learn about manufacturer's/distributors partnerships

√ Examine distribution channels

√ Request information on users group

√ Determine desktop videoconferencing strategy

Ask: Characteristics of the system

√ Is the system personal or room system product?

√ Is the product software-only or software and hardware?

√ Integrated into single package or component-based?

√ If component based is integration also proposed?

√ If PC is required for operation, who provides PC?

√ Product's dimensions and its weight (if hardware)

√ State system's environmental requirements.

√ What documentation is provided with the product?

Ask: Proprietary Algorithms Offered

√ Name compression algorithms employed for video?

√ Name compression algorithms employed for audio?

√ Backward compatibility with earlier algorithms?

√ Picture resolutions offered with each algorithm?

√ Average fps across range of network speeds offered?

√ Maximum fps across range of speeds offered?

√ Echo cancellation algorithms offered?

Ask: Standards Compliance

√   Ask about compliance with H.32X Recs. individually

√   Does product offer 4CIF/CIF/QCIF or QCIF only?

√   Average/maximum fps at various H.320 bandwidths?

√   T.120/T.130 compliance?

√   Does the product support JPEG and MPEG? How?

Ask: Network-Specific Issues

√   What network arrangements does product support?

√   Can system transmit images using only POTS lines?

√   What network interfaces are required for configuration as proposed?

√   Who supplies interface equipment and cables?

√   Does proposal offer turnkey installation including network connections?

√   For what network speed is the codec optimized?

√   How is the transmission speed changed? Describe.

√   What LAN interfaces are available for the product?

√   With which ITU-T network-oriented standards does the product comply?

√   Does the product offer a built-in inverse multiplexer?

√   If so, describe it. Does it comply with BOnDInG?

Ask: Videoconferencing System Features

√   Examine system documentation during procurement.

√   Compare system features relative to how well-documented they are for each product considered.

√   What type of system interface is provided with the system?

√   Is the control unit a push button-type device, a touch-screen based unit, a PC keyboard, wireless remote or electronic tablet and stylus based?

√   Can the document camera be controlled from the primary system control unit or operator's console?

√   If the system's primary control unit is not a PC, is a PC keyboard also used to activate or change any system features that a conferee would commonly need during a conference? Which ones?

√   Does the product offer camera presets?

√   Does it offer far-end camera control?

√   Does the product offer picture-in-picture? In a dual-monitor arrangement what can be seen in the window? A single-monitor arrangement?

√ Can audio-only conferees be part of conference?

√ How many audio-only conferees can be included?

√ How are conferences set up? Describe process.

√ Can frequently-dialed numbers be stored? How many speed dialing numbers can be stored? How are they activated?

√ Do codecs automatically handshake to determine compatible algorithm?

√ Can the system be placed in a multipoint mode, and support a multipoint conference?

√ Does the product offer on-line help?

√ Is it backward compatible with codec manufacturer's previous products? Describe fully.

√ Does the product offer continuous presence?

√ Does the product offer scheduling software?

√ Does the product offer applications sharing? What OSs (e.g., MacOS, Linux)?

Ask: Audio considerations

√ List the names of proprietary audio algorithms offered and describe them.

√ What range of frequencies is encoded with the above?

√ Is bandwidth assignment flexible? How are adjustments made?

√ What method is used to eliminate echoes?

√ If echo cancellation is used, is a burst of noise used to acquaint the system with the room? Or, does unit use voices to train acoustic line-side echo cancellers?

√ What is the echo cancellation system's convergence time measured in milliseconds?

√ What is the echo cancellation system's tail length measured in milliseconds?

√ What is the echo cancellation system's AERLE rating measured in dB?

√ How many and what type of microphones are provided?

√ Is audio input provided via a speakerphone or external microphone?

√ Where are speakers located on system?

√ Are microphones unidirectional or omnidirectional?

√ Are lapel, ceiling or wireless microphones included in system price?

√ Examine products to see how well audio is synched with video particularly in a desktop videoconferencing application.

Ask: Camera considerations

√ Is the camera color and does it offer pan/tilt/zoom?

√ What is the room camera's sensitivity (in lux)?

√ What is the range of viewing angles of the room camera using the zoom?

√ Is a wide-angled lens provided? Can it be ordered optionally if not?

√ Does camera offer automatic gain control (AGC)?

√ Does camera offer white balance?

√ What type of auxiliary camera is used (light and CCD ratings, etc)?

Ask: Monitors

√ What size monitor is proposed? Single or dual?

√ What resolution does monitor offer?

√ What is the dot pitch and how does it correlate with the monitor's resolution?

Ask: Graphics Support and Subsystems

√ How is document transfer supported?

√ What document stand is included, is it backlit for slides, how large can documents be?

√ What standard resolutions are supported for still images?

√ Are high-resolution graphics supported? What resolutions?

√ How are high-resolution graphics compressed and formatted?

√ Do audiographics systems comply with the ITU-T's T.120 Recommendation?

√ Does product offer built-in ability to create and store images as slides (T.120)?

√ Is application sharing supported (T.130)?

√ Can participants electronically annotate shared images in real time (T.120)?

√ Can users at different sites interactively share, manipulate and exchange still images (T.120)?

√ Does system offer very-high-resolution graphics (including CAD support)?

√ Is graphics subsystem an integral component or is it an outboard product?

√ If external and PC-based, how does the graphics sub-system connect to the codec?

√ Can system support the need for photorealism? (X-rays, medical imaging applications).

√ Is a flatbed scanner included in price? (This is an application-driven requirement).

√ If scanner is included what is the size of the bed?

√ What is the scanner's color depth (4-bit grayscale, 24-bit color, etc.)?

Ask: Multipoint Control Unit

√ List proprietary video/ audio algorithms supported.

√ Are conference ports are universal?

√ State the MCU maximum port size. With how many ports is it proposed? Describe the expansion process.

√ Describe how the product can be upgraded to a more fully-featured model.

√ How can MCUs be cascaded to expand capacity?

√ Does cascading consume ports? Does it create delays in end-to-end transmission? Explain.

√ Describe MCU in terms of network interfaces (e.g., public, private, LAN).

√ What network speeds are supported?

√ How many simultaneous conferences can take place at any given speed and how are the MCU bridging capabilities subdivided?

√ Does system offer a director/chairperson control mode of operation? A voice-activated mode? A rotating mode? Other? Explain.

√ In voice-activated mode, how does the system prevent loud noises from diverting the camera?

√ Does the product offer full support of the ITU-T H.231/H.243 Recommendation? Explain fully.

√ How does the MCU provide graphics support and what limitations exist?

√ Does the MCU support G.729a audio?

√ Does the MCU operate in meet-me, dial-out and hybrid arrangements?

√ After a conference is arranged, state all limitations in terms of the user's ability to add conferees or change transmission speeds.

√ Are conference tones provided to signal a conferee's entry-to and exit-from a conference?

√ Can audio-only participants be included?

√ Does the system provide for voice and video encryption and if so, which methods are used?

√ How is the MCU managed? Provide detailed information on the administrative subsystem.

√ Can the MCU store a database of sites and the codecs installed at those locations as well as network information?

√ How does the network operator configure the MCU for a multipoint conference?

√ Can the administrative console be used to configure, test and diagnose problems with the MCU? Can it keep an event log and account for use?

√ What type of conference scheduling package is provided? Can conferences be scheduled in advance? Does the MCU automatically configure itself? Can it dial out to participating sites? Does the MCU reservation system allow the user to control videoconference rooms or is it used just to schedule bridge ports?

√ Is there an additional cost for the scheduling package?

Ask: Security-Related Features

√   Is AES encryption supported?

√   Are H.233/234 and H.235v3 supported?

√   How does encryption differ between point-to-point and multipoint conferences?

√   Is the operator's console password-protected?

Ask: Miscellaneous Questions

√   If system is sold through a distributor or value-added reseller, can you call the manufacturer directly?

√   Who will install the equipment? Do they have a local presence? Who will provide technician and end-user training?

√   What is the length of warranty and terms of service?

√   Where are spare parts stocked and what response time is guaranteed when the system fails? What are the terms of the second year service agreement? Cost? How is defective equipment or software repaired or replaced? Is immediate replacement in possible? In what time frame?

F

INSTALLATION PLANNING CHECKLIST

Physical space considerations

√   Acceptable and convenient location

√   Will excess capacity be brokered? Plan for it.

√   Dimensions adequate for group size

√   Minimal windows

√   Proximity to communications demarcation

√   Anteroom or next-group gathering space

√   Videoconferencing coordinator's work space

√   Sufficient electrical power and additional outlets

√   Conduit, ducts, etc. for concealing cables

√   Signage directing conferees to room

√   Sufficient privacy to meet application's needs

√   Security (locking door, card-key, punch-coded entry)

Interior Design

√   Carpeting

√   Wall color and treatments

√   Drapes, curtains or blinds

√   Facade wall and shelving (boardroom applications)

√   Table type, shape, surface and placement

√   Spectator seating for

√   Chairs and upholstery

√   Whiteboards

√   Clocks showing times at local and primary remote sites

√   Table sign identifying site and organization

Approaches to ambient noise control

√   HVAC system improvements

√   Acoustic panels and heavy draperies

√   Replacement, rearrangement or addition of lighting fixtures

√   Double-glazed windows

Network considerations

√   Network installed and tested

√   Necessary cables on hand

√   CSU/DSU or NT-1 installed and tested

√   Network documented at completion of project

Videoconferencing peripherals

√   Cameras (auxiliary)

√   Monitors (auxiliary)

√   Audio bridging

√   Still-video (graphics) components such as document cameras

√   Presentation graphics subsystems

√   PCs for file access and file sharing applications

√   Electronic whiteboards

√   Fax machines and copiers

√   DVD burner & player?

√   RGB projection system interfaces (front or rear)

√   Interfaces to CAD and/or scientific equipment (scopes, etc.)

√   I-MUX installed and tested

√   Multipoint conferencing unit installed and tested

Promotional and Usage-Oriented Preparation

√   Select personnel in remote locations to support system

√   Develop and publish videoconferencing usage policy

√   Determining chargeback policy

√   Determining scheduling system

√   Address scheduling issues (prioritization, bumping, etc.)

√   Develop instruction sheet for scheduling

√   Develop etiquette tips for point-to-point and multipoint conferences

√   Develop tips for setting up inter-company conferences

G

PERSONAL CONFERENCING
PRODUCT EVALUATION CRITERIA

Implementation

√ Client/server, server-only or client-only?

√ Point-to-point or multipoint?

√ Standards-based or proprietary?

√ Is the system offered as a cross-platform solution?

√ Product family? What capabilities are included in the family?

√ Do all applications in the product family have a similar look and feel?

√ Is the product offered in multiple languages?

Platforms Supported

√ Which Microsoft Windows versions

√ Which Mac OS versions

√ UNIX (Linux, Solaris, BSD, other)

Hardware Configuration

√ What class of processor is required (minimum / recommended)?

√ Processor clock speed (minimum / recommended)?

√ How much RAM is required (minimum / recommended?

√ How much free space on the hard drive is required?

√ What number & type slots (ISA, EISA, PCI) does the system require?

Hardware Supplied with Product

√ V.90 modem

√ Camera

√ Sound card

√ Microphone

√ Speakers

√ Video capture board

√ Hardware-level codec
√ NT-1 for ISDN BRI
√ All required cables

Networks and Protocols Supported

√ POTS
√ Circuit-switched digital ISDN BRI
√ Other circuit-switched (T-1, E-1, ISDN PRI)
√ Multirate ISDN (Nx64, H0, H11, H12)
√ IP networks (TCP or UDP)

Frame Rates, Data Transfer Rates

√ Range of operating data rates?
√ Optimal data rate?
√ Average frame rate at optimal data rate?
√ Maximum frame rate at optimal data rate?
√ Can users control bandwidth allocation between voice, video & data?

Standards Compliance

√ H.320 compatible?
√ H.264
√ H.320-based multipoint support?
√ H.321 ATM network support?
√ H.323 supported?
√ H.324 supported? If so, is V.90-compliant modem included?
√ H.323-compliant gateway between H.320 and packet network?

Picture Viewing and Resolution

√ What is the maximum / minimum size of viewing window?
√ Can the user scale the viewing window(s) using a mouse?
√ What video resolutions are supported (SQCIF, QCIF, CIF, 4CIF, other)?
√ Can the user adjust the picture resolution?
√ Can users make on-the-fly trade-offs between frame rate and resolution?
√ Does the product provide a local video window for self-view?
√ Can users control the viewing window screen quadrant?
√ Does the product offer far-end camera control?

Making and Receiving Calls

√ Does the product offer "voice-call first" (to add a video connection)?

√ Does an on-line directory permit "scroll & click" to place a video call?

√ Is an incoming caller's ID captured and displayed?

√ Is an incoming caller's number automatically stored for callback (ISDN)?

Audio Sub-system

√ In-band or out-of-band audio?

√ Are microphone and speakers included?

√ Is microphone built in to the camera?

√ Is a headset or handset included?

√ Audio/video synchronization (subjective evaluation of)

√ Audio-delivery quality (subjective evaluation of microphone)

√ Audio-receive quality (subjective evaluation of speakers)

Camera Features

√ Does the camera offer pan/tilt/zoom?

√ Can camera swivel in order to act as a document camera?

√ Choices of focal lengths?

√ Color or black and white?

√ Does the camera offer brightness, contrast, color, and tint adjustments?

√ Does the camera include an audio/video shutter for privacy?

Data Collaboration / Graphics / Document Conferencing Features & Support

√ T.120 compliance?

√ Document conferencing / data collaboration?

√ Application viewing / screen sharing?

√ Bit-mapped whiteboard (clipboard) capabilities?

√ File transfer?

√ Can the user control the flow of files during a file transfer? If so, how?

√ Can conferees prevent network floods during multipoint file transfers?

√ Chat (message pad) features?

√ Annotation tools (markers, highlighters, pens, etc.)?

√ Drawing tools (inclusion and sophistication of)?

√ Pointing tools?

√ Slide show capabilities?

√ Screen-capture and storage as "photo"?

√ Are full- or partial-screen video "freeze frame" snapshots offered?

√ What format are snapshots stored in?

Intranet / Internet Adaptations

√ Is the client browser-based?

√ RSVP client capabilities?

√ UDP (vs. TCP) sessions?

√ Does the system support the User Location Service (IP-network)?

√ Is a user automatically logged onto the ULS when they sign on to their computer?

Multipoint Capabilities

√ Does the system support multipoint conferencing? How many parties?

√ What conference management features are provided?

√ Multipoint Control Unit (bridge) required? Is it included?

√ Multipoint file transfers supported?

√ Multipoint capabilities—broadcast one-to-many

√ Multipoint capabilities—can all desktops interact with one another?

√ Multipoint capabilities—maximum number of simultaneous conferees?

√ Password protection?

√ Conference attendance control features

√ Standards supported

√ When one first signs on to a conference, does the system display the number of meeting participants and the names of the other sites?

Miscellaneous Features

√ Does the product offer traditional "voice" features (do-not-disturb, call forward, call waiting and hold)?

√ Does the product offer video messaging? What compression method is used to compress audiovisual files?

√ Context-sensitive on-line help or application "assistant" to guide user?

√ Audio-only callers included in conference?

√ Call center features?

√ MPEG video file playback from video sources?

Cost

√ Cost per desktop

√   Cost of server

√   Concurrent-use licensing cost

√   Other miscellaneous costs

Installation and Performance Tests

√   Ease-of-installation, as rated by microcomputer-oriented trade press & surveys?

√   Trial installation. How easy is the product to install?

√   Does the product include a quick install feature?

√   Does the package also include an uninstall feature?

√   Is documentation adequate?

√   Does the system minimize the transfer of non-essential data (e.g., mouse movements) in performance tests?

√   How fast can the system transfer files?

√   Did you encounter any bugs that crashed the application during performance tests?

Supplier Support / Commitment to Customer's Success

√   What support (applications, installation, testing) does the supplier provide?

√   On screen diagnostics and messages provided?

√   Are application and installation notes offered on line?

√   Does the manufacturer offer low-cost (no cost) upgrades to allow the installation to keep pace with product evolution?

√   Does the manufacturer publish a list of known bugs and product shortcomings and provide information on when and how these problems will be corrected?

√   Does the supplier provide a test-center that you can call when you are trying to get your application working?

GLOSSARY OF TERMS

— 0 ...9—

3-D

A way to visually describe an image using height, width and depth components so that the object appears to have physical depth in relation to its surroundings. 3-D modeling is the process of defining the shape and related characteristics of objects that can later be rendered in 3-D form.

4CIF

The ITUT's H.263 coding and compression standard specifies a common intermediate format to provide resolutions that are four times greater than that of CIF. Support of 4CIF in H.263 enables the standard to compete with higher bit-rate coding schemes such as MPEG. 4CIF specifies 576 non-interlaced luminance lines, that each contain 704 pixels. Support of 4CIF in H.263 is optional. 4CIF is also referred to as Super CIF, which was defined in Annex IV of H.261 in 1992.

16CIF

As is true with 4CIF, the ITU-T's H.263 standard specifies, but does not mandate, support of 16 CIF, a picture resolution that is composed of 1152 non-interlaced luminance lines, that each contain 1408 pixels. At this resolution, H.263 can provide resolutions about twice as good as NTSC television.

— A—

A/D conversion

Analog to digital conversion. A/D converters accept a series of analog voltages and describe them via a series of discrete binary-encoded values that approximate the original signal. This process is known as digitizing or sampling. D-to-A converters reverse the process.

absorption loss

The attenuation of an optical signal within a transmission system, specified as dB/km.

access	A method of reaching a carrier or a network. In the world of wide area networking, access channels (which may be copper, microwave, fiber) carry a subscriber to a carrier's point of presence (POP). In the world of local area networking access methods are used to mediate the use of a shared bus.
access method	The technique and protocols that govern how a communications device uses a local area network (LAN). The IEEE's 802 standards 802.3 through 802.12 specify access methods for LANs and MANs.
acoustic echo canceller	An AEC is used to eliminate return echoes in an acoustically-coupled tele-meeting. AEC's are used in full-duplex audio arrangements in which all microphones are active at all times. This situation causes an increase in ambient noise that an AEC is designed to mediate.
acoustics	The qualities of an enclosed space that define how sound is transmitted, its clarity and how the original signal will be distorted.
active video lines	The lines that convey information in a television signal, e.g., all except for those that occur in the horizontal and vertical blanking intervals.
additive color	Direct light that is visible directly from the source: the sun, light bulbs, video monitors. The wavelengths of direct light can be viewed in three primary colors, red, greed and blue (RGB). Combinations of these three frequencies (for that is what colors are) result in most perceivable color variations.
additive primaries	By definition, three primary colors result when light is viewed directly as opposed to being reflected: red, green and blue (RGB). According to the tri-stimulus theory of color perception, blending some mixture of these three lights can adequately approximate all other colors. This theory is harnessed in color television and video communications.
addressable	The ability of a device to receive communications over a network whereby the unique destination of the device can be specified. Typically, an address is a set of numbers (such as a telephone number) that allows a message to be intercepted and interpreted for

purposes of an application.

ADPCM　　CCITT Recommendation G.721. Adaptive Differential Pulse Code Modulation. A technique for converting an analog signal into digital form. It is based on standard sampling at 8 kHz and generates a 32 Kbps output signal. ADPCM was extended in G.726, which replaces both G.721 and G.723, to allow conversion between 64 Kbps PCM and 40, 32, 24 or 16 Kbps channels.

ADSL　　Asymmetrical Digital Subscriber Line. A method of sending high-speed data over existing copper-wire twisted pair POTS lines. ADSL, developed by Bellcore and deployed by the telephone companies, uses a modulation technique known as discrete multitone (DMT) to transmit multimegabit traffic more slowly upstream than downstream. ADSL will not work over portions of the network that attenuate signals above 4 kHz. It also can not be used where there is bridged taps and cross-coupled interference.

affine map　　A function that identifies similar frequency patterns in an image and uses one to describe all that are similar.

AIN　　Advanced Intelligent Network. A digital network architecture based on out-of-band signaling that maximizes the intelligence, efficiency, and speed of the PSTN. AIN relies on databases that store vast amounts of data about network nodes and end-points and which are accessed across a packet-switched network that is separate from the one that carries customer traffic. AIN allows moment-to-moment call routing, automatic number identification (ANI), customer call-control, and more.

algorithm　　A computational procedure that includes a prescribed set of processes for the solution of a problem in a finite number of steps; the underlying numerical or computational method behind a code or process. Algorithms are fundamental to image compression (both motion and still), because they allow an information-intensive file or transmission to be squeezed to a more economical size.

389

alias	Unwanted signals generated during the A-to-D conversion process. Aliasing is typically caused by a sampling rate that is too low to faithfully represent the original analog signal in digital form; typically, this occurs at a sampling rate that is less than half the highest frequency to be sampled.
aliasing	A subjectively disturbing distortion in a video signal that manifests in different ways depending on the cause. When the sampling rate interferes with the frequency of program material, aliasing takes the form of artifact frequencies known as sidebands. Spectral aliasing is caused by interference between two frequencies such as the luminance and chrominance signals and appears as herringbone patterns, wavy lines where straight lines should be, and loss of color fidelity. Temporal aliasing is caused when information is lost between line or field scans. It appears when a video camera is focused on a CRT. The lack of scanning synchronization produces an annoying flicker on the receiving device's screen.
amplifier	A device that receives an input signal in wave form (analog) and gives a magnified signal as an output.
amplify	To increase the magnitude of a voltage or a waveform in order to increase the strength of the signal.
amplitude	The magnitude of a waveform or voltage. Greater amplitude results when waves are set in motion with greater force. The term amplitude is also used to describe the strength of a signal.
amplitude modulation	AM. A method of changing a signal by varying its height or amplitude in order to superimpose it on a carrier wave. Used to impress radio waves (audio or video) onto a carrier in analog transmissions.
analog	Representations of numerical values by physical variables such as voltage and amplitude. Analog signals are continuously varying; indeed, depending on the precision with which they are sampled/measured, they can vary infinitely. By this we mean that each sample can produce a value that corresponds to the unique magnitude of the variable. An analog signal is one that uses electrical

	transmission methods to duplicate an original waveform, and thereby capture and convey these unique magnitudes.
analog transmission	A method of sending signals whereby the transmitted signal is analogous to the original signal. Sending a stream of continuously varying electrical waves represents the original sine wave.
animation	The process used to link a series of still images to create the effect of a motion sequence.
Annex A	(To Recommendation H.261). Inverse Transform Accuracy specification that defines the maximum tolerable error thresholds for the DCT.
Annex B	(To Recommendation H.261). Sets forth a Hypothetical Reference Decoder.
Annex C	(To Recommendation H.261). Specifies the method by which the video encoder and decoder delays are established for a particular H.261 implementation.
ANSI	The *American National Standards Institute*. A non-governmental industry organization that develops and publishes voluntary standards for the US ANSI has published standards for out-of-band signaling, for voice compression, for network performance, and for various electrical and network interfaces.
antenna	An aerial or other device that collects and radiates electromagnetic energy.
API	*Application Programmer Interface* A set of formalized software calls and routines, which can be referenced by an application program to access underlying network or other services.
application	An application is software that performs a particular useful function for a user—e.g., a spreadsheet tool, a word processing facility. Examples include word processing, spreadsheets, distance learning, document conferencing, and telemedicine.
application sharing	A collaborative conferencing feature that provides personal conference participants with read/write access to an application, even when one or more of these participants does not have the application

running at their desktop. In application sharing, one user launches and controls the application.

application viewing
In personal conferencing, the ability of one system to screen-slave off another system. Every keystroke or mouse movement made by the user who runs the application can be seen by the user at the other end, even though he/she is not running the application and has no control over it.

architecture
The design guidelines, physical and conceptual organization, and principles that describe how a system or network will support an activity. Architecture discusses scalability, security, topology, capacity and other high-level attributes.

artifacts
Spurious effects introduced to a signal that result from digital signal processing. These effects manifest as jagged edges on moving objects and flicker on fine horizontal edges.

ASCII
American Standard Code for Information Interchange, a digital coding scheme that is capable of representing 256 (text) characters. ASCII is a 7-level code for asynchronous character transmission over a network. It is a universal code.

ASIC
Application-Specific Integrated Circuit. A chip designed for a specific application or purpose.

aspect ratio
The ratio of the width to the height of an image or video displayed on a monitor. NTSC and PAL television uses an aspect ratio of 4 wide to 3 high, which is expressed 4:3.

asymmetrical compression
Techniques in which the decompression process is not the reverse of the compression process. Asymmetrical compression is more processing-intensive on the compression side so that video images can be easily decompressed at the desktop or in applications in which sophisticated codecs are not cost effective.

asynchronous
Lacking in synchronization. A method of transmitting data over a network using a start bit at the beginning of a character and a stop bit at the end. The time interval between characters may be of varying lengths. In video, a signal is asynchronous

	when its timing differs from that of the system reference signal.
ATM	Asynchronous Transfer Mode (also known as cell relay). ATM provides a single network interface for audio, video, image and text with sufficient flexibility for handling these different media types. The ATM transport technique uses a multiplexing scheme in which data are divided into small but fixed-size units called cells. Each cell contains a 48-byte information field and five-bytes of header information for a total cell size of 53-bytes. Although it is a packet switching technique, ATM can achieve the integration of all types of traffic, including those that require isochronous service.
ATSC	The Advanced Television Systems Communications. This group was formed by the Joint Committee on Inter-Society Coordination (JCIC) to establish voluntary technical standards for advanced television systems, including HDTV. In April 1995, the ATSC approved the Digital Television Standard for HDTV Transmission.
ATSC A/53	The digital television standard for HDTV transmission proposed by the Grand Alliance and approved by the Technical Subgroup of the Federal Communications Commission (FCC) Advisory Committee. The standard specifies the HDTV video formats, the audio format, data packetization, and RF transmission. New television receivers will be capable of providing high-resolution video, CD quality multi-channel sound, and ancillary data delivery to the home.
attenuation	The decrease in the amplitude of a signal. In video communications this usually refers to power loss in electromagnetic signals between a transmitter and the receiver during the process of transmission. Thus, the received signal is weaker or degraded when compared to the original transmission.
ATV	Advanced TV. Any system of distributing TV programming that results in better video and audio quality than that offered by the NTSC standard. ATV is based on digital signal processing and

transmission. HDTV can be considered one type of ATV but systems can also carry multiple pictures of lower quality.

audio

In video communications, electrical signals that carry sounds. The term is also describes sound recording and transmission systems —speech pickup systems, transmission links that carry sounds, amplifiers.

audio bridge

Equipment that mixes multiple audio inputs and feeds back composite audio to each station after it removes the individual station's input. This equipment may also be called a mix-minus audio system.

auto focus

In a camera, a device for measuring the distance of the lens from a given object is included to automatically set the lens-film distance. In videoconferences when there is almost no set-up time and subjects may be moving around from time to time this is particularly valuable.

auto iris

A process of correlating aperture size to the amount of light entering the camera. Auto Iris produces unpredictable quality in video production because white backgrounds or clothing will cause a camera to close down the lens when a person's face would be the desired gauge for the f-stop. Although it is a good feature in a videoconferencing camera, auto iris is not as effective as manual adjustment of the camera's iris in video production.

AVI

Audio Video Interleaved. The filename extension for compressed video usually used under Microsoft Windows. AVI decompression usually takes place in software. AVI compression works on key frames to achieve the maximum possible entropy through redundancy elimination. After key frames are intra-frame compressed AVI then constructs subsequent delta frames by recording only interframe differences. AVI competes with MPEG-1 although MPEG-1 produces higher-quality video.

AVT

Audio Visual Terminal. A term used in the ITU-T's H.320 specification. It refers to a videoconferencing implementation that can deliver an audio and video signal.

AWG American Wire Gauge, a standard measuring technique used for non-ferrous conductors (copper, aluminum). The lower the AWG the thicker the wire; 22 AWG cable is thicker than 26 AWG cable.

— B—

B Blue (as in RGB).

B channel The ISDN circuit-switched bearer channels, capable of transmitting 64 Kbps of digitized information.

B frame In MPEG, the B frame is a video frame that is created using bi-directional interframe compression. Computationally demanding, B frames are created by using I frames and P frames. Through bi-directional encoding the P (predictive) frame, which is created by using a past frame as a model, is compared an I (intraframe coded) frame: a frame that has had the spatial redundancy eliminated from it, without reference to other frames. Using interpolation the codec uses hints derived by analyzing past and predicted events to develop a "best-guess" present frame.

back porch The portion of a video signal that contains color burst information and which occurs between the end of the horizontal synch pulse and the start of active video.

back projection When a projector is placed behind a screen (as it is in television and videoconferencing applications) it is described as a back projection system. The viewer sees the image via the transmission of light as opposed to reflection used in front projection systems.

backbone network A transmission facility used to interconnect distribution networks of typically lower speed. A backbone network often connects major sites (hubs). From these sites, spoke-like tail circuits (spurs), emanate and, in turn, often terminate in minor hubs.

bandwidth A term that defines the information carrying capacity of a channel—its throughput. In analog systems, it is the difference between the highest frequency that a channel can carry and the lowest, measured in hertz. In digital systems the unit of measure of bandwidth is bits per second.

bandwidth-on-demand

The ability to vary the transmission speed in support of various applications, including videoconferencing. In videoconferencing applications, an inverse multiplexer or I-MUX takes a digital signal that comes from a codec and divides it into multiple 56- or 64 Kbps channels for transmission across a switched digital network. On the distant end, a compatible I-MUX recombines these channels for the receiving codec, and, therefore, ensures that, even if the data takes different transmission paths, it will be smoothly recombined at the receiving end.

BAS

Bit-rate Allocation Signal. Used in Recommendations H.221 and T.120 to transmit control and indication signals, commands and capabilities.

baseband

In a Local Area Network (LAN) context, this means a single high-speed information channel available to and shared by all the terminals or nodes on the network. Because there is sharing of this resource, means have to be provided to control access to the channel and to minimize information "collisions" and distortions caused by more than one terminal transmitting at the same time. Different types of LANs use different access methods to avoid collisions. Baseband LANs present a challenge to companies that wish to put video over their networks because video requires isochronous service (i.e., the delivery of information is smoothly timed).

Baseline Sequential JPEG

The most popular of the JPEG modes. It employs the lossy DCT to compress image data as well as lossless processes based on DPCM. The "baseline" system represents a minimum capability that must be present in all Sequential JPEG decoder systems. In this mode, image components are compressed either individually, or in groups. A single scan pass completely codes a component or group of components. If groups of components are coded, the data is interleaved; it allows color images to be

	compressed and decompressed with a minimum of buffering.
Basic Rate ISDN	See BRI.
BBC	British Broadcasting Corporation, formed in 1923 as the monopoly radio and later television, broadcaster in the United Kingdom. Also used as an abbreviation of background color cancellation.
Bell Operating Company	Any of the 22 regulated telephone companies that were "spun off" from AT&T during divestiture. The BOCs (or regional bell operating companies—RBOC) are grouped into RBHCs—Regional Bell Holding Companies such as Nynex, BellSouth and others.
Bellcore	An abbreviation for Bell Communications Research. Bellcore is a resource for software engineering and consulting that created many public network architectures for the Regional Bell Holding Companies (RBHCs) over the years. Formed to take the place of Bell Labs, which, after divestiture, severed all formal ties with the BOCs, it was owned by all seven RBHCs until the fall of 1996. At that time the RBHCs sold it to Science Applications International Corporation (SAIC), a company that specializes in government consulting for the Defense Department's Advanced Research Projects Agency (DARPA) and other federal customers.
B-frame	A mandatory MPEG picture coding technique that provides bi-directional interframe compression and which uses interpolation to predict a current frame of video data based on a past frame and a "future" frame.
binary	A method of coding in which there are only two possible values, 0 and 1, for a given digit. Each binary digit is called a "bit."
binary large objects	BLOBs. Events on a network caused by the transmission of bit-intensive images that cause bottlenecks.
B-ISDN	Broadband ISDN. The ITU-T is developing the B-ISDN standard, incorporating the existing ISDN switching, signaling, multiplexing and transmission

standards into a higher-speed specification that will support the need to move different types of information around the public switched network.

bit

Binary Digit. The basic signaling unit in all digital transmission systems.

bit plane

The memory used to represent, on a VDT, one bit per pixel. Multiple bit planes can be introduced to produce deeper color and, as the number of bit planes increase, so does the color resolution. One bit plane yields two colors (monochrome). Two yields four colors (00, 01, 10, 11), four can describe 16 colors, and so on.

bit rate

The number of bits of information transmitted over a channel in a given second. Typically expressed bps.

bit-block transfer

Bit-BLT. The movement of an associated group of pixels around on a screen. When a window is opened or moved around on a PC or X-terminal a Bit-BLT occurs.

bitmap

The total of all bit planes used to represent a graphic. Its size is measured in horizontal, vertical and depth of bits. In a one-bit (monochrome) system there is only one bit plane. As additional planes are added, color can be described. Two bit planes yield four possible values per pixel, eight yield 256, and so on.

black level

The lowest luminance level that can occur in video or television transmission and which, when viewed on a monitor, appears as the color black.

blanking interval

Period during the television picture formation when the picture is suppressed to allow the electron gun to return from right to left after a line (horizontal blanking) or from top to bottom after each field (vertical blanking).

blanking pulses

The process of transmitting pulses that extinguish or blank the reproducing spot during the horizontal and vertical retrace intervals.

block

In H.261, a block consists of 8 pixels by 8 pixels. It is the lowest element in the hierarchical video multiplex structure, which, at the top of the hierarchy, includes the picture, then a group of blocks, then a macroblock and individual blocks that comprise the macroblock. A block can be of two types, luminance or color difference.

blue
One of three additive primaries and B in the RGB.

BNC
Refers to Bayonet Neill-Concelman. A twist-lock connector widely used for the connection of video cables.

board
Boards consist of a flat backing made of insulating material and inscribed with conductive circuits etched on their surface. A fully-prepared circuit board is meant to be permanently affixed into a system as opposed to a module that is meant to slide in although the terms are now used interchangeably.

BOC
See Bell Operating Company. Also referred to as Regional Bell Operating Company or RBOC.

BOnDInG
Bandwidth On Demand Interoperability Group. This consortium of over 30 vendors developed the standard for inverse multiplexing that carries their name. Version 1.0 of the standard, approved in August 1992, describes four modes of inverse multiplexer interoperability. It allows inverse multiplexers from different manufacturers to subdivide a wideband signal into multiple 56- or 64 Kbps channels, pass these individual channels over a switched digital network and recombine them into a single high-speed signal at the receiving end.

bps
The speed at which bits are transmitted over a communications medium; in other words, the number of bits that pass a given point in a communications line in one second. The term "bps" is also used more generically to describe the information-carrying capacity of a digital channel.

branch
Part of a cable television distribution system. Branches in the network are analogous to tree limbs that attach to a main trunk. Branches provide separate locales or communities with cable television service. Branch and tree systems are being replaced with fiber-optic distribution systems in which the cable television head-end is connected via fiber optics

to a local hub.

BRI Basic Rate Interface. In ISDN there are two interfaces, the BRI and the PRI or Primary Rate Interface. The BRI offers two circuit-switched B (bearer) channels of 64 Kbps each and one packet-switched D (delta) channel that is used for exchanging signals with the network. Known in Europe as the Basic Rate Access or BRA.

bridge A bridge connects three or more conference sites so that they can simultaneously communicate. In video communications, bridges are often called MCUs, multipoint conferencing units. In IEEE 802 parlance, a bridge is a device that interconnects LANs or LAN segments at the data-link layer of the OSI model to extend the LAN environment physically.

brightness The luminance portion of a television or video signal.

broadband The term applied to networks that have bandwidths significantly greater than that found in telephony networks. Broadband systems can carry a large number of moving images or a vast quantity of data simultaneously. Broadband techniques usually depend on coaxial or optical cable for transmission. They utilize multiplexing to permit the simultaneous operation of multiple channels or services on a single cable. Frequency division multiplexing or cell relay techniques can both be used in broadband transmission.

broadcast To send information to two or more receiving devices simultaneously. The term originated in farming in which it referred to the scattering of seeds. Now it is used to describe the transmission of radio and television signals.

broadcast quality In the US this corresponds to the NTSC system's 525-line, 30 fps, 60 fields-per-second audio-video delivery system. It is also a subjective concept, used to describe an audiovisual signal that delivers quality that appears to be approximately as good as that of television.

broadcasting A means of one-way, point-to-multipoint transmission. For our purpose, we will consider this word to have two meanings. First it is the relaying of audio/visual information across the frequency spectrum where it propagates in free space and is picked up by properly equipped antennas. Second, it is the placement of information on digital networks (LAN, MAN or WAN) which can support many different applications including cable television.

BTV   See business television.

buffer   A storage reservoir designed to hold digital information in memory. Used to temporarily store data when the circuit used to transmit it is engaged or when differences in speed are involved.

burst   To send a group of bits in data communications, typically in a baseband transmission scheme. A color burst is used for synchronization in the NTSC standard for color television.

bursty data   Information that flows in short intense data groupings (often packets) with relative long silent periods between each transmission burst.

bus   A common path shared by multiple input and output devices. In the computer world a bus can be the short cable link between terminals networked in an office; in the world of circuitry a bus can be a thin copper wire on a printed circuit board. In video production, there are program buses that determine what is sent to the record deck, preview buses that allow a video source to be shown on a separate monitor, and mixing buses which work with special effect generators and which allow separate video signals to be combined.

business television   Point-to-multipoint videoconferencing. Often refers to the corporate use of video for the transmission of company meetings, training and other one-to-many broadcasts. Typically incorporates satellite transmission methods and is migrating from analog to digital modulation techniques. Also known as BTV.

B-Y   One of the color signals of a color difference video signal—the blue minus luminance signal. The formula for deriving B-Y is -.30R, -.59G and -.89B.

byte

A group of eight bits usually the smallest addressable unit of information in a data memory storage unit. Also known as an octet.

—C—

cable

A number of insulated metallic conductors or optical fibers assembled in a group and covered by a flexible protective sheath. Sometimes used in a slang sense to refer to cable television.

Cable Act of 1984

An Act passed by Congress that deregulated most of the CATV industry including rates, required programming and municipal fees. The FCC was left with virtually no jurisdiction over cable television except among the following areas: (1) registration of each community system prior to the commencement of operations; (2) ensuring subscribers had access to an A-B switch to permit the receipt of off-the-air broadcasts as well as cable programs; (3) carriage of television broadcast programs in full without alteration or deletion; (4) non-duplication of network programs; (5) fines or imprisonment for carrying obscene material; and (6) licensing for receive-only earth stations for satellite-delivered via pay cable. The FCC could impose fines on CATV systems violating the rules. The Cable Reregulation Act of 1992 superseded this Act.

cable modems

Cable modems are external devices that link PCs to cable television systems' coaxial networks to provide broadband connectivity. They work by modulating the Ethernet data that comes out of a PC, and converting it to a specific frequency to send it over the cable network. The cable modem also receives and demodulates incoming data, and re-converts it into Ethernet format. To the PC, a cable modem looks and acts like an Ethernet-based connection.

Cable Reregulation Act

Reregulation Bill 1515 that passed Congress in October of 1992, that forced the FCC to regulate cable television and cable television rates. A lengthy and extremely complex set of rules in which the FCC defined allowable monthly rates for Basic service, Expanded Basic service, equipment and installation. Rates must now conform to these FCC benchmarks and can be reduced if too high. Another provision of the Act requires cable television companies to sell cable programming to DBS operators and owners of home satellite dishes. The Act places a huge regulatory burden on the understaffed FCC. President Bush vetoed it in his last months of office but Congress overrode the veto.

camcorder   Cameras and video recorder systems packaged as a whole that permanently integrate camera, recorder and microphone components. Camcorders are used for remote production work and consumer activities.

Cameo   Macintosh-based personal videoconferencing system announced by Compression Labs in January of 1992. Developed jointly with AT&T and designed to work over ISDN lines and, most recently, Ethernet LANs. The Cameo transmits 15 fps of video and requires an external handset or headset for audio transmission.

camera   In video, an electronic (or in the past electromechanical) device used to convert visual images into electrical impulses. The camera scans an image and describes the light that is present using an optical system and a light-sensitive pick-up tube.

carrier   A term used to refer to various telephone companies that provide local, long distance or value added services; alternately, a system or systems whereby many channels of electrical information can be carried over a single transmission path.

carrier wave   A single frequency that can be modulated by another wave that contains information. Thus, the information contained in the second wave form is superimposed on the carrier for the purpose of transmitting it.

cathode ray tube Developed by a German Karl Ferdinand Braun, the CRT is a glass picture tube, narrow at one end and wide at the other. The narrow end contains a negative terminal called a cathode The cathode emits a stream of electrons. These electrons are focused or beamed with a "gun" to "paint" an image on a luminescent screen at the wide end. The inside of the wide end is coated with phosphors that react to the electron beam by lighting up, thus creating a picture. CRTs are used in TV receivers, oscilloscopes, PC monitors, and video displays. In video cameras, they are part of the scanning mechanism.

CATV

Community Antenna Television. Developed in 1958, this technology was first used to carry television programming to areas where television service was not available. The term is now used to refer to cable television; which is a method of distributing multi-channel television signals to subscribers via a broadband cable or fiber optics networks. Early systems were generally branch-and-tree types, with all programs transmitted to all subscribers, who used a channel selection switch to indicate which program they wanted.

CBR

Constant Bit Rate. A feature offered with isochronous service and required for real-time interactive video and voice communications.

CCD

Charge coupled device. Used in cameras and telecines as an optical scanning mechanism. It consists of a shift register that stores samples of analog signals. An analog charge is sequentially passed along the device by the action of stepping voltages and stored in potential wells formed under electrodes. The charge is moved from one well to another by the stepping voltages.

CCIR

Comité Consultatif International Radio-communications. An organization, part of the United Nations, that sets technical standards for international television systems as part of its responsibilities. The CCIR is now known as the ITU-R.

CCIR Rec. 656

The international standard that defines the electrical and mechanical interfaces for digital TV that operates under the CCIR-601 standard. It defines the serial and parallel interfaces in terms of connector pinouts as well as synchronization, blanking and multiplexing schemes used in these interfaces.

CCIR Rec. 601	An internationally agreed-upon standard for the digital encoding of component color television derived from the SMPTE RP125 and EBU 324E standards. It uses a 4:2:2 sampling scheme for Y, U and V with luminance sampled at 13.5 MHz and chrominance (U and V components) sampled at 6.75 MHz. After sampling, 8-bit digitizing is used. The particular frequencies set forth in the standard were chosen because they work for both 525/60 (NTSC) and 625/50 (SECAM and PAL) television systems. In the US the system specifies that 720 pixels be displayed on 243 lines of video and that 60 interlaced fields be sent per second. Chrominance channels are sent with 360 pixels on 243 lines, again at 60 fields/second. CCIR Recommendation 601 is used in professional digital video equipment.
CCIS	Common Channel Inter-Office Signaling. In this scheme, which is used for ISDN, the signaling information is carried out-of-band over a special packet-switched signaling channel.
CCITT	Abbreviation of Comité Consultatif International Téléphonique et Télégraphique, an organization that sets international telecommunications standards. The CCITT is now called the International Telecommunications Union's Telecommunications Standardization Sector or ITU-T.
CCTV	Closed circuit television. Typically used in security and surveillance applications and usually based on slow-scan technology.
CD	A high-capacity optical storage device that measures 4.75-inch in diameter and which contains multimedia or audio-only information. Originally developed for sound, CD technology was quickly seen as a storage medium for large amounts of digital data of any type. The information on a CD is digitally encoded in the constant linear velocity (CLV) format, which replaced the older CAV (constant angular velocity) format.
CD-i	Philips Compact Disc-interactive specification, which embraces the same storage concept as CD-ROM and Audio CD, except that CD-i also stores compressed

full-motion video.

CD-ROM
Compact Disc Read-Only Memory. A standard used to place any type of digital data onto a compact disc.

CDV
Compression Labs Compressed Digital Video, a compression technique used in satellite broadcast systems. CDV is the technique used in CLI's SpectrumSaver system to compress a NTSC or PAL analog TV signal so that it can be transmitted via satellite in as little as 2 MHz of bandwidth.

CellB
A Sun Microsystems Computer Corporation-proprietary video compression encoding technique developed by Michael F. Speer.

cell relay
The process of transferring data by dividing all transmissions (voice, video, text, image, etc.) into 53-byte packets called cells. A cell has 48 bytes of information and 5 bytes of address. The objective of cell relay is to develop a single high-speed network based on a switching and multiplexing scheme that works for all data types. Small cells favor low-delay, a requirement of isochronous service.

CELP
Code-Excited Linear Prediction, a low-bit audio encoding method, a low-delay variation of which is used in the ITU-T's G.728 compression standard.

channel
A physical transmission path along which signals can be sent, e.g., a video channel.

charge-coupled device
CCD (full name Interline Transfer Charge-Coupled Device or IT CCD). CCDs are used as image sensors in an array of elements in which charges are produced by light focused on a surface. They are specialized semiconductors, based on MOS technology and consist of a rectangular array of hundreds of thousands of light-sensitive photo diodes (pixels). Light from a lens is focused onto the pixels, and thereby releases electrons (charges) which accumulate in the photo diodes. The charges are periodically dumped into vertical shift registers and moved by charge-transfer so they can be amplified.

chip
An integrated circuit. The physical structure upon which circuits are fabricated as components of systems such as memory systems, and coding and

decoding systems.

chroma
The color information in a television or video signal composed of hue and saturation.

chromaticity
The quality of light, in terms of its color, as defined by its wavelength and purity. Chromaticity charts describe this combination of hue and saturation, independent of intensity. The relative proportion of R, G and B determines the color perceived.

chrominance
The combination of hue and saturation that, taken together with luminance (brightness), define color.

CIE
Commission Internationale de l'Eclairage, an international body that specifies colors based on their frequencies.

CIF
Common Intermediate Format, an optional part of the ITU-T's H.261 and H.263 standards. CIF specifies 288 non-interlaced luminance lines, that each contain 352 pixels and 144 chrominance lines that contain 176 pixels. CIF is to be sent at frame rates of 7.5, 10, 15 or 30 per second. When operating with CIF, the number of bits that result cannot exceed 256 K bits (where K equals 1024).

Cinepak
A proprietary software-based compression method developed by Radius for use on Apple Macintosh computers. Cinepak video is sent at 15 fps, with a 240-pixel high by 320-pixel wide resolution.

circuit
In telecommunications, pair of channels, which together provide bi-directional communications. A circuit includes the associated terminal equipment at the carrier's switching center.

circuit switching
The process of establishing a connection for the purpose of communication in which the full use of the circuit is guaranteed to the parties or devices that are exchanging information. After the communication has ended, the connection is released for use by others.

CISSP
(ISC)_ grants the *Certified Information Systems Security Practitioner* designation to information systems security practitioners for passing a rigorous CISSP examination and subscribing to the (ISC)_ Code of

Ethics. Additional information is available at http://www.isc2.org/welcome.html.

CIVDL   Collaboration for Interactive Visual Distance Learning. A collaborative effort by 10 leading US universities that uses dial-up videoconferencing technology for the delivery of engineering programs.

clear channel   The characteristic of a digital transmission path in which the circuit is entire bandwidth is available for information exchange. This differs from channels in which part of the channel is reserved for signaling, control or framing bits.

CLI   Compression Labs, Incorporated, San Jose, California is one of the foremost codec manufacturers in the world. CLI was the developer of the first "low-bandwidth" codec in the US, VTS 1.5. This codec was one of the first two codecs (along with one from NEC from Japan) able to compress full-motion video to 1.5 Mbps transmission speeds.

client   A service-requesting program in a client/server computing environment that solicits support (service/resources) from a server, using a network to convey its request. The client provides the important resources required by a user to interface with a server. The term 'client' has, however, strayed from this strict definition to become a catchall phrase. A client today is often assumed to be a front-end application that offers user-friendly GUI tools for such actions as setting up conferences, adding additional participants, opening applications, copying files, and storing files.

clock   An oscillator. A PC's CPU clock regulates the execution of instructions. Clocks are also used to create timing reference signals in digital networks and systems for the purpose of synchronization.

CO   Central office. A CO can be one of many types of switching systems, either analog or digital, which connect subscriber lines to other lines and network trunks on a circuit-switched basis. The two most common in the US are AT&T's 5ESS and Nortel's DMS-100, both of which are digital.

coaxial cable	A cable with one central conductor surrounded by an insulator that is, in turn, surrounded by a cylindrical outer conductor and covered by an outer insulation sheath. The insulator next to the central conductor is typically polyethylene or air and the outer conductor is typically braided copper. Coaxial cables are used to carry very high-frequency currents with low attenuation; often in cable television networks.
codec	A sophisticated digital signal-processing unit that takes an analog input and converts it to digital on the sending end. At the receiving end, another codec reverses the by reconverting the digital signal back to analog. Codec is a contraction of code/decode (some experts in the video industry assert it also stands for ompress/decompress).
codec conversion	The back-to-back transfer of an analog signal from one codec into another codec in order to convert from one proprietary coding scheme to another. The analog signal, instead of being displayed to a monitor, is delivered to the dissimilar codec where it is re-digitized, compressed and passed to the receiving end. This is obviously a bi-directional process. Carriers offer conversion service.
color	That which is perceived as a result of differing qualities of the light reflected or emitted. Humans see color via the additive process in direct light (e.g., television in which the primary colors are red, green, and blue) and the subtractive process in reflected light (e.g., books, in which the primary colors are yellow, magenta, cyan, and black). The three basic color components are hue, saturation, and brightness.
color burst	A few cycles (8 to 12) of sub-carrier frequency that serves as a color synch signal and communicates the proper hues to a video monitor or television. The color burst is part of an NTSC or PAL composite video signal. It provides a reference for the demodulation of the color information. The absence of color burst indicates black and white video or television.
color depth	The number of distinct colors that can be represented

by a piece of hardware or software. The number of bits that are used to describe a color determine the system's color depth.

color difference signal

The first step in encoding the color television signal. Subtracting the luminance information from each primary color forms the color difference signals: red, green or blue. Color difference conventions include the Betacam format, the SMPTE format, the EBU-N10 format and the MII format. Color difference signals are NOT component video signals—these are, strictly speaking, the pure R, G and B waveforms.

color monitor

CRT that works on the principle of the additive primary colors of red, green and blue. The phosphors of these monitors are tinted with these hues so that they glow in unique colors when excited with electrons beamed by an electron gun. The phosphor dots inside the visible face of the screen are organized in tightly grouped trios of red, green and blue; each trio is a pixel.

color shift

The unwanted changing of colors caused when too few bits are used to express a color.

color space

The three properties of brightness, saturation and hue can be pictured as a three-dimensional color space. The center dividing line or brightness column is the axis where no color exists at all. Hues (colors) form circles around this axis. There is also a horizontal axis that describes the amount of saturation. Highly saturated colors are closest to the center and less saturated colors are arranged toward the outer edges.

color subcarrier

The NTSC color subcarrier conveys color information and has a frequency of 3.579545 MHz. Color saturation is conveyed via signal amplitude and hue (tint) is conveyed via signal phase.

color timing

The synchronization of the burst phase of two or more video signals to ensure that no color shifts occur in the picture.

comb filter

An electrical filter that separates the chroma (color) and luma (brightness) components of a video signal into separate parts. It does this by passing some frequencies and rejecting others in between. Using a comb filter reduces artifacts but also causes some resolution loss in the picture. S-Video permits a video signal to bypass a comb filter, and thereby results in a better image.

common carrier   A telecommunications operating company that provides specific telephony services.

compact disc   CD. Information is stored on a CD's surface so that when it is scanned the fluctuations in the surface create two states: on and off. See CD-ROM.

companding   Like 'codec' or 'modem,' companding is a contraction, in this case, combining the words compressing and expanding. It refers to the reduction of the dynamic range of an audio or video signal in which the signals are sampled and transformed into non-linear codes.

component video   Transmission and recording of color television with luminance and chrominance (red, green and blue picture components) treated as separate signals. Component video is not a standard but rather a technique that yields greater signal controls and image quality. Hue and saturation (chrominance) are considered a single component, as is luminance (brightness), which is recorded at a higher frequency than chrominance; this makes it possible to exceed 400 lines of resolution. Component video is also known as Y/C. In component video, synchronization information may be added with the G signal; it can also be a separate signal.

composite video   A color television signal in which the chrominance signal is a sine wave that is modulated onto the luminance signal that acts as a subcarrier. This is used in NTSC and PAL systems.

compression   The process of reducing the information content of a signal so that it occupies less space on a transmission channel or storage device and a fundamental concept of video communications. An uncompressed NTSC signal requires about 90 Mbps of throughput, greatly exceeding the speed of all but the fastest and shortest of today's networks. Squeezing the video information can be accomplished by reducing the quality (sending fewer frames in a second or displaying the information in a smaller window) or by eliminating redundancy.

conditional frame replenishment	A process of compression in which only the changes that are present in the current video frame, when compared to the past video frame, are transmitted.
conferencing	The ability to meet over distance in which meetings can include both visual and audible information. Typically videoconferencing systems incorporate screens that can show the faces of distant-end participants, graphics, close-ups of documents or diagrams and other objects.
connection	A path that is established between two devices and which provides reliable stream delivery service.
content	In the context of video, the information object or objects that are packaged for playback by a viewer.
continuous presence	A technique used in video processing and transmission in which the sending device combines portions of more than one video image and transmits those separate images in a single data stream to a receiver or receivers. The receiver displays these multiple images on a single monitor where they are arranged side-by-side or stacked vertically. Continuous presence images can also be displayed on multiple monitors.
contone	Continuous tone. Used to describe the resolution of an image, particularly photographic-quality images.
contrast	The range of light-to-dark values of an image that are proportional to the voltage differences between the black and white levels of the signal.
convergence	The trend, now that media can be represented digitally, for historical distinctions between the boundaries of key industries to blur. Companies from consumer electronics, computer and telecommunications industries are forming alliances and raiding each other's markets. Convergence will be accelerated with the coming of the much-heralded U.S. information superhighway.

CPU

Central processing unit: the chip in a microcomputer or printed circuit board in a mainframe or minicomputer in which calculations are performed.

cross connect

The equipment used to terminate and manage communications circuits in a premises distribution system. Jumper wires or patch cords are used to connect station wiring to hardware ports of various types.

CSMA/CD

Carrier Sense Multiple Access with Collision Detection. A baseband LAN access method in which terminals "listen" to the channel to detect an idle period during which they can begin the transmission of their own messages. They might also hear a collision occur when two devices attempt to send information across the same channel at the same time.

CSU

Channel Service Unit. A customer-provided device, a CSU provides an interface between the customer and the network. The CSU ensures that a digital signal enters a communications channel in a format that is properly shaped into square pulses and precisely timed. It also provides a physical and electrical interface between the data terminal equipment (DTE) and the line.

CU-SeeMe

A free videoconferencing program (under copyright of Cornell University and its collaborators) available to anyone with a Macintosh or Windows and a connection to the Internet. CU-SeeMe allows a user to set up an Internet-based videoconference with another site anywhere in the world. By using a reflector, multiple parties at different locations can participate in a CU-SeeMe conference, each from his or her own desktop computer.

– D–

D-channel

In an ISDN network the D-channel is a signaling channel over which the carrier passes packet-switched information. The D channel can also support the

	transmission of low-speed data or telemetry sent by the subscriber.
D1	Digital Tape Component Format. The CCIR 601-approved digital standard for making digital video recordings. It records each component separately and employs the YCbCr coding scheme.
D2	Digital Tape Composite Format. A digital system that is considerably less costly than D1. It records a composite signal in an 8-bit digital format that is derived by sampling the video signal at a rate of four times the frequency of the subcarrier.
D2-MAC	One of two European formats for analog HDTV.
DACs	Digital-analog converters.
DAI	DMIF-Application Interface (DAI) is an expression (a protocol message) of the functionality provided by DMIF.
data	The much-misused term that indicates any representation, such as characters or analog quantities, to which meaning is or can be attached or assigned. Data is the plural of datum, which means "given information" or "the basis for calculations." In general usage, however, data usually means characters (letters, numbers, and graphics) that can be stored or transmitted over various types of telecommunications networks. Until recently, voice and video signals were not considered data but now that they are being converted to a digital format they are being referred to as data.
data compression	Reducing the size of a data file by reducing unnecessary information, such as blanks and repeating or redundant characters or patterns.
Data Over Cable Service Interface Specifications	A standard interface for cable modems, the devices that manipulate signals between cable TV operators and customer computers or TVs. DOCSIS specifies modulation schemes and the protocol for exchanging bi-directional signals over cable. Now known as *CableLabs Certified Cable Modems,* the ITU ratified DOCSIS 1.0 in March 1998.
dB	Decibel. One-tenth of a Bell and a logarithmic

measure of electric energy. A decibel expresses the ratios between voltages, sound intensities, currents, the power (amplitude) of sound waves, and so on.

DBS Direct broadcast satellite, a transmission scheme used for program delivery, most generally entertainment. There are several DBS providers; none of the systems that they use are, however, compatible. These systems provide downstream speeds of 400 Kbps to 30 Mbps or higher and upstream speeds of 28 Kbps or higher.

DC Direct Current.

DCE Data communications equipment. The network side of an equipment to network connection with DTE, data terminal equipment that plugs into DCE, which provides a means of network connection.

DCT See Discrete Cosine Transform.

DCT coefficient An expression of a pixel block's average luminance, as used in DCT. The value for which the frequency is zero in both the horizontal and vertical directions.

decode A process that converts an incoming bitstream, that consists of digitized images and sounds, into a viewable and audible state.

decryption Decryption reverses an encryption process to return an encrypted transmission into its original form. Decryption applies a special decoding algorithm (key) to the encrypted exchange; any party without access to the proper key required for decryption can not receive the transmission in an intelligible form.

dedicated access A leased, private connection between a customer's equipment and a telephone company location, most often that of an IXC.

dedicated leased line A transmission circuit leased by one customer for exclusive use around the clock. See also private line.

delay The time required for a signal to pass between a sender and a receiver; alternately the time required for a signal to pass through a device or conductor.

demultiplex The process of separating two or more signals previously combined for transmission over a shared channel. Multiplexing merges multiple channels onto

	one channel prior to transmission; de-multiplexing separates them again at an appropriate network node. Often shortened to *DeMUX*.
depth of field	The range of distances from the camera in which objects around a focal point (the distance from the surface of a lens or mirror to its subject)) will be in focus. Use of a smaller lens aperture increases depth of field.
DES	Data encryption standard developed by the National Bureau of Standards and specified in the FIPS Publication 46, published in January 1977. Generally replaced by AES.
desktop	A term used to refer to a desktop computer or workstation used by an individual user.
Dense wavelength division multiplexing	Dense WDM transmits up to 32 OC-48 (2.5Gbps) signals over a single fiber, and offers the potential to transmit trillions of bits per second over one fiber.
dial-up	The ability to arrange a switched connection, whether analog or digital, by entering a terminating address such as a telephone number, in order that the call can be routed by the network. Differs from point-to-point services that can be used only to communicate between two locations.
digital	Information that is represented using codes that consist of zeros and ones (binary coding). Binary digits (bits) can be either zeros or ones and are typically grouped into "words" of various lengths—8-bit words are called bytes.
Digital Access Cross Connection System	DACS. A switch/multiplexer that permits DS0 cross connection from one T-1 transmission facility to another.
Digital Signal Hierarchy	A TDM multiplexed hierarchy used in telephone networks. DS0, the lowest level of the hierarchy, is a single 64 Kbps channel. DS-1 (1.544 Mbps) is 24 DS0s. DS-2 (6.312 Mbps) is four DS-1 signals multiplexed together; DS-3 (44.736 Mbps) is seven DS-2 signals multiplexed together. At the top of the hierarchy is DS-4, which is six DS-3 signals and which requires a transmission system capable of handling a

	274.176 Mbps signal.
Digital Signal Processor	See DSP.
digital transmission	The conveyance over a network of digital signals by means of a channel or channels that may assume in time any one of a defined set of discrete values or states.
digitizing	The process of sampling an analog signal so that it can be represented by a series of bits.
digitizing tablets	Graphics systems used in conjunction with videoconferencing applications. Using a special stylus or electronic "pen" a meeting participant can write on the tablet and the message can be viewed by the distant end and, if desirable, stored on a PC. Photos and text can also be annotated electronically. These devices are unsettling to use, however, because no image appears on the tablet, thus it is difficult to orient the letters.
direct broadcast satellite	The use of satellite to broadcast directly to homes or businesses. Subscribers are obliged to purchase and install a satellite dish. DBS service originated in Japan, which is composed of many islands and which has a harsh geography that includes mountains, rivers, valleys and ridges that made it very difficult to plan and execute a terrestrial broadcasting CATV system.
directional couplers	In cable systems, multiple feeder cables are coupled with these devices to match the impedance of cables.
Discrete Cosine Transform	DCT. A pixel-block based process of formatting video data in which it is converted from a three-dimensional form to a two-dimensional form suitable for further compression. In the process, the average luminance of each block or tile is evaluated using the DC coefficient. Used in the ITU-T's H.261 and H.263 videoconferencing compression standards and the ISO/ITU-T's MPEG and JPEG image compression recommendations.
Discrete Wavelet Transform	DWT. Based on same principles as DCT, this method segregates the spectrum into waves of different lengths and then processes all frequencies to retain the image's sharp lines, which are partially lost in the

DCT process.

distance learning The incorporation of video and audio technologies into the educational process so that students can attend classes and training sessions in a location distant from that where the course is being presented.

dithering In color mapping, dithering is a method of representing a hue, not available in the color map, by intermingling the pixels of two colors that are available and letting the eye-brain team average them together into a single perceived median color.

Divestiture In early 1982, AT&T signed a Consent Decree and thereby agreed to spin off the 22 local Bell Operating Companies (BOCs). These were grouped into seven Regional Bell Holding Companies that managed their business, coordinated their efforts, and provided strategic direction. Restrictions were placed on both the BOCs and AT&T. The US Department of Justice stripped AT&T of the Bell name (except in their Bell Labs operation), the authority to carry local traffic, and the option to discriminate in favor of their former holdings. BOCs were awarded Yellow Pages publishing and allowed to supply local franchised-monopoly services, but were not allowed to provide information services or to manufacture equipment. They could carry calls only within local access and transport areas (LATAs). This agreement changed the composition of telephone service in the US when it became effective on, January 1, 1984.

document camera A specialized camera that is mounted on a long adjustable neck for taking pictures of still images—pictures, graphics, pages of text and objects—for manipulation such as a video conference.

downlink The communications path from a satellite to an earth station.

DMIF Delivery Multimedia Integration Framework is a session protocol for the management of multimedia streaming over generic delivery technologies. It is similar to FTP except that, where FTP returns data, DMIF returns pointers toward (streamed) data.

DOCSIS See *Data Over Cable Service Interface Specifications.*

DPCM	Differential Pulse Code Modulation is a compression technique that sends only the difference between *what was* (a past frame of video information) and *what is* (a present frame). DPCM requires identical codecs on each, the transmitting and receiving ends to predict, from a past frame of pixels, what the present frame will be. The transmitting frame, after computing its prediction, compares the actual to its speculation and sends information on the difference. In turn, the receiving codec interprets these differences (called errors) and makes adjustments to the present video frame.
driver	A driver is software that provides instructions for reformatting or interpreting software commands for transfer to and from peripheral devices and the central processing unit (CPU). Many printed circuit boards require a software driver in order for the other PC components to work correctly. In other words, the driver is a software module that drives the data out of a specific hardware port. Video drivers may be required for desktop video.
DS-0	Digital service, level zero or DS-zero. A single 64 Kbps channel in a multiplexing scheme that is part of the North American and European digital hierarchy and which results from the process of digitizing an analog voice channel through the application of time division multiplexing, pulse code modulation and North American or European companding.
DS-1	(T-1). A multiplexing scheme that is part of the North American digital hierarchy and which specifies how to subdivide 1.544 Mbps of bandwidth into twenty-four 64 Kbps channels using time division multiplexing, pulse code modulation and North American companding. Europe has a similar multiplexing hierarchy that produces a 2.048 Mbps signal (E-1/CEPT).
DS-3	Also called T-3. A multiplexing scheme that is part of the North American digital hierarchy and which specifies how to subdivide 45 Mbps of bandwidth into 28 T-1 (1.544 Mbps) carrier systems—a total of

672 channels. The techniques for accomplishing this include time division multiplexing, pulse code modulation and North American companding.

DSP
Digital signal processor. A specialized computer chip designed to perform speedy and complex operations on digitized waveforms. Useful in processing sound and video signals.

DSU
Data service unit. A device used to transmit digital data on digital transmission facilities. It typically interfaces to data terminal equipment via an RS-232-C or other terminal interface, connecting this device to a DSX-1 (digital system cross connect) interface.

DSX
Digital Signal Cross-Connect, a panel that connects digital circuits to allow cross-connections via a patch and cord system.

DTE
Data terminal equipment. The equipment side of an equipment to network connection with DCE, data communications equipment connecting the DTE to the network for the purposes of transmission.

dual 56
Combination of two 56 Kbps lines (usually switched 56) to yield a 112 Kbps channel used for low-bandwidth videoconferencing. Dual 56 allows for direct dialing of a videoconference call and can be obtained from IXCs or LECs.

DVC
Digital Video Cassette. A DVC is a storage medium based on a _-inch-wide tape made up of metal particles. The DVC source is sampled at a rate similar to that of CCIR-601 but additional chrominance subsampling (4:1:1 in the NTSC 30 kHz mode) provides better resolutions. When the NTSC 30-fps signal is encoded, the image frame resolution is 720 pixels by 480 lines with 8 bits used for each pixel.

DVD
DVD (Digital Versatile Disk) is a type of CD. It has storage capacity of 17 gigabytes, which is much higher than CD-ROM's 600 Mbytes and a higher data delivery than that of CD-ROM. DVD uses MPEG and Dolby compression algorithms to achieve its storage capacity.

DVE
Digital Video Everywhere was developed by InSoft Incorporated as a core software architecture for open

systems platforms. It features a hardware-independent API for running multimedia collaborative and conferencing applications across LANs, WANs, and TCP/IP-based networks.

DVI Digital Video Interactive. A proprietary compression and transmission scheme from Intel. Compression is asymmetric, requiring relatively greater amounts of processing power at the encoder than at the decoder. DVI played an important role in the PC multimedia market.

DWDM See *Dense Wavelength Division Multiplexing.*

— E—

earth station An antenna transmitter or receiver that accepts a signal from a satellite and may, in turn, be capable of transmitting a signal to a satellite.

EBIT A three-bit integer that indicates the number of bits that should be ignored in the final data octet of a RTP H.261 packet.

EBU European Broadcasting Union, an organization that developed technical recommendations for the PAL television system.

echo The reflection of sound waves that results from contact with non-sound-absorbing surfaces such as windows or walls. Reflected signals sound like a distorted and attenuated version of the original sound that, in videoconferencing, would primarily be speech. Echoes in telephone and videoconferencing applications are caused by impedance mismatches, points where energy levels are not equal. In a four-wire to two-wire connection, the voice signal moving along the four-wire section has more energy than the two-wire section can absorb; consequently, the excess energy bounces back along the four-wire path. When the return-delay approaches 500 ms, speakers will hear their own words transmitted back at them.

echo cancellation A process that uses a mathematical process to predict an echo and remove that portion of the signal from an audio waveform to eliminate acoustical echo.

echo modeling	A mathematical process whereby an echo is created from an audio waveform and, subsequently subtracted from that form. The process involves sampling the acoustical properties of a room, calculating approximately what form an echo might take, and then removing that information from the signal.
echo suppression	The insertion of mild attenuation in audio transmit and/or receive signal paths. Used to reduce annoying echoes in the audio portion of a videoconference, an echo suppresser is a voice-activated "on/off" switch that is connected to the four-wire side of a circuit. It silences all sound when it is on by temporarily deadening the communication link in one direction. Unfortunately, echo suppression clips the remote end's new speech as it stops the echo.
EIA	Electrical Interface Association, a standards-setting group in the US that defines standards for interfaces including jacks and other network connections.
EISA	Extended Industry Standard Architecture. The independent computer industry's alternative to IBM's Micro-Channel data bus architecture that IBM used in its PS/2 line of desktop computers. EISA is a 32-bit bus or channel that expands on the original Industry Standard Architecture (ISA) 16-bit channel. EISA capabilities are particularly important when a machine is being used for processor-intensive applications.
electromagnetic spectrum	The range of wavelengths that includes light, physical movement of air (sound) or water, radio waves, x-rays, etc. These wavelengths propagate throughout the entire universe.
electromagnetic waves	Oscillations of electric and magnetic forces that produce different wavelengths and which include light, radio, gamma, ultraviolet, and other forms of energy.
electron beam	A stream of electrons focused on a phosphorescent screen and fired from a "gun" to create images. Deflecting it from magnetic coils or plates so that it hits a precise location on the screen focuses the beam.

electrons	Negatively charged particles that, along with positively charged protons, allow atoms to achieve a neutral charge.
encoding	The process through which media content is transformed for the purposes of digital transmission.
encryption	The conversion, through the application of an algorithm, of an original signal into a coded signal in order to secure it from unauthorized access. Typically the process involves the use of "keys" that can unlock the code. The most common encryption standard in the US Bureau of Standard's is the DES (data encryption standard) which enciphers and deciphers data using a 64-bit key specified in the Federal Information Processing Standard Publication 46, published in January, 1977.
enhanced standard	Standards enhancement is a common practice of videoconferencing codec manufacturers. They begin with an algorithm that is compliant with a formal standard (typically the ITU-T's H.320 algorithm). However, they add capabilities that are transparent to dissimilar products that run the standard, but which improve implementation when operating exclusively in their proprietary environment.
entrance facilities	In a premise distribution system, entrance facilities are the point of interconnection between a building wiring system and a telecommunications network outside the building.
entropy	A measure of the information contained in any exchange. Entropy is the goal of nearly every compression technique. If information is duplicated, the excess amount (that portion that is over and above what is necessary to convey) is redundant. The goal is to remove the redundant portion of whatever is being conveyed (for our purposes, motion, video, audio, text, and still images). The remainder is entropy-information.
envelope	The variations of the peaks of the carrier that contain the video signal information.
envelope delay	The characteristics of a circuit that result in some simultaneously transmitted frequencies reaching

their destination before others.

ESS AT&T's term for Electronic Switching System, a central office switch designed for "class five" or "exchange office" operations. The 5ESS is one of the more common CO switches in the US and, for that matter, the world.

ether The medium that, according to one theory, permeates all space and matter and which transmits all electromagnetic waves.

Ethernet The original CSMA/CD LAN as invented by Xerox in 1973 and standardized by Digital Equipment Corporation, Intel and Xerox. Also known as an IEEE 802.3 network, Ethernet is most commonly implemented as 10Base-T, a LAN based on unshielded twisted pair and arranged in a star topology. In Ethernet, all terminals are connected to a single common highway or bus.

Ethernet switch A device that examines an Ethernet packet to determine which port it should exit. It conserves bandwidth by sending packets only to destination ports rather than broadcasting them (as in the case of a simple hub) to all ports. Ethernet switches provide virtual LAN segments between end-stations. Like telephone switches, they employ circuit-switching to support parallel Ethernet conversations. Video communications benefit from the resulting bandwidth conservation.

ETSI European Telecommunications Standards Institute, a group charged with devising Europe-wide telecommunications standards. This group issues Common Technical Regulations (CTRs) some of which pertain to video communications.

exchange A telephone company's switching center or wire center where subscriber lines terminate at a central location and are switched to other lines and to trunks.

Exchange Microsoft's electronic messaging server.

— F—

fade

The gradual reduction in the received signal that results in the disappearance of a picture to black or, in the case of fade in, the gradual introduction of light to an all-black image.

fast packet multiplexing

The combination of TDM, packet and computer intelligence to allow multiple digital signals to share a high-speed network. Fast packet multiplexing assumes a clean network and, therefore, does not buffer information, but rather moves it along, and assumes it will arrive with little or no degradation. The two most common forms of fast packet multiplexing are cell relay and frame relay.

FCC

Federal Communications Commission, a US regulatory body established by Congress in 1934.

FDDI

Fiber Distributed Data Interface. An ANSI standard for a 100 Mbps token-ring-based fiber-optic LAN. It uses a counter-rotating token ring topology and is compatible with the physical layer of the OSI model.

FDDI II

Emerging ANSI standard that incorporates both circuit and packet switching over fiber optics at 100 Mbps. Not compatible with the original FDDI standard.

FDM

Frequency Division Multiplexing. A method of transmitting multiple analog signals on a single carrier by assigning them to separate and unique frequency bands and then transmitting the entire aggregate of all frequency bands as a composite.

fiber optics cable

A core of fine glass or plastic fibers that has extremely high purity and is surrounded by a cladding of a slightly lower refractive index. Light or infrared pulses that carry coded information signals are injected at one end. They pass through the core using a method of internal reflection or refraction. Attenuation is low; pulses travel as far as 3,500 feet or more before needing regeneration.

field

A normal television image is comprised of interlaced fields—each field contains one-half of a video frame's information. Each field carries half the picture lines. In the NTSC video standard 60 fields/30 frames are sent in a second and in the European PAL and SECAM systems 50 fields/25 frames are sent in a second.

field interlacing

In television, the process of creating a complete video frame by dividing the picture into two halves with one containing the odd lines and the other containing the even lines. This is done to eliminate flicker.

field of view

The focal length combined with the size of the image area light is focused on. Field of view is measured in degrees and is not dependent on the distance from a subject.

field sequential system

The first broadcast color television system, approved by the FCC in 1950. It was later changed to the NTSC standard for color broadcasting.

file

A set of related records treated by a computer as a complete unit. Retrieving information from one computer memory storage facility to another is called a file transfer.

FireWire

A high-speed serial bus that Apple Computer and Texas Instruments developed that allows for the connection of up to 63 devices. Also known as the IEEE 1394 standard, it provides 100, 200, 400, 800, 1600, and 3200 Mbps transfer rates. FireWire supports isochronous data transfer, and thereby guarantees bandwidth for multimedia operations. It supports hot swapping and multiple speeds on the same bus. FireWire is used increasingly for attaching video devices to the computer.

filter

An electrical circuit or device that passes a selected range of energy while rejecting all others that are not in the proper frequency range.

firmware

Programs or data that are stored on a semiconductor memory circuit, often a plug-in board. Firmware-stored memory is non-volatile.

FITL

Fiber in the loop. A Bellcore technical advisory, number 909, Issue 2, which addresses the ability of telephone companies to provide video services and delivery using optical fiber cable.

FlexMux

The synchronized delivery of streaming data may require the use of different QoS schemes as it traverses multiple public and private networks. MPEG defines the delivery-layer FlexMux multiplexing tool to allow grouping of Elementary Streams (ES) with low multiplexing overhead. It may be used to group ES with similar QoS requirements, to reduce the number of network connections, or to reduce end-to-end delay. Use of the FlexMux tool is optional, and the FlexMux layer may be empty if the underlying TransMux instance provides adequate functionality.

flicker

An unwanted video phenomenon that results when the screen refresh rate is too slow or when the two interlaced fields that make up a video frame are not identically matched. Flicker is also known as jitter and sometimes as jutter. In the early days of television, interlacing two video fields to create a single video frame was used to combat flicker and eliminated it fairly well at rates over 40 fields per second. Flicker is not a problem with non-interlaced video display formats (those used for computer monitors).

fluorescent lights

Used for illumination in many corporate and public settings. These lights produce spectral frequencies of a less balanced nature than incandescent lights, and cause problems in a videoconference or video production process, and may cause 60 Hz flicker.

footprint

The primary service area covered by a satellite where the highest field intensity is normally in the center of the footprint, with intensity reducing toward the outer edges.

format

In television the specific arrangement of signals to compose a video signal. There are many different ways of formatting a video signal: NTSC, PAL, SECAM, component video, composite video, CD-I, QuickTime and so on.

forward motion vector

Used in motion compensation. A motion vector is a physical quantity with both direction and magnitude; a course of motion in which pixels are the objects that are moving. A forward motion vector is derived from a video reference frame sent previously.

forward prediction   A technique used in video compression. Specifically, compression techniques based on motion compensation in which a compressed frame of video is reconstructed by working with the differences between successive video frames.

fps   Frames per second. The number of frames contained in a single second of a moving series of video images. 30 fps is considered to be 'full-motion' video in Japan and the US, while 25 fps is considered to be full-motion video in Europe.

fractal compression   An asymmetrical compression technique that shrinks an image into extremely small resolution-independent files and stores it as a mathematical equation rather than as pixels. The process starts with the identification of patterns within an image, and results in collection of shapes that resemble each other but that have different sizes and locations. Each shape-pattern is summarized and reproduced by a formula that starts with the largest shape, repeatedly displacing and shrinking it. Patterns are stored as equations and the image is reconstructed by iterating the mathematical model. Fractal compression can store as many as 60,000 images on one CD-ROM and competes with techniques such as JPEG, which uses DCT to drop redundant information. One disadvantage of fractal compression is that it requires considerable processing power. JPEG is much faster but fractal compression is more efficient; it squeezes information into smaller files. Applications using fractal compression center on desktop publishing and presentation creation.

fractal geometry   The underlying mathematics behind fractal image compression, discovered by two Georgia Tech mathematicians, Michael Barnsley and Alan Sloan.

Fractal Image Format   FIF. A compression technique that uses on-board ASIC chips to look for patterns. Exact matches are rare; the basis of the process is to find close matches using a function known as an affine map.

fractional T-1	FT-1 or fractional T-1 refers to any data transmission rate between 56 Kbps and 1.544 Mbps. It is typically provided by a carrier in lieu of a full T-1 connection and is a point-to-point arrangement. A specialized multiplexer is used by the customer to channelize the carrier's signals.
frame	An individual television, film or video image. There are either 25 or 30 frames per-second sent with television, 24 are sent in moving picture films. A variable number, typically between 8 and 30, are sent in videoconferencing systems, depending on the transmission bandwidth offered.
frame buffer	Memory used for holding the data for a single and complete frame (or screen) of video. Some systems have enough memory to store multiple screens of data. Frame buffers are evaluated in terms of how many bits are used to represent a pixel. The more bits that are used, the "deeper" the color. The greater the number of buffers used to store captured video frames the higher the possible frame rate.
frame dropping	The process of dropping video frames to accommodate the transmission speed available.
frame grab	The ability to capture a video frame and temporarily store it for later manipulation by a graphics input device.
frame grabber	A PC board used to capture and digitize a single frame of NTSC video and store it on a hard disk.
frame rate	The number of frames that are sent in a second, and the equivalent of fps. NTSC video has a frame rate of 30 fps. PAL and SECAM send frames at a rate of 25 per second and motion picture (film) images are delivered at a frame rate of 24 per second.
frame store	A system capable of storing complete frames of video information in digital form. This system is used for television standards conversion, computer applications that incorporate graphics, video walls and various video production and editing systems.
freeze frame	A single frame from a video segment displayed motionless on a screen. Also, a method of

transmitting video images in which less than one or two frames are sent in any given second. Sometimes known as slow-scan, still video or captured frame video. When freeze frame video is viewed, the viewer sees successive images refreshing a scene but they lack a sense of continuous motion.

frequency

The number of times that a periodic function or oscillation repeats itself in a specified period of time, usually one second. The unit of measurement of frequency is typically hertz (Hz) which is used to measure cycles or oscillations per second.

frequency interleaving

The process of putting hue and color saturation information into the vacant frequency spectrum via a process of offsetting that chrominance spectrum exactly so that its harmonics are made to fall precisely between the harmonics of the luminance signal.

frequency Modulation

FM. A method of passing information by altering the frequency of the carrier signal.

front porch

With reference to a composite video signal, the front porch is the portion of the signal that occurs between the end of the active video on each horizontal scan line and the beginning of the horizontal synch pulse.

full-motion video

Generally refers to broadcast-quality video transmissions in which the frame rate is 30 per second in North American and Japan and 25 per second in Europe.

– G–

G

Green, as in RGB. The G signal sometimes includes synchronization information.

G.711

CCITT Recommendation entitled, "Pulse Code Modulation (PCM) of Voice Frequencies." G.711 defines how a 3.1 kHz audio signal is encoded at 64 Kbps using Pulse Code Modulation (PCM) and either mu-law (US and Japan) or A-law (Europe) companding.

G.721

CCITT Recommendation that defines how a 3.1 kHz audio signal is encoded at 32 Kbps using Adaptive Differential Pulse Code Modulation (ADPCM).

G.722	CCITT Recommendation that defines how a 7.5 kHz audio signal is encoded at a data rate of 64 Kbps.
G.723	ITU-T Recommendation entitled, "Dual Rate Speech Coder for Multimedia Communications Transmitting at 5.3 and 6.4 Kbps." G.723 is part of the H.323 and H.324 families
G.728	ITU-T Recommendation for audio encoding using Low Delay Code Excited Linear Prediction (CELP). The bandwidth of the analog audio signal is 3.4 kHz whereas after coding and compression the digitized signal requires a bandwidth of 16 Kbps.
G.729	Coding of speech at 8 Kbps/s using conjugate-structure algebraic-code-excited linear-prediction (CS-ACELP). Part of the ITU-T's H.323 standard for videoconferencing over non quality-of-service guaranteed LANs.
gain	An increase in the strength of an electrical signal; measured in decibels.
gamma	A display characteristic of CRTs defined in the following formula: light = volts $\wedge$ gamma. Gamma values range in CRTs, with most ranging between 2.25 and 2.45.
gate	A digital logic component whose output state depends on the states of the logic signals presented to its inputs.
GBIC	Gigabit Interface Converter
generic	Applicable to a broad range of applications, i.e., application independent.
Genlock	Short for generator locking device, a genlock is a device that enables a composite video system (e.g., a television) to accept two signals simultaneously. A genlock holds on to one set of signals while it processes a second set. This allows the combination of signals (video and graphics) into a single stream of video. Traditionally, genlock allowed multiple devices (video recorders, cameras, etc.) to be used together with precise timing so that they captured a scene in unison.
geostationary satellite	A satellite whose orbital speed is matched or

synchronized with respect to the earth's rotation so that the satellite remains fixed, relative to a specific point on the earth's surface.

geosynchronous orbit   A satellite orbit that is synchronous with a point on the earth's surface. An orbit, approximately 22,500 miles above the earth's equator, where satellites circle at the same speed as the rotation of the earth, and thereby appear stationary to an earth bound observer.

ghost   A duplicate shadowy image set off slightly from the primary picture image.

GIF   Graphical Interchange Format, a commonly-used graphics file format for transferring image files over a network.

giga   A prefix denoting a factor of 10^9, abbreviated GHz.

Gigabit Interface Converter   An interface for attaching network devices to fiber-based transmission systems (e.g., Fibre Channel or Gigabit Ethernet). GBICs convert serial electrical signals to serial optical signals and vice versa. GBIC modules contain ID and system information that a switch can use to assess network device capabilities, and are hot-swappable.

GOB   Group of Blocks. The ITU-T's H.261 Recommendation process divides a video frame into a group of blocks (GOB). At CIF resolutions there are 12 such blocks and at QCIF there are three. A GOB is made up of 12 macroblocks (MB) that contain luminance and chrominance information for 8448 pixels. A GOB relates to 176 pixels by 48 lines of Y and the spatially-corresponding 88 pixels by 24 lines of CB and CR. Each GOB is divided in 33 macroblocks, in which the macroblocks are arranged in three rows that each contain 11 macroblocks.

GPT   A codec manufacturer. GEC and Plessy combined their efforts to create an early video codec that depended on BT's compression algorithm. GPT codecs are prevalent in Europe.

Grand Alliance   The merger of four competitive HDTV systems, previously proposed to the FCC as individual standards, into a collaborative venture that is backed by all the supporters of the original four separate systems. The companies that comprise the Grand Alliance include AT&T, the David Sarnoff Research Center, General Instrument, MIT, North American Philips, Thomson Consumer Electronics, and Zenith Electronics. The new HDTV standard was approved in late 1996 after a compromise between the computer and broadcasting industries was reached.

graphical user interface (GUI). A "point and click" computing environment that depends on a pointing device such as a mouse to evoke commands and move around a screen as opposed to a standard computer keyboard.

graphics Artwork, captions, lettering, and photos used in programs, or the presentation of data of this nature, in a video communications system.

graphics accelerator A video adapter with a special processor that can enhance performance levels during graphical transformations. A CPU can often become bogged down during an activity of this type. The accelerator allows the CPU to execute other commands while it takes care of the compute-intensive graphics processing.

graphics coprocessor A programmable chip that speeds video performance by carrying out graphics processing independently of the computer's CPU. Among the coprocessor's common abilities are drawing graphics primitives and converting vectors to bitmaps.

gray scale A range of luminance levels with incremental brightness steps from black to gray to white. The steps generally conform to a logarithmic relationship.

groupware A term for software that runs on a LAN and allows coworkers to work collaboratively and concurrently. Groupware is now being enhanced with video capabilities and many of the new desktop conferencing products offer capabilities commonly associated with groupware.

GSM See Global System for Mobile Communications

guardband Unused radio frequency spectrum between television channels.

GUI See Graphical user interface.

— H—

H.221 The framing portion of the ITU-T's H.320 Recommendation that is formally known as "Frame Structure for a 64 to 1920 Kbps Channel in Audiovisual Teleservices." The Recommendation specifies synchronous operation in which the coder and decoder handshake and agree upon timing. Synchronization is arranged for individual B channels or bonded H0 connections.

H.222 ITU-T Recommendation, ratified in July of 1995, that specifies "Generic coding of moving pictures and associated audio information. H.222.0 is entitled, "MPEG-2 Program and Transport Stream" and H.222.1 is entitled, "MPEG-2 streams over ATM."

H.223 Part of the ITU-T's H.324 standard that specifies a control/multiplexing protocol, it is formally called "Multiplexing protocol for low bitrate multimedia communication." Annexes, passed in February 1998, deals with video packetization with H.263, and with multimedia mobile communications over error-prone channels.

H.225 Part of the ITU-T's H.323 Recommendation. H.225 establishes specific messages for call control such as signaling, registration and admissions as well as the packetizing and synchronizing of media streams. Annex I addresses call signaling protocols and media stream packetization for packet-based multimedia communications systems.

H.226 Part of the ITU-T's H.32x Recommendation, H.226 establishes channel aggregation protocols for multilink operation on circuit-switched networks.

H.230 A multiplexing Recommendation that is part of the ITU-T family of video interoperability Recommendations. Formally known as "Frame-synchronous Control and Indication Signals for Audiovisual Systems," the Recommendation specifies how individual frames of audiovisual information are

	to be multiplexed onto a digital channel.
H.231	ITU Recommendation, formally known as "Multipoint Control Unit for Audiovisual Systems Using Digital Channels up to 2 Mbps," H.231 was added to the ITU-T's H.320 family of Recommendations in March 1993. It specifies the multipoint control unit used to bridge three or more H.320-compliant codecs together in a multipoint conference.
H.235	Formerly known as H.SECURE, H.235 was ratified in February 1998 to establish security and encryption protocols for H.323 and other H.245-based multimedia terminals. H.235 is not itself a method of encryption and security, but establishes a methodology for leveraging such protocol families as IPSec (that, in turn, leverages security association and key distribution protocols such as ISAKMP and Oakley).
H.242	Part of the ITU-T's H.320 family of video interoperability Recommendations. Formally known as the "System for Establishing Communication Between Audiovisual Terminals Using Digital Channels up to 2 Mbps," H.242 specifies the protocol for establishing an audio session and taking it down after the communication has terminated.
H.243	ITU-T "System for Establishing Communication Between Three or More Audiovisual Terminals Using Digital Channels up to 2 Mbps."
H.245	Part of the ITU's H.324 protocol that defines control of communications between multimedia terminals. Its formal Recommendation name is "Control protocol for multimedia communication."
H.246	ITU recommendation to establish a method for interworking of H-Series multimedia terminals with H-Series multimedia terminals and voice/voiceband terminals on the PSTN (more accurately, GSTN, or General Switched Telephone Network) and ISDN.
H.247	ITU recommendation to establish a method for multipoint extension for broadband audiovisual

communication systems and terminals.

H.261
The ITU-T's Recommendation that allows dissimilar video codecs to interpret how a signal has been encoded and compressed, and to decode and decompress that signal. The standard, formally known as "Video Codec for Audiovisual Services at Px64 Kbps," it also identifies two picture formats: the Common Intermediate Format (CIF) and the Quarter Common Intermediate Format (QCIF). These two formats are compatible with all three television standards: NTSC, PAL and SECAM.

H.262
ITU recommendation for using the MPEG-2 compression algorithm for compressing a videoconferencing transmission.

H.263
H.263 is the ITU-T's "Video Coding for Low Bit Rate Communication." It refers to the compression techniques used in the H.324 Recommendation and in other ITU-T recommendations, too. H.263 is similar to H.261 but differs from it in several ways. Its advance negotiable options enable implementations of the Recommendation that employ them to achieve approximately the same resolution quality as H.261 at half the data rate. H.263 uses half-pixel increments for motion compensation (optional in both standards) while H.261 uses full-pixel precision and specifies a loop-filter. In H.263, portions of the data stream hierarchy have been rendered optional. This allows the codec to be configured for enhanced error recovery or, alternatively, for very low data rate transmission. There are, in H.263, four options designed to improve performance. They are advance prediction, forward and backward frame prediction, syntax-based arithmetic coding and Unrestricted Motion Vectors. H.263 supports not only H.261's optional CIF and mandatory QCIF resolutions but also SQCIF (128x96 pixels), 4CIF (704x576 pixels) and 16CIF (1408x1152 pixels).

H.310
H.310 is an ITU-T draft standard for broadcast (HDTV) quality video conferencing. It describes how MPEG-2 video can be transmitted over high-speed ATM networks. The standard includes subparts such

as H.262 (MPEG-2 video standard), H.222.0 (MPEG-2 Program and Transport Stream), H.222.1 (MPEG-2 streams over ATM) and a variety of G.7XX audio compression standards. H.310 takes H.320 to the next generation of networks (broadband ISDN and ATM).

H.320 An ITU-T standard formally known as "Narrow-band Visual Telephone Systems and Terminal Equipment." H.320 includes a number of individual recommendations for coding, framing, signaling and establishing connections (H.221, H.230, H.321, H.242, and H.261). It applies to point-to-point and multipoint videoconferencing sessions and includes three audio algorithms, G.721, G.722 and G.728.

H.321 H.321 adapts H.320 to next-generation topologies such as ATM and broadband ISDN. It retains H.320's overall structure and some of its components, including H.261 and adds enhancements to adapt it to cell-switched networks.

H.322 H.322 is an enhanced version of H.320 optimized for networks that guarantee Quality of Service (QoS) for isochronous traffic such as real-time video. It will be first used with IEEE 802.9a isochronous Ethernet LANs.

H.323 H.323 extends H.320 to Ethernet, Token-Ring, and other packet-switched networks that do not guarantee QoS. It will support both point-to-point and multipoint operations. QoS issues will be addressed by a centralized gatekeeper component that lets LAN administrators manage video traffic on the backbone. Another integral part of the spec defines a LAN/H.320 gateway that will allow any H.323 node to interoperate with H.320 products. In addition to H.320's H.261 video codec H.323 also H.263, a more sophisticated video codec. Also in the family are H.225 (specifies call control messages); H.245 (specifies messages for opening and controlling channels for media streams); and the G.711, G.722, G.723, G.728 and G.729 audio codecs.

H.324 H.324 defines a multimedia communication terminal that operates over POTS lines. It can incorporate H.261 or H.263 video encoding. The H.324 family includes H.223, for multiplexing, H.245 for control, T.120 for audiographics, and the V.90 modem specification.

H.332   ITU Recommendation for large multipoint H.323 conferences (defined as "loosely coupled conferences").

H.450.1   ITU Recommendation that defines a generic functional protocol for the support of supplementary services in H.323.

H.450.x   ITU Recommendation that defines specific supplementary services to H.323 that leverage H.450.1.

H0   The switched 384 Kbps dialing standard as defined by an ITU-T Recommendation.

H11   The switched 1.544 Mbps dialing standard as defined by an ITU-T Recommendation.

H12   The switched 1.920 Mbps dialing standard as defined by an ITU-T Recommendation.

handshake   The electrical exchange of predetermined signals by devices in preparation of a connection. Once completed, the transmission begins. In video communications, handshake is a process by which codecs interoperate by seeking out a common algorithm.

hard disk   A sealed mass storage unit that allows archival of large amounts of data and smaller amounts of video or audio information.

hardware   The mechanical, magnetic and electronic components of a system. Examples are computers, codecs, terminals, and scanners.

harmonic distortion   A problem caused by the production of nonlinearities in a communications channel in which harmonics of the input frequencies appear in the output channel.

harmonics   In periodic waves, the component sine waves that are integer multiples—exponents—of the fundamental frequency.

HD	*MAC* One of the two analog HDTV standards the EC's Council of Telecommunications Ministers promoted, and which was abandoned in favor of digital formats that were subsequently proposed.
HDTV	High Definition Television. TV display systems with approximately four times the resolution of standard television systems. Systems labeled as HDTV typically offer at least 1,000 lines of resolution and an aspect ratio of 16:9.
header	A string of bits in a coded bit stream that is used to convey information about the data that follows.
head-end	The originating point in a cable television system where all the television and telecommunication signals are assembled for transmission over a broadband cable system to the subscribers. Signals may be generated from studios, received from satellites or conveyed via landline or microwave radio trunks. Head-end sites can be located in the same building as the cable operator's business headquarters or they can be sited close to satellite receiving dishes. They are generally equipped with amplifiers or signal regenerators.
hertz	Hz. The unit of measurement used in analog or frequency-based networks named after the German physicist Heinrich Hertz. One hertz equals one cycle-per-second.
high frequency	Electromagnetic waves between 3 and 30 MHz.
high-pass filter	A device that attenuates the frequencies below a particular cut-off point, a high-pass filter is useful in removing extraneous sounds such as hum and microphone boom.
holography	The recording and representation of three-dimensional objects by interference patterns formed between a beam of coherent light and its refraction from the subject. The hologram is the holographic equivalent of a phonograph. A three-dimensional image is stored which, if broken apart, can still reconstruct a whole image, though at reduced definition. Holograms require laser light for their

creation and, in many cases, for their viewing.

horizontal

H. In television signals, the horizontal line of video information that is controlled by a horizontal synch pulse.

horizontal blanking interval

The period of time during which an electron gun shuts off to return from the right side of a monitor or TV screen to the left side in order to start painting a new line of video.

horizontal resolution

Detail expressed in pixels that provide chrominance and luminance information across a line of video information.

host

A computer that processes data in a communications network.

hub

A network or system signal distribution point where multiple circuits convene and are connected. Some type of switching or information transfer can then take place. Switching hubs can also be used in Ethernet LAN environments in an arrangement whereby a LAN segment might support only one workstation. This relieves congestion through a process called micro-segmenting.

hue

The attribute by which a color may be identified within the visible spectrum. Hue refers to the spectral colors of red, orange, yellow, green blue and violet. A huge variety of subjective colors exist between these spectral colors.

Huffman encoding

A lossless, statistically based entropy-coding compression technique. Huffman encoding is used to compress data in which the most frequently occurring code groups are represented by shorter codes and rarely occurring code groups are represented by longer codes. The idea behind Huffman encoding is similar to that of Morse code in that short, simple codes are assigned to common characters and longer codes are assigned to lesser-used characters. Huffman coding, used in H.261 and JPEG, can reduce a test file by approximately 40%.

hybrid fiber/coax

Hybrid fiber-coax (HFC) is being used by the cable companies to provide local loop service. HFC uses fiber links from the central site to a neighborhood hub. Existing coax cable connects the hub to several hundred nearby homes. Once installed, HFC provides subscribers with telephone service over the cable and, in addition, interactive broadband signaling, which can support videoconferencing over the cable network.

— I—

IANA   See Internet Corporation for Assigned Names and Numbers

ICCAN   See Internet Corporation for Assigned Names and Numbers

I frame   Intraframe coding, as specified by the MPEG Recommendation in which an individual video frame is compressed without reference to other frames in the sequence. Also used in the JPEG compression method. I-frame coding generates much data and is used when there are major scene changes or as a reference frame to eliminate accumulated errors that result from interframe coding techniques.

I-Signal   In the NTSC color system, the I signal represents the chrominance on the orange-cyan axis.

IEC   International Electrotechnical Commission. Also synonymous with IXC—interexchange carrier.

IEEE   Institute of Electrical and Electronics Engineers; the organization that evolved from the IRE—the Institute of Radio Engineers.

IEEE 802 standards   Various IEEE committees that developed LAN physical layer standards and other attributes of LANs and MANs. IEEE 802.9 specifies the isochronous Ethernet communications protocol, which adds 6 Mbps of bandwidth to the same twisted pair copper wire that carriers 10 Mbps Ethernet. This additional bandwidth takes the form of 96 ISDN circuit switched B channels. IEEE P 802.14 is a protocol for "Cable-TV Based Broadband Communications Network," another IEEE 802.X standard that has relevance for video communications.

IETF	Internet Engineering Task Force. The standards body that adopted the MIME protocol for sending video-enabled e-mail and other compound messages across TCP/IP networks and one of two working bodies of the Internet Activities Board.
image	An image, for the purposes of this book, refers to a complex set of computer data that represents a picture or other visual data. Images can offer high-resolution (in other words, they can be composed of many pixels-per-square-inch that causes them to have a photographic quality) or low-resolution, (crude animation, sketches and other pictures that contain a minimal number of pixels-per-square-inch.
image bit map	Digital representation of a graphics display image as a pattern of small units of information that represents one or more pixels.
imaging	The process of using equipment and applications to capture visual representations and transmit them over telecommunications networks. Imaging can be used to move and store medical X-rays, for engineering and design applications that allow engineers to develop 3-D images of a products or components for the purpose of design-refinement, and to store large quantities of documents.
impedance	The ratio of voltage to current as it is measured along a transmission circuit.
in-band signaling	Networks exchange information between nodes through the use of signaling. In-band signaling uses the communications path to exchange such information as request for service, addressing, disconnect, busy, etc. In-band signaling has been largely replaced by out-of-band signaling, an example of which is the ITU-T's Signaling System Number 7 or SS7.
input	The data or signals entering a computer or, more generally, the signal applied to any electronic device or system.
input selector	A switch or routing device that is used to select video inputs in a digital picture manipulator or keying device.

integrated circuit	An electronic device *IC*'s are made by layering semiconductive materials and packaging both active and passive circuits into a single multi-pin chip.
intelligent peripheral	In the AIN, the intelligent peripheral or IP collects information from designated "triggers" and issues SS7 requests based on that information.
inter-LATA	Communications that cross a LATA boundary and, therefore, must be carried by an IXC in accordance with the MFJ.
interactive	Action in more than one direction, either simultaneously or sequentially. In interactive video there is a bi-directional interplay between two or multiple parties, this is different from television, which is a send-only system that does not allow the receiver to respond to the signal.
interactivity	The ability of a video system user to control or define the flow of information, including the capability of communicating in return on a peer-to-peer basis.
interexchange carrier	IXC or IEC. The long distance companies in the US that provide inter-LATA telephony and communications services. This concept is going away as a result of the passage of the Telecommunications Deregulation and Competition Act of 1996.
interface	A shared boundary or point common to two or more systems across which a specified flow of information passes. If interfaces are carefully specified, it should be possible to plug in, switch on, and operate equipment purchased from different sources. *Interface* can also refer to a circuit that converts the output of one system or subsystem to a form suitable to the next system or subsystem.
interference	Unwanted energy received with a signal that scrambles it or otherwise degrades it.
interframe coding	A compression technique used in video communications in which the redundancies that can be found between successive frames of video information are removed. This type of compression is also referred to as temporal coding.

interlace	A technique for "painting" a television monitor in which each video frame is divided into two fields with one field composed of the odd-numbered horizontal scan lines and the other composed of the even-numbered horizontal scan lines. Each field is displayed on an alternating basis—this is called interlacing. It results in very rapid screen refreshment and is done to avoid visual flicker. Interlacing is not used on computer monitors and is not required as part of the US HDTV standard.
International Standards Organization	A non-treaty standards-making body that, among other things, helped to develop the MPEG and JPEG standards for image compression. See ISO.
International Telecommunications Union	One of the specialized agencies of the United Nations that is composed of the telecommunications administrations of roughly 113 participating nations. Founded in 1865, it was invented as a telegraphy standards body. It now develops international standards for interconnecting telecommunications equipment across networks. Known as the CCITT until early 1994, the ITU played a big role in developing audiovisual communications standards through its Telecommunications Standardization Sector (see ITU-T). In this book we refer to standards as CCITT Recommendations if they were ratified before 1994 and ITU-T Recommendations if they were adopted after 1994.
Internet	The Internet is a vast collection of computer networks throughout the world. Originally developed to serve the needs of the Department of Defense's Advanced Research Projects Agency, the Internet has become a common vehicle for information exchange, commerce, and communications between individuals and businesses.
Internet Corporation for Assigned Names and Numbers	ICANN, the successor to the Internet Assigned Names Authority (IANA) is a not-for-profit organization with an international Board of Directors that oversees the operations of the necessary central coordinating functions of the Internet.
Internet reflector site	A multipoint technique used by Internet-based

videoconferencing. A reflector is a server that bounces signals back to all parties of a multipoint connection, allowing any number of users to conference with each other.

internetworking   The ability of LANs to interconnect and exchange information. Bridges and routers are the devices that connect these LANs with each other, with WANs, and with MANs. Users can, therefore, exchange messages, access files on hosts other than those on their own LAN, and access and utilize various inter-LAN applications.

interoperability   The ability of electronic components produced by different manufacturers to communicate across product lines. The trend toward embracing standards has greatly furthered the interoperability process.

interpolation   A compression technique with which a current frame of video is reconstructed using the differences between it and past and future frames.

intra-LATA   A connection that does not cross over a LATA boundary, and that regulated LECs are allowed to carry on an end-to-end basis. Recently passed legislation may make this term obsolete.

intraframe coding   A compression technique whereby redundancies within a video frame are eliminated. DCT is an intraframe coding method. Interframe coding is also called spatial coding.

inverse multiplexer Equipment that receives a high-speed input and breaks it up for the network into multiple 56- or 64 Kbps signals so that it can be carried over switched digital service. The I-MUX supports the need for dialing over this network. It also provides the synchronization necessary to recombine the multiple channels into a single integrated transmission at the receiving end. This is necessary because, once the transmission has been subdivided into multiple channels, some parts of the transmission may take a different and longer path to the destination than others. This results in a synchronization problem even though the difference may be measured in milliseconds. The job of the I-MUX is to make sure that the channels are recombined into a cohesive single transmission.

IP　　Internet Protocol. Defined in RFC 791, the Internet Protocol is the network layer for the TCP/IP protocol stack. It is a connectionless, best-effort packet-switched delivery mechanism. By connectionless we mean that IP can carry data between two systems without first requiring their connection. IP does not guarantee the data will reach its destination; neither can it break large packets down into smaller ones for transmission or recover lost or damaged packets. For that it relies on the Transmission Control Protocol.

ISDN　　Integrated Services Digital Network. ITU-T standard for a digital connection between user and network. A multiplicity of services, voice, data, full-motion video and image can be delivered. Two interfaces are defined, Basic Rate Interface, and Primary Rate Interface. Basic Rate Interface (BRI) provides two 64 Kbps circuit-switched bearer (B) channels for customer data and one 16 Kbps delta (D) or signaling channel that is packet-switched. The Primary Rate Interface (PRI) differs between the US and Europe: in Europe it provides thirty 64 Kbps B channels and two 64 Kbps D channels (one for signaling and the other for synchronization) and in the US there are twenty-three B channels and one 64 Kbps D channel, which is used for synchronization and signaling.

ISO　　International Standards Organization. Founded in 1946, ISO has approximately 90 member countries, including the US's ANSI and its British equivalent, the British Standards Institution (BSI). ISO members include the national standards bodies of most industrialized countries. The ISO, in conjunction with the ITU-T, continues to develop video communications standards. Their joint efforts have produced the Joint Photographic Experts Group (JPEG) and the Motion Picture Experts Group (MPEG) family of encoding techniques.

isochronous　　Channels that are capable of transmitting timing information in addition to the data itself, and which thereby allow the terminals at each end to maintain exact synchronization with each other (the term is pronounced 'I-SOCK-ron-us').

IsoENET An approach to providing isochronous service over Ethernet. It derives an additional 96 switched B channels from 10Base-T. National Semiconductor and IBM developed it.

ITCA International Teleconferencing Association, a professional association organized to promote the use of teleconferencing (audio and video conferencing). Located in Washington D.C.

ITFS Antenna System Instructional Television Fixed Service. A type of local distance learning system that provides one-way over-the-air television service at microwave frequencies. These frequencies are reserved for educational purposes. The signals can be received only by television installations equipped with a converter that changes the signals to NTSC.

ITU One of the specialized agencies of the United Nations, the International Telecommunications Union was founded in 1865, before telephones were invented, as a telegraphy standards body.

ITU-R Formerly the United Nation's CCIR, the ITU-R sets international standards that relate to how the radio spectrum is allocated and utilized.

ITU-T The International Telecommunications Union's Telecommunications Standardization Sector. The ITU-T (formerly the CCITT) is the telephony standards-setting arm of the ITU. The ITU-T developed the H.320, H.323 and H.324 protocol suites that define how videoconferencing codecs interoperate over various types of networks.

IXC Interexchange carrier, long distance service providers in the US that provide inter-LATA service.

—J—

Jitter Random signal distortion, a subjective effect caused by time-base errors. Jitter in a reproduced video image moves from left to right and causes an irregular. Jitter can be controlled by time-base correction.

JPEG   Joint Photographic Experts Group. The joint ITU-T/ ISO standard for still image compression.

— K—

K   A prefix that denotes a factor of one thousand or 10^3. This abbreviation is used frequently when discussing computer networks and digital transmission systems. In computer parlance, K stands for 1024 bytes as in 64K bytes memory. In transmission systems the K stands for 1000—e.g., 64 Kbps means 64 thousand bits per second bandwidth.

K-Band   In satellite communications a frequency band that ranges between 10.9 and 36 GHz.

Kbps   Kilobits per second. The transport of digitized information over a network at a rate of 1,000 bits in any given second.

kHz   Kilohertz. One thousand Hertz or cycles.

kilobyte   1,024 bytes of data.

kiosk   A small structure, open at one or more sides located in a public place. Kiosks today can be designed and equipped with motion video-enabled and multimedia displays to enable even the least computer-literate people to access information on a particular topic or product. Typically they use point-and-click or single-press methods for video-enabled information retrieval.

Ku-Band   In satellite communications systems, this frequency band ranges between 10.9 and 11.7 GHz for Direct Broadcast Satellite (DBS) service.

— L—

LAN   Local Area Network. A computer network, usually within a single office complex and increasingly within a single department, that connects workstations, servers, printers and other devices, and thereby permits resource sharing and message exchange.

LAN segmentation   Splitting one large LAN into multiple smaller LANs. This technique is used to keep LANs from becoming congested with multimedia and desktop video applications.

laser
Light Amplification by the Stimulated Emission of Radiation. A device that produces optical radiation both in the range of visible light wavelengths and outside this range. Lasers are used in optical fiber networks to transmit signals through a process of oscillation.

LATA
Local Access and Transport Areas. The areas within which the Bell Operating and independent telephone companies can provide transport services. The Telecommunications Deregulation Act of 1996 is changing this distinction as sufficient competition in local access is achieved. The FCC is gradually allowing RBOCs and IXCs to compete outside of their historic constraints.

latency
Latency refers to the transmission delays encountered in a packet-switched network. Latency refers to the tendency of packet-switched networks to slow down when they become congested. It is lethal for data types such as voice and video that require constant bit rates to ensure that these time-sensitive data types arrive with minimal and predictable delays.

lavaliere microphone
A small clip-on microphone, popular because it is unobtrusive and maintains a fixed distance to the speaker's mouth.

layer
Layering is an approach taken in the process of developing protocols for compatibility between dissimilar products, services and applications. In the seven-layer OSI model, layering breaks each step of a transmission between two devices into a discrete set of functions. These functions are grouped within a layer according to what they are intended to accomplish. Each layer communicates with its counterpart through header-defined interfaces. The flexibility offered through the layering approach allows products and services to evolve while accommodating changes made at the layer level.

leased line
A transmission facility reserved from a communications carrier for the exclusive use of a

subscriber. See private line.

LEC	Local Exchange Carrier. LECs include the Bell Operating Companies and independent telephone companies that provide the subscriber local loop.
Lempel-Ziv-Welch compression	LZW, a data compression method named after its developers. LZW techniques are founded on the notion that a given group of bytes can be compressed by substituting an index phrase.
lens	An optical device of one or more elements in an illuminating or image forming system such as the objective of a camera or projector.
level	The intensity of an electrical signal.
line	A facility between a subscriber and a telecommunications carrier, generally a single analog communications path physically composed of twisted copper wire.
line of sight	Some spectrum-based transmission systems need an unobstructed path between the transmitter and receiver. The ability for the receiver to see the sender is described as having a clear line of sight.
lip synch	The techniques used to maintain a precise coordination between the delivery of sound and the facial movements associated with speech. Lip synch is required because it takes much longer to process the video portion of a signal than the audio portion (the video part contains much more information and takes longer to compress). To mitigate the delay, a codec incorporates an adjustable audio delay circuit to delay-equalize sounds with faces, and allow talking-head images to look natural.
LiveBoard	LiveWork's group-oriented electronic whiteboard that allows users in different locations to interactively share information. A wireless pen is used to create documents or images viewed and edited in real-time over POTS lines. LiveBoard can also display full-motion color video and provide audio for mixed media applications.
local loop	In telephone networks, the lines that connect customer equipment to the switching system in the

central office.

local multipoint distribution service (LMDS)
Broadband wireless local loop service that offers two-way digital broadcasting and interactive service over a special type of microwave.

logarithm
An exponential expression of the power to which a fixed number or base must be raised in order to produce a given number. Logarithms are usually computed to the base of 10 and are used for shortening mathematical calculations—e.g., 10^3 = 1,000.

Logarithmic Quantization Step Encoding
A digital audio encoding method that yields 12-bit accuracy using only eight-bit encoding. It is a technique used in CCITT G.711 audio.

loop filter
H.261 codecs break an original video field into 8-x-8 pixel blocks, then filters and compresses each block. A loop filter is used to separate the images' spatial frequencies so that they can be uniquely compressed.

lossless compression
Techniques that, when they are reversed, yield data identical to the original.

lossy compression
A compression technique that, when reversed, contains less information than the original image. Techniques that, when reversed, do not yield data that is identical to the original. Lossy compression techniques are good at compressing motion video sequences and achieve much higher compression ratios than lossless techniques.

low-pass filter
A device that attenuates the frequencies above the cut-off point and which is used in sound synthesis and digital audio sampling.

LSI
Large Scale Integration. Refers to degree of miniaturization achieved in manufacturing complex integrated circuits.

luma
The brightness signal in a video transmission.

lumen
A measure of light emitted by a source.

luminance
The information about the varying light intensity of an image produced by a television or video camera. Also called brightness.

Lux	A contraction of luminance and flux and a basic unit for measuring light intensity. A Lux is approximately 10 foot candles.
LWZ	Lempel-Ziv-Welch coding developed by three mathematicians in the 1970s. LWZ looks at repetitive bit combinations and represents the most commonly occurring sequences with abbreviated codes.

— M—

MAC	Multiple Analog Component, one of the original European HDTV formats.
MAC	In data networking, media access controllers are the interface between an application and an output device.
macroblock	In H.261 encoding, a macroblock is made up of six 8x8-pixel blocks of which two contain chrominance information and four contain luminance information components. A macroblock contains 16x16 pixels-worth of Y (luminance) information and the spatially-corresponding 8x8 pixels that contain CB and CR information.
MAN	Metropolitan Area Network.
Mbps	Megabits per second or approximately one million bits per second.
MCU	See multipoint conferencing (or control) unit.
media	Air, water, space and solid objects through which information travels. The information that is carried through all natural media takes the form of waves.
Media Access Control	MAC. The network protocol that controls access to a LAN's bandwidth and could include techniques that would reserve bandwidth for isochronous communications.
megabyte	1,048,576 bytes—2^{20}.
Megastream	British Telecom's brand name for a digital system that offers the customer thirty 64 Kbps channels. Used in high-bandwidth video conferencing systems, whether point-to-point or dial-up (using the ITU-T's H0 dialing standard).
memory	A digital store for computer data that commonly

	consist of integrated circuits. There are two primary types: read-only memory or ROM and random-access memory or RAM.
Merged Gateway Control Protocol	A protocol for controlling media gateways. It is the combination of Level 3's IPDC and Bellcore's and Cisco's SGCP (Simple Gateway Control Protocol).
mesh topology	A networking scheme whereby any node can communicate directly with any other node.
MFJ	Modified Final Judgment. The out-of-court settlement that broke up the Bell System, severing the Bell Operating Companies from AT&T.
MGCP	See Merged Gateway Control Protocol.
microsegmenting	The process of configuring Ethernet and other LANs with a single workstation per segment using hubs and inexpensive wiring. The goal is to remove contention from Ethernet segments to guarantee enough bandwidth for desktop video and multimedia. With each segment having access to a full 10 Mbps of Ethernet bandwidth, users can avail themselves of applications that incorporate compressed video.
microwave	Radio transmission above 1 GHz used for transmitting communications signals, including video pictures, between various sites using dish aerials.
middleware	Middleware is a layer of software that provides the services required to link distributed pieces of an application across a network, typically in a client/server environment. Middleware is transparent, hiding the fact that components of an application are distributed. It helps developers to overcome the difficulties of interfacing application components into heterogeneous network environments.
MIME	Multipurpose Internet Mail Extension. Developed and adopted by the Internet Engineering Task Force, this applications protocol is designed for transmitting mixed-media files across TCP/IP networks.
mixing	The process of combining separate sound and/or visual sources to make a smooth composite.

modulation	Alteration of the amplitude, frequency or phase of an analog signal by a different frequency in order to impress a signal onto a carrier wave. Also can be used in digital signaling to make multiple signals share a single channel. In the digital realm, this is generally achieved by dividing a channel into time slots, into which each separate signal contributes bits in turn.
moiré	In a video image, a wavy pattern caused by the combination of excessively high frequency signals; the mixing of these signals results in a visible low frequency that looks a bit like French watered silk, after which it is named.
monitor	A precision display for viewing a picture. The word is now often applied to any TV set with video (as opposed to RF) input. A receiver-monitor is a set that is both a conventional TV used for viewing broadcast television but which also accepts video input from a computer. Also refers to computer displays.
monochrome	Reproduction in a single color, normally as a black and white picture.
motion compensation	An interframe coding technique that examines statistics of previous video frames to predict subsequent ones. In a motion sequence each pixel can have a distinct displacement from one frame to the next. From a practical point of view, however, adjacent pixels tend to have the same displacement. Motion compensation uses motion vectors and rate-of-change analysis to exploit inter-frame similarities. An image is divided into rectangular blocks and a single displacement vector to describe the movement of all pixels within the block.
Motion JPEG	A compression technique that applies JPEG compression to each frame of a video clip.
motion vectors	In the H.261 Recommendation this optional calculation can be included in a macroblock. Motion vector data consists of one code word for the horizontal information followed by a code word for the vertical information. The code words can be of variable length and are specified in the H.261 standard. Performing motion vector calculations

places more demand on a processor and many systems do not include this capability, which is optional for the source coder. If a source encoder performs the motion compensation processing, the H.261 Recommendation requires that the decoder use it during decompression.

motion video capture board   A device designed to capture, digitize and compress multiple frames of video for storage on magnetic or optical storage media.

MPEG   Motion Picture Experts Group JTC1/SC29/WG11 standard that specifies a variable rate compression algorithm that can be used to compress full-motion image sequences at low bit rates. MPEG is an international family of standards that are not used for videoconferencing, but are more generally used for video images stored on CD-ROM or video servers and retrieved in a computer or television application. MPEG compresses YUV SIF images. It uses the same DCT algorithm used in H.261 and H.263 but uses 16 by 16 pixel blocks for motion compensation.

MPEG-1   Joint ISO/IEC recommendation known as "Coding of Moving Pictures and Associated Audio for Digital Storage Media at up to about 1.5 Mbps." MPEG-1 was completed in October 1992. It begins with a rather low image resolution; about 240 lines by 352 pixels per line. This image is compressed at 30 frames per second. It provides a digital image transfer rate up to about 1.5 Mbps and compression rates of about 100:1. The MPEG-1 specification consists of four parts: system, video, audio, and compliance testing.

MPEG-2   The MPEG-2 is Committee Draft 13818, and can be found in documents MPEG93/N601, N602 and N603. This draft was completed in November 1993. Its formal name is "Generic Coding of Moving Pictures and Associated Audio." MPEG-2 is targeted for use with high-bandwidth applications aimed at digital television broadcasts. It specifies 720-by-408 pixel resolution at 60 fields-per-second with data transfer rates that range from 500 Kbps to more than 2 Mbps.

MPEG-4	"Very Low Bitrate Audio-Visual Coding." An 11/94 Call for Proposals by ISO/IEC resulted in a Working Draft specification in 11/96.
MS-DOS	A computer operating system developed by Microsoft and originally aimed at the IBM PC.
multicasting	Conferencing applications that typically use packet-switched transmission to broadcast a signal that can be received by multiple recipients, all of whom are listening on a single multicasting address.
multi-mode fiber	Optical fiber with a central core of a diameter sufficient to permit light-pulses to zig-zag from side to side as well as to travel straight down the middle of the core. Step-indexed fibers have a sudden change of refractive index where cladding meets core. Signals are reflected back at this boundary. Graded index fibers have a gradual change of refractive index with radial distance from the center of the core, pulses are refracted back by this change.
multimedia	According to the Defense Information Systems Agency: "Two or more media types (audio, video, imagery, text, and data) electronically manipulated, integrated, and reconstructed in synchrony." Generally, multimedia refers to these media in a digital format for it is the digitization of voice and video information that is lending much of the power to multimedia communications.
Multiple Systems Operator	When one cable company runs multiple cable systems, it creates what is referred to as a Multiple Systems Operator. MSOs benefit from economies of scale in areas of equipment procurement, marketing, management and technical expertise. Local decisions are left to the individual cable company operators.
multiplex	A method of transmitting multiple signals onto a single circuit in a manner so that each can be recovered intact.
multiplexer	Electronic equipment that allows multiple signals to share a single communications circuit. Multiplexers are often called "muxes." There are many different types of multiplexers including T-1 and E-1 that use time division multiplexing, statistical and frequency

456

	division multiplexers, etc. Some use analog transmission schemes, some digital. A compatible demultiplexer is required to separate the signal back into its components.
multiplexing	The process of combining multiple signals onto a single circuit using various means.
multipoint control unit	A device that bridges together multiple inputs so that more than three parties can participate in a videoconference. An MCU uses fast switching techniques to patch the presenters or speaker's input to the output ports representing the other participants. An ITU-T H.320-compliant MCU must meet the requirements of H.231 and H.243.
Multirate ISDN	A Bandwidth-on-Demand scheme in which telephone customers access, and subsequently pay for, ISDN bandwidth on an as-needed basis.
multisensory	Involving more than one human sense.

— N—

N-ISDN	Narrowband ISDN. Another name for conventional ISDN that defines bandwidths between 64 Kbps and 2.048 Mbps.
narrowband	Networks designed for voice transmission (typically analog voice), but which have been adapted to accommodate the transmission of low-speed data (up to 19.2 Kbps). Trying to get the existing narrowband public switched telephone network to transmit video is a significant challenge to the local telephone operating companies.
narrowcast	A cable television or broadcast program aimed at a very small segment of the market. Typically these are specialized programs likely to be of interest to a relatively limited audience.
network	Interconnected computers or other hardware and software between which data can be transferred. Network transmission media can vary; it can be optical fiber, metallic media, or a wireless arrangement.
Nipkow disc	A disc with holes that, when rotated, permit some

light reflected from a screen to hit a photo tube. This creates a minuscule current flow; it is proportionate to the light in the image being reproduced. This was the basis of early television as pioneered by John Logie Baird in England in the 1920s. Paul Nipkow, a German physicist invented it, in 1884.

NLM
: Network Loadable Module. An application that resides in a Novell NetWare server and coexists with the core OS. NLMs are optimized to this environment and provide superior service to applications that run outside the core.

node
: A concentration point in a network where numerous trunks come together at the same switch.

noise
: Any unwanted element accompanying program material. This can take the form of snow, random horizontal streaks or large smears of varying color.

NT-1
: Network Termination 1, an ISDN standards-based device that converts between the U-Interface and the S/T-Interface. The NT-1 has a jack for the U-Interface from the wall and one or more jacks for the S/T Interface connection to the PC or videoconferencing system, as well as an external power supply

NTSC
: National Television Standards Committee, the body formed by the FCC to develop television standards in the US. NTSC video has a format of 525 scan lines. Video is sent at a rate of 30 frames per second in an interlaced format, in which two fields comprise a frame (60 Hz). The bandwidth required to broadcast this signal is 4 MHz. NTSC uses a trichromatic color system known as YIQ. The NTSC signal's line frequency is 15.75 kHz. Finally, the color subcarrier has a frequency of 3.58 MHz.

NTT
: Nippon Telephone and Telegraph.

Nx384
: N-by 384. The ITU-T's approach to developing a standard algorithm for video codec interoperability that was expanded into the Px64 or H.261 standard, and approved in 1990. It was based on the ITU-T's H0 switched digital network standard.

Nyquist's Theorem
: A formula that defines the sampling frequency in analog to digital conversions that states that a signal

must be sampled at twice its bandwidth in order to be digitally characterized without distortion. For a sine wave, the sampling frequency must be no less than twice the maximum signal frequency.

— O—

OC	Optical carrier, as defined in the SONET specification. OC-1 or Optical Carrier Level 1 is defined as 51.84 Mbps. OC-3 is defined as 155.52 Mbps.
octet	An eight-bit datum representing one of 256 binary values. Also known as a BYTE
operating system	A fixed program used for operating a computer. An OS accepts system commands and manages space in memory and on storage media.
optical disk drives	Peripheral storage disks. Used in video communications to store and play back motion image sequences, often accompanied by sound.
optical fiber	Very thin glass filaments of extremely high purity onto which light from lasers or LEDs is modulated in such a way that the transmission of digitally encoded information takes place at very high speeds. Fiber optics systems are capable of transmitting huge amounts of information. See fiber-optic cable.
OSI	Open Systems Interconnection. An international standard that describes seven layers of communication protocols that allows dissimilar information systems and equipment to be interconnected.
out-of-band signaling	Network signaling (addressing, supervision and status information) that traverses a network using a separate path, channel or network than the information (call or transmission) itself. One well-known type of out-of-band signaling is the ITU-T's Signaling System #7.
output devices	Electrical components such as monitors, speakers, and printers that receive a computer's data.

— P—

P frame
Predictive framing, as specified in the MPEG Recommendation. Pictures are coded by predicting a current frame using a past frame. The picture is broken up into 16x16 pixel blocks. Each block is compared to a the block that occupies the same vertical and horizontal position in a previous frame.

packet
A unit of digital data that is switched as a unit. A packet consists of some number of bits (either a fixed number or a variable number), such as those that serve as an address. The packet can be sent over a packet switching network by the best available route (switched virtual circuit) or over a predetermined route (permanent virtual circuit). In a switched virtual circuit or TCP/IP networking environment packets from a single message may take many different routes. At the destination they are reunited with other packets that comprise the communication, re-sequenced and presented to the recipient in their original form.

packet switching
A technique for transmitting data in which the message is subdivided into smaller units called packets. These packets each carry the destination address and sequencing information. Packets from many users are then collated on a single communications channel.

PAL
Phase Alternate Line. The television standard used in most of Western Europe except France. PAL is not a single format however; variations are used in Australia/New Zealand, China, Brazil, and Argentina. PAL uses an image format composed of 625 lines. Frames are sent at a rate of 25 per second and each frame is divided into two fields to accommodate Europe's 50 Hz electrical system. PAL uses YUV, a trichromatic color system.

pan
To pivot a camera in a horizontal direction.

parallel communications
A multi-wire or bus system of communications. The bits that make up the information are transmitted separately, each assigned to a separate wire in the cable. Parallel communications are used frequently in computer-to-printer outputs in which the data must

be transferred very rapidly but over a short distance. The opposite method of transmission is serial in which information travels over a single channel a bit at a time.

passband	A range of frequencies transmitted by a network or captured by a device.
Pay-Per-View	PPV. A type of CATV or DBS service whereby a subscriber notifies a provider that he or she is interested in receiving a program and is billed for the specific service on a one-time basis.
PC	A self-contained desktop or notebook computer that can be operated as a standalone machine or connected to other PCs or mainframes via telephone lines or a LAN.
PCM	Pulse Code Modulation. A method of converting an analog signal to a digital signal that involves sampling at a specified rate (typically 8,000 per second), converting the amplitude of the sample to a number and expressing the number digitally.
pel	Picture element. Generally synonymous with pixel, the term pel is used in the analog world of television broadcasting. As television moves toward the digital world of HDTV the term pixel is being embraced by the broadcasting industry.
peripheral equipment	Accessories such as document cameras, scanners, disks, optical storage devices that work with a system but are not integral to it.
persistence	1) The time taken for an image to die away. 2) Ability to maintain a session despite inactivity or interruption.
phase	The relationship between the zero-crossing points of signals. A full cycle describes a 360-degree arc; a sine wave that crosses the zero-point when another has attained its highest point can be said to be 90 degrees out of phase with the other.
phase modulation	The process of shifting the phase of a sine wave in order to modulate it onto a carrier.
Phoneme	An element of speech such as, "ch," "k," or "sh," that is used in synthetic speech systems to assemble words for audio output.

phosphor	Substance that emits visible light of specific wavelengths when irradiated, as, for example, by an electron beam in a cathode ray tube or by ultra-violet radiation in a fluorescent lamp.
photo diode	A basic element that responds to light energy in a solid-state imaging system. It generates an electric current that is proportional to the intensity of the light falling on it.
picture-in-picture	PIP. A video display mode in which a small video image is superimposed on a quadrant or smaller area of a video screen. This small image or "window" can be opened in order to display a second video input on a monitor. This is a particularly valuable feature in a single monitor videoconferencing system; it allows the near-end to view themselves in a small window while simultaneously seeing the far end.
Picturephone	AT&T's video-telephone that was introduced at the 1964 World's Fair in Flushing Meadow, New York. The device incorporated a camera that was mounted on top of a 5.25" x 4.75" screen. Since audio signals were transmitted separately from video signals, the system required a transmission bandwidth of 6.3 Mbps and, therefore, could not use the PSTN. It never became popular.
pixel	The smallest element of a raster display. A picture cell with specific color and/or brightness.
POP	Point of presence. Typically the closest location of a carrier to a subscriber and the point at which that subscriber will be supplied with service from that carrier.
port	A multi-wire input/output to a computer or other electronic component. The number of ports determines the number of simultaneous users, the number of peripheral devices one can simultaneously connect with, or a combination of both.
POTS	Plain Old Telephone Service. Analog narrowband telecommunications service designed for transmitting voice calls.
prediction	A type of motion compensation in which a compressed frame of video is reconstructed by a receiving codec by

comparing it to a preceding frame of video and making assumptions about what it should be.

Presence	Ability to identify the location of a communication recipient and the type of device that recipient might be using at a given moment (e.g., landline phone, mobile phone, or desktop computer).
PRI	The ISDN Primary Rate Interface. In the US the PRI consists of 23 64 Kbps bearer or "B" channels and one 64 Kbps delta or "D" channel. In Europe, the PRI-equivalent is known as Primary Rate Access or PRA. The PRA consists of 30 64 Kbps B channels and two 64 Kbps D channels.
prism	Crystalline bodies with lateral faces that meet at edges that are parallel to each other. A prism is used to refract light, often dispersed into component wavelengths.
private line	A telephone line rented for the exclusive use of a business and which connects two locations on a point-to-point basis. Communications can only be exchanged between these two locations, as opposed to switched services in which calls can be placed to any other addressable location with compatible service.
private network	A collection of leased, private circuits that link multiple locations of an enterprise or organization. May be used to transmit different combinations of voice, data and video signals between sites over the point-to-point lines that exist between sites.
proprietary	Systems that use techniques and processes that are closely-guarded trade secrets and which, because of these highly-individual approaches to achieving a result, cannot interoperate with other manufacturers' equipment without a great deal of difficulty.
protocol	A standard procedure agreed upon by regulating agencies, companies, or standards-setting bodies to regulate transmission and, therefore, to achieve inter-communications between systems or networks.
PSTN	Public Switched Telephone Network. The conventional voice telephony network as provided by Local Exchange Carriers.

PTT	Post Telephone and Telegraph company. A generic term for the telephony providers in Europe and other non-US countries.
public room	A commercial facility that offers video communications services for rent.
Px64	Pronounced "P times 64." Another (early) name for the ITU-T's H.261 codec.

—Q—

Q-Signal	In the NTSC color system, the Q signal represents the chrominance on the green-magenta axis.
Q.920	One of the two ITU-T Recommendations in which the Link Access Protocol-D (LAPD) is formally specified.
Q.921	Along with Q.920, this ITU-T Recommendation specifies the LAPD protocol, an OSI layer-2 protocol. The connection-oriented service establishes a point-to-point D channel link between a customer and a carrier's switch. Data is sent over the link as Q.921 packets.
Q.931	Whereas the D channel is packet-switched, B channels are circuit-switched. They are established, maintained and torn down using ITU-T Recommendation Q.931 (also known as ITU-T I.451). In Q.931, the ITU-T specified how 'H' channels—multiple contiguous channels on T-1 or E-1/CEPT frames—can be bound together and switched across an ISDN network.
QCIF	Quarter Common Intermediate Format, a mandatory part of the ITU-T H.261 compression standard that requires that non-interlaced frames of NTSC video be sent with 144 luminance lines and 176 pixels. When operating in the QCIF mode the number of bits that result from encoding a single image cannot exceed 64 K bits where K equals 1024.
QoS	Quality of Service (QoS) defines a level of performance in a communications network. For example, video traffic requires low delay and jitter; data traffic does not, but will burst and thereby impede the transmission of voice traffic. Any channel that transports both must, therefore, reserve and execute preferential treatment for video.

quality reduction	One way of compressing data in which some compromises are made in picture resolution, frame rate and image size in order to accommodate the bandwidth available for transmission.
quantization	Quantization is one step in the process of converting an analog event into its digital representation. It involves the division of a continuous range of signal values into contiguous parts, a unique value being assigned to each part. Prior to quantization the signal is sampled at regular intervals (typically at a rate that is least twice its highest frequency). The instantaneous amplitude of each sample is expressed as one of a defined number of discrete levels. These levels can then be assigned a digital value.
quantization matrix	A set of values used by a de-quantizer. The H.261 quantization matrix specifies 64 8-bit values.
QuickTime	Apple's video compression solution developed for the Macintosh product line. Decompression can be accomplished using nothing more than standard computer hardware although QuickTime accelerator cards with hardware decompression chips allows much better results.

— R—

RAM	See random access memory.
Random Access Memory	RAM. A storage receptacle that is volatile, i.e., that it only retains information until a computer is turned off.
raster	The pattern of horizontal lines that form the image scanning of a television system.
raster scan	The process of continually scanning both vertically and horizontally using an electron beam to create moving images on a television screen.
RBOC	Regional Bell Operating Company. This term is used in a general sense today to refer to any of the original 22 Bell Operating Companies that were spun off parent AT&T as part of the MFJ (pronounced R-BOK).
RCA	Radio Corporation of America. First large commercial enterprise to wholeheartedly pursue television in the

US under the direction of David Sarnoff.

RCA Jack

A physical interface commonly used in audio and video equipment (often colored red, white, or yellow).

Real-time

The time that elapses when events occur naturally. This term describes computing and electronic applications in which there is no perceived delay in the transmission of an interactive communication.

Real-time compression

Capture and compression of video in one step so that it can be written to disk or other storage media.

Real-Time Transport Protocol

RTP adds a layer to IP to address problems caused when interactive multimedia communications are transported over TCP/IP networks. Packet-switched networks were designed for data in which latency is tolerable. In voice and video communications, a constant bit rate is required for smooth delivery. RTP's approach is to give low-latency connection-oriented communications a higher priority than connectionless data. RTP also addresses the need for multicasting and compression over the Internet and other TCP/IP-based networks. TCP is a reliable transport, but its efforts on behalf of reliability are not ideal for video communications. TCP re-sends packets of data that have not been properly received. In video communications, packets that do not arrive on time are no longer useful. RTP does not resend missed packets, but notifies a server that a packet was missed. If the missed packets threshold reaches a specified level, RTP can request additional bandwidth allocation.

redundant

In the context of compression, redundant information is that which does not change over time (temporal) or space (spatial). Temporal redundancy manifests in a sequence of video frames in which successive frames contain the same information as those that preceded them. This information is, thus, redundant; e.g., not necessary to convey. Rather, it is only necessary to convey the areas of the frame that *did* change for that information is critical to the recipient's accurate interpretation of the new frame. Spatial redundancy manifests within a frame of video in which a group of similar pixels are clustered. The goal of video compression is to eliminate redundancy that, in turn, reduces the data rate required for transmission.

reference frames
In video compression reference or intra-coded frames are sent periodically to eliminate the accumulated errors generated through the process of interframe coding.

refresh
To recharge the cells of a volatile digital memory so that their contents are not lost. Alternately, to "paint" a new picture on the screen of a VDT is to refresh the screen.

repeater
A device used to reprocess and amplify a weak signal in a transmission system. Used where the distance between the sender and receiver is too great for the signal to travel between the two without being boosted along the way. Radio signals, as they travel through free space, are attenuated; their power is gradually lost. Thus repeaters are needed to interpret the original signal and amplify it. They also often amplify background noise and interference in the process.

reproducing spot
In television broadcasting, images are sent as a series of signals that describe a series of scan lines. A black and white television receiver captures these signals and reproduces the image on the screen using a *reproducing spot* that writes to the screen, varying in intensity in accordance with the instantaneous amplitude variations of the received signal.

Request in Progress Message
H.323v3 feature that allows the receiver of a request to specify, when a message can not be processed before the initial timeout value expires, that a new timeout value should be applied.

resolution
The ability of an optical or video system to produce separate images of objects very close together and, hence, to reproduce fine detail.

retrace
The return of an electron beam to its starting place to prepare for sending a new field of information.

retrace interval
The process of moving a scan line back to the left-most position in order to begin a new scan.

RF	Radio frequency. Electromagnetic signals that travel from radios or other sources through free space.
R-Y	A designator used to name one of the color signals in the color difference video signal, with a formula of .70R, -.59G, -.11G.
RFI	Radio frequency interference. Often manifests in television signals as vertical bars moving slowly across a picture. It can also take the form of diagonal or vertical lines or a pattern of horizontal bars.
RGB	Red, green, blue. Color television signals are oriented as three separate pictures: red, green, and blue. Systems can merge them or, for maximum quality, leave segregated as *component video signals.*
RIP	See Router Information Protocol
RIP	See Request in Progress Message
roadrunner	A digital multi-node network switching system, manufactured by ACTION/Honeywell. It allowed voice, data and video to share point-to-point digital bandwidth.
roll-about	A totally self-contained videoconferencing system that includes a codec, monitor(s), audio system, and network interfaces. These systems can, in theory, be moved from room-to-room. In fact, they are not portable because they include electronic equipment that is not durable enough for a great deal of jostling about and, moreover, is quite heavy.
router	Equipment that facilitates the exchange of packets between autonomous networks (LANs and WANs) of similar architecture. Connection is made at OSI layer 3 (network layer). A router responds to packets addressed to it by either a terminal or another router. Routers move packets over a specific path or paths based on the packet's destination, network congestion and the protocols implemented on the network.
Routing Information Protocol	Hop-count based interior gateway routing protocol. RIP is one of the oldest and, perhaps, still the most prevalent routing protocol.
RS-170A	A specification for adding color to the NTSC television signal. It uses the YIQ method to add hue and

saturation to a luminance subcarrier.

RS-232-C
An EIA interchange specification specifying electrical characteristics and pin allocations in a 25-pin connection cable. The equivalent of the international V24 standard.

RS-366
An EIA interface standard for autodialing.

RS-449
A physical layer specification that defines a 37-pin data connector designed for high-speed transmissions. The signal pairs are balanced and each signal pin has its own return line instead of a common ground.

RSVP
Resource-Reservation Protocol. An IETF specification that allows users of conferencing over the Internet to reserve bandwidth at a given time.

RTCP
Real-Time Transport Control Protocol. A companion to RTP that will allow users to provide feedback on how an RTP session is working; how well packets are being delivered, etc.

RTP
See Real-Time Transport Protocol.

rubber bandwidth
Coined by I-MUX manufacturer Ascend, the term refers to an ability to support applications that require varying speeds by dividing a signal up into 56- or 64-Kbps chunks and sending them over a switched digital network.

run-length encoding
Compression method that works at the bit level to reduce the number of repeating bits in a file. Sending only one example, followed by a shorthand description of the number of times it repeats, indicates a string of identical bits.

— S—

S-Video
Super Video (as opposed to composite). S-Video is a hardware standard that defines the physical cable jack used for video connections. The connector takes the form of a 4-pin mini plug when separate channels are provided for luminance and chrominance. Also known as Y/C Video, S-Video is used in the S-VHS and Hi8 videotape formats.

sampling
The process of measuring an analog signal in time

slices so that it can be quantized and encoded in a digital format. The sampling interval, per Nyquist's Theorem, must be at least twice the highest frequency contained in the analog input. Thus, in an analog signal in which frequencies of 4,000 hertz are possible, the sampling rate must be 8,000 per second.

satellite A man-made object designed to orbit the earth. Used to receive and retransmit telecommunications signals that originate at an earth station.

satellite delay In satellite transmissions a signal must travel from the earth to a transponder placed in geosynchronous orbit, a distance of 22,300 miles above the equator. Radio waves, which travel at approximately the speed of light (186,000 miles per second) take about.24 seconds to make the round trip, and thereby induce perceptible delay in satellite communications that are particularly annoying in speech and interactive video. transmissions.

satellite receiver A microwave antenna that is capable of intercepting satellite transmitted signals, lowering the frequency of those signals and retrieving the modulated information.

saturation In color reproduction, the spectral purity or intensity of a color. Adding white to a hue reduces its saturation.

SBIT A three-bit integer that indicates the number of bits that should be ignored in the first data octet of a RTP H.261 packet.

scan converter A device that converts computer images to a television display format. Used to convert the scan rates of VGA or SVGA to NTSC, PAL or SECAM scan rates.

scan line During scanning, a number of nearly horizontal passes are made across an image and the light recorded by the image is converted into an electrical signal. Each pass across the image is called a scan line.

scanner A device used to capture and digitize images. A scanner reads the light and dark areas of an image and converts the light into a coded signal.

scanning The process of analyzing the brightness and colors of an image, using a camera or other collection device, according to a predetermined pattern. In monitors, the number of scanning lines per frame gives an indication of the quality of the picture transmitted and displayed.

SCIF
: Super CIF. A standard video exchange format made up of a square of four CIF images. SCIF specifies 576 lines, each of which contains 704 pixels. Also known as 4CIF.

screen
: Devices that emit or reflect light to a viewer. Examples include the silver sheets of beaded vinyl upon which motion pictures or slides are projected, conventional CRTs, LCD displays, and conventional televisions.

SDP
: See Session Description Protocol.

SECAM
: Sequential Couleur Avec Memoire. The color television system that offers 625 scan lines and 25 interlaced frames per second. It was developed after NTSC and PAL and is used in France, the former Soviet Union, the former Eastern bloc countries and parts of the Middle East. Two versions of SECAM exist: horizontal SECAM and vertical SECAM. In this system, frequency modulation is used to encode the chrominance signal.

Security policy
: An information security policy as an explicit set of prescribed practices that fosters protection of the enterprise network and of information assets.

security video
: Video systems that are used to observe the events in a distant location, nearly always in a receive-only arrangement. Many such systems are slow-scan and use inexpensive cameras to capture images. The video signal. Typically not digitized or compressed.

serial communication
: Used to connect computers with modems, printers and networks. Networks based on serial communication typically use the EIA's RS-232-C interface in which the pins that carry the electrical signal are arranged in a pattern that is understood by both the sending and receiving device.

serial interface
: Refers to data being transferred between a computer and some type of peripheral device, often a printer, in a series of bits in which each bit is sent one at a time as part of a sequence.

server	In a network, a server is the device that host applications and allocates the activities of shared support resources. Servers store applications, files, and databases; they provide access to printers and fax machines, and they permit clients to access external communications networks and information systems.
Session Description Protocol	Associated with SIP, SDP is a Protocol for conveying information about streams in multimedia sessions to enable recipients of a *session description* to participate in a session. In a Multicast session, SDP is a means to communicate the timing and existence of a session, and to convey sufficient information to enable recipients to join and participate in the session.
Session Initiation Protocol	Defined in IETF RFC 2543, SIP is an application-layer control (signaling) protocol for creating, modifying, and terminating sessions with one or more participants. Sessions can be Internet multimedia conferences, Internet telephone calls, or multimedia distribution. Members in a session can communicate via multicast, a mesh of unicast relations, or a combination of these.
set-top box	In a CATV environment, the cable terminates in the subscriber's location on a decoder or set-top box, which is connected directly to the television receiver. Set-tops vary greatly in complexity. Older models merely translate the frequency received off the cable into a frequency suitable for the television receiver; newer models are addressable and offer a unique identity much like a telephone. Interactive set-tops allow subscribers to communicate interactively.
seven-layer model	See OSI (Open Systems Interconnection) model.
shadow mask	In a color monitor, this component is placed between the electron guns and the screen to ensure that the electron beams are restricted to their targeted dots.
shutter	A device that cuts off light in an optical instrument.
SIF	Source input format. Used in MPEG, and not to be confused with the Common Intermediate Format (CIF) used in the ITU-T's H.32X family of videoconferencing standards.

signal-to-noise	A ratio used to describe the clarity or degradation of a circuit. Signal-to-noise is measured in decibels. An acceptable signal-to-noise ratio is generally 200:1 or 46 dB.
SIP	See Session Initiation Protocol.
simplex	A circuit capable of transmissions in only one direction.
sine wave	The smooth, symmetrical curve that is created by natural vibrations that occur in the universe.
single mode fiber	Optical fiber with a central core of such small dimension that only a single mode of transmission is possible. Light travels straight down the middle of the fiber without zigzagging from side to side. Single mode fibers transmit high bit rates over long distances.
slow-scan	Television or video scanning system in which a frame rate lower than normal is used to reduce the bandwidth required for transmission. Often associated with the transmission of video frames over the public switched telephone network.
SMATV	Satellite Master Antenna Television. A distribution system that feeds satellite signals to hotels, apartments, etc. Often associated with pay-per-view.
SMDS	Switched Multimegabit Digital Service. A public networking service based on the IEEE's 802.6 specification.
smear	The undesirable blurring of edges in a compressed image, often caused by the DCT that tends to eliminate the high-frequency portions of an image that represent sharp edges.
SMPTE	Society of Motion Picture and Television Engineers, the authors of the SMPTE time code used in video film and A/V production.
SMPTE Time Code	An 80-bit time code used in video editing. Developed by the SMPTE.
snow	Random noise or interference appearing in a video picture as white specs.

software	The programs, procedures, rules and routines used to extend the capabilities of computers.
SONET	Synchronous Optical Network.
sound waves	Naturally occurring oscillations that are created when objects vibrate at a certain frequency, and thereby compress the air molecules around them and cause these molecules to seek equilibrium by decompressing outward. During decompression dynamic energy is expended, thereby causing neighboring groups of molecules to compress. This starts a chain reaction. Compressed air is pushed in all directions from the source. If the resulting waves are between 50 and 20,000 Hz humans can hear them as sounds.
spatial	Compression/coding that is performed on a single frame of video data as opposed to temporal (time) coding that is performed by comparing successive video frames to each other.
spectrum	A range of wavelengths or frequencies of acoustic, visible or electromagnetic radiation. Also refers to a display of visible light radiation that is arranged in order of wavelength.
SpectrumSaver	Compression Labs compression system designed to digitize and compress a full-motion NTSC or PAL analog TV signal so that it can be transmitted via satellite in as little as 2 MHz of bandwidth.
splitter	In CATV networks, multiple feeder cables are attached to splitters that match the impedance of the cables.
square wave	A term typically used to refer to a digital signal comprised of only two voltage values, that represent on and off (or one and zero).
SQCIF	H.263, a scalable version of H.261, Sub-QCIF, a 128-by-96 pixel image resolution format. SQCIF was also added to the H.261 recommendation, to allow low-resolution picture coding.
S/T-Interface	The ISDN Subscriber/Termination (S/T) Interface uses two pairs of wires to deliver an ISDN signal from a wall jack to an ISDN adapter or other ISDN equipment.
standards conversion	The process of adapting a television or video production from one set of standards (e.g., NTSC) to

	another (e.g., PAL). This is necessary when programming that was developed for one part of the world (e.g., North America), is viewed in another (e.g., Europe).
star topology	A network in which there is an individual connection between a switching point (or in the CATV world, a head-end) and a device or subscriber.
store-and-forward	Systems in which messages (including video-enabled messages) are stored at an intermediate point for later retrieval.
streaking	In video, shifting brightness across an image at the vertical location of a bright object.
streaming data	Any transmission with a time component, such as video and audio signals.
Study Group	Groups of subject experts that are appointed by the ITU-T, an international standards-setting body that is part of the United Nations. Study Groups draw up Recommendations that are submitted to the ITU-T at the next plenary assembly for adoption.
subcarrier	A band of frequencies superimposed onto a main carrier frequency. In color television the color information is carried on the same subcarrier as the luminance signal by modulating it onto the upper frequencies of the luma waveform.
subjective evaluation	During the process of comparing videoconferencing systems, the portion of the comparison that is in the *eye-of-the-beholder* as opposed to being quantifiable.
subscriber	A customer of a carrier, service provider or utility.
subtractive color	Color information that results from reflection or refraction of light. Reflected wavelengths correspond to the colors that people attribute to those objects. The subtractive primaries are magenta, cyan, and yellow. In contrast, additive color results from wavelengths of radiation emitted by light sources. Additive primaries are red, blue and green (RGB).
Super CIF	A video format defined in the 1992 Annex IV amendment of the ITU-T's H.261 coding standard. Also known as 4CIF in the ITU-T's H.263 standard, it

is comprised of 704 by 576 pixels.

super trunks Large diameter coaxial or fiber-optic cables used to carry a signal from a head-end to various hub sites in a CATV network.

Super VGA SVGA. A higher resolution version of the VGA standard that allows applications to use video resolutions of 1,024-by-768 pixels.

switch A device that establishes, monitors, and terminates a connection between devices connected to a network.

Switched 56 Switched 56 is a legacy service that enables dialing and transmitting of digital information up to 56 Kbps in a manner similar to placing an analog telephone call. The service is billed like a voice line—a monthly charge plus a cost for each minute of usage. Any switched 56 offering can connect with any other offering, regardless of carrier.

switching The process of establishing a connection between an input and an output. Switching occurs in many parts of a network, such as at a PBX, at the local exchange, at an IXC's office, or in X.25 or frame relay networks. It allows a subscriber to establish communication with multiple parties by sending an address to the switch which, in turn, attempts to make a connection.

switching hub See *Ethernet Switch*

symmetrical Techniques in which the decompression techniques are an exact reverse of the compression techniques. Used in Real-Time interactive video applications.

synch pulse Oscillator circuits generate the horizontal and vertical sweep signals in a television receiver; they are controlled, in precise synchrony, by the movement of the scanning element at the camera. To achieve horizontal control, the camera sends synch pulses along with the luminance information. To control the vertical sweep oscillator, a synch pulse is sent at the end of each field.

synchronization The process of controlling two or more systems so that they establish a time-based relationship with each other. Clocks are used to ensure precision in the sending and receiving of bits or other signals;

characters are spaced by time, not bits.

system bus

Analogous to the nervous system in the human body, the system bus serves to interconnect the microprocessor with memory, storage devices, and input/output hardware.

— T—

T-1

Also known as DS-1 or T-carrier. High-speed digital transmission system characterized by a bit rate of 1.544 Mbps and subdivided, via time division multiplexing, into 24 channels. Each channel has a bit rate of 64 Kbps. Frames of data are created 8,000 times a second by combining each channel's 8-bit time slot into a group of 192 bits, a synchronization bit is added to the frame that makes it 193 bits wide. These 193-bit frames are sent 8,000 times a second and received by a channel bank or T-1 multiplexer.

T.120

The ITU-T's "Transmission Protocols for Multimedia Data," a data sharing/data conferencing specification that lets users share documents during any H.32x videoconference. Like H.32x specifications, T.120 is an umbrella Recommendation that includes a number of other Recommendations. Data-only T.120 sessions can be held when no video communications are required, and the standard also allows multipoint meetings that include participants using different transmission media. The mandatory components of T.120 include recommendations for multipoint file transfer and shared-whiteboard implementation.

T.121

The T.120 family member that is formally known as "Generic Application Template." This template encompasses those operations that are common to most T.120 application protocols.

T.122

ITU-T Recommendation that is part of the T.120 family and which is known as "Multipoint Communication Service for Audiographics and Audiovisual Conferencing Service Definition." T.122 defines the multipoint services available to an applications developer while T.125 specifies the data protocol used to implement these services.

T.123   ITU-T Recommendation that is part of the T.120
               family and which is known as "Protocol Stacks for
               Audiographic and Audiovisual Teleconference
               Applications." This protocol covers public switched
               telephone networks, ISDN, circuit switched digital
               networks, packet-switched digital networks, Novell
               NetWare IPX (via a reference profile) and TCP/IP.

T.124   ITU-T Recommendation that is part of the T.120
               family and which is known as that allows "Generic
               Conference Control."

T.125   ITU-T Recommendation that is part of the T.120
               family and which is known as "Multipoint
               Communication Service Protocol Specification."

T.126   ITU-T Recommendation that is part of the T.120
               family and which is known as "Multipoint Still Image
               and Annotation Protocol." This standard defines a
               protocol for annotated shared whiteboard applications
               and still image conferencing. It uses services provided
               by T.122 and T.124. Remote pointing and keyboard
               data exchanges are covered, and therefore allow
               terminals to share applications, even if they are
               running on different platforms or operating systems.

T.127   ITU-T Recommendation that is part of the T.120
               family and which is known as "Multipoint Binary File
               Transfer Protocol." This allows participants in an
               interactive data conference to exchange binary files,
               and provides a means to distribute and retrieve one or
               more such files simultaneously.

T.128   T.128 - Audio Visual Control for Multipoint
               Multimedia Systems. Now replaced with the T.130
               family of ITU-T Draft Recommendations.

T.130   The ITU-T's Draft Recommendation for "Real Time
               Architecture for Multimedia Conferencing. Provides
               an overview of how T.120 data conferencing will work
               in conjunction with H.32x videoconferencing.

T.131   The ITU-T's Draft Recommendation for "Network-
               Specific Mappings." Defines how Real-Time audio and
               video streams should be carried over different types of
               networks including B-ISDN and ATM when used in

conjunction with T.120 data conferencing.

T.132	The ITU-T's Draft Recommendation for "Real Time Link Management." Defines how real time audiovisual streams can be routed between diverse multimedia endpoints.
T.133	The ITU-T's Draft Recommendation for "Audio Visual Control Services." Defines how to control the source and link devices associated with real time information streams.
T.84	ISO-JPEG standard.
T.RES	The ITU-T's Draft Recommendation for "Reservation Services." Defines how terminals, MCUs and reservations systems will interact to reserve a conference. Part of the T.130 family.
T.TUD	The ITU-T's Draft Recommendation for "User Reservation." Describes how to transport user defined bit streams between conferencing end-points. Part of the T.130 family.
T-3	See DS-3.
talking head	The portion of a person that can be seen in the typical business-meeting-style videoconference; the head and shoulders. Because there is little motion in a talking head image (most is in facial expression) it is easy to capture with compressed video.
tap	In CATV systems the subscriber drop connects to feeder cables via taps; passive devices that isolate the feeder cable from the drop.
tariff	The terms and conditions of telecommunications or transmission service. Tariffs generally require approval by a regulatory body such as the FCC, PUCs or Often (UK).
TCP/IP	Transmission Control Protocol/Internet Protocol. A *de facto* standard and a set of internetworking protocols originally developed by the Department of Defense. Connects dissimilar computers across networks and is, therefore, widely used in the private sector. TCP/IP protocols work in the third and fourth layers of the OSI model to guarantee the delivery of data even in

very congested networks.

TDM	Time Division Multiplexing. A technique for interleaving multiple voice, data and video signals onto a single carrier by assigning each signal its own separate time slot during which it can place a segment of its digitally-encoded transmission. At the distant end, the signals are separated so that each discrete signal can be re-oriented and recombined as an entire communication or message.
Telco	A generic name generally used to refer to local exchange carriers; telephone companies providing local exchange service.
telecommunications	The art and science of applying services and technologies in order to enable communications over distance. Uses technologies such as radio, terrestrial and cable-based services, wireless transmission and optical fiber networks.
Telecommunications Competition and Deregulation Act of 1996	Passed by the 104th Congress and signed by President Clinton in February, 1996, the Telecommunications Competition and Deregulation Act of 1996 was a rewrite of the Communications Act of 1934. The stated goal of the act was to promote competition and reduce regulation to secure lower prices and higher quality, more robust, services for American telecommunications consumers by encouraging the rapid deployment of new telecommunications technologies.
telecommuting	The process of commuting to work electronically rather than physically.
teleconferencing	The use of telecommunications links to provide audio or audio/video/graphics capabilities. These systems allow distant workgroups or individuals to meet, and thereby reduce the administrative and opportunity costs of holding in-person meetings involving people from many different cities or countries.
telemedicine	The practice of using videoconferencing technologies to diagnose illness and provide medical treatment over a distance. Used in rural areas where health care is not readily available and to provide medical services to prisoners, among other applications.

telephony	The convergence of telephone and computer functionality. Specifically, the process of converting voice and other sounds into electrical signals that can be transmitted by wire, radio or fiber; stored and reconverted to audio upon receipt at the distant end.
telewriting-based terminal equipment	An older term that refers to devices that use digitizing tablets and monitors to enable the sharing of text and graphics in video- and teleconferences. These devices were more evident in the very early 1990s. They used voice-grade circuits or required low data rates for transmission, typically about 9.6 Kbps. They were the predecessor to document conferencing systems.
temporal coding	Compression that is achieved by comparing successive frames of video over time to eliminate redundancies between them.
tile	See tiling. An individual pixel block that becomes visible on the viewer's monitor. Tiles, together, appear as a series of mosaic squares.
tiling	Video effect in which the image appears as pixel blocks. Can be caused by compression when the sampling rate or bandwidth is not adequate to fully describe the original image.
time-division multiplexing	A way of enabling a single bearer signal to carry more than one information channel simultaneously. This is achieved by sharing the common transmission path on a cyclical basis. Each channel takes a turn using the path by entering the source data into a "time slot." Common TDM systems can support 24 channels in the US and 32 in Europe.
time slot	In TDM or TDSW, the slice of time that belongs to a digitized communication whether it be voice, data or video communication.
transcoding	Conversion from one format to another. This may be from one digital format into another, or between dissimilar formats such as from digital frames into analog video or from text into speech.
transform coding	A technique of video compression that requires VLSI chips with hundreds of thousands of gates. See DCT for a description of one type of transform coding.

TransMux	MPEG-4 specifies an interface to the *TransMux* (Transport Multiplexing) models, the layer that offers transport services for matching the requested QoS. Any suitable existing transport protocol stack (e.g., RTP in UDP/IP, AAL5 in ATM, or Transport Stream in MPEG-2) over a suitable link layer may become a specific TransMux instance.
transponder	A microwave repeater mounted on a satellite and used to receive communication from an uplink and retransmit it at a different frequency on a downlink.
tree and branch	A cable television arrangement in which large capacity cables called super trunks radiate from the head-end to various sectors of the franchise area to hub sites. From these hub sites, lower-capacity cables branch out into the surrounding districts to be split and split again until cable runs along every street. A tapping point outside each residence enables the final connection to the residence.
tuner	A piece of equipment or portion of a circuit used to select one signal or channel from amongst many signals or channels.
twisted pair	Two insulated copper wires twisted at regular intervals and normally covered by a protective outer sheath composed of PVC.

— U—

U-Interface	The U-Interface carries ISDN formatted signals over a single pair of wires between a subscriber's location and a telephone company's central office.
UDP	User datagram protocol; an unreliable connectionless transport protocol that sends un-sequenced packets across a packet-switched network. UDP is defined in RFC 768.
UHF	Ultra High Frequency. The spectrum band that occupies frequencies from 470 to 890 MHz, within which television channels 14 through 83, are transmitted. CATV systems cannot transport these frequencies so they are converted for the purposes of cable transmission.

ULS	User Location Service. A white-pages directory that allows users of Internet videoconferencing and personal conferencing tools to refresh IP addresses dynamically and to determine who is logged on and available to have a conference.
unicast	Application of conferencing, usually over packet-switched networks, where only one user or site receives data. Contrast this with multicast in which data is received by more than one user or site.
up-link	The portion of a communications link used for the transmission of signals from a satellite earth station to satellite transponders in space.
upstream transmission	The ability for a CATV subscriber to communicate interactively with the head-end or a hub-site.
USB	Universal Serial Bus is a hardware interface for peripherals such as keyboards, mice, scanners, and telephony devices. USB supports MPEG-1 and MPEG-2 digital video. USB allows for 1–127 attached devices, and allows devices to use its full 12 Mbps bandwidth or a 1.5 Mbps subchannel (it also allows for hot-swapping). Low-power devices that might normally require a separate AC adapter can be powered through the cable because the USB bus distributes 0.5 amps of power through each port.
UTP	Unshielded Twisted Pair.

— V—

V.35	An ITU-T standard interface between a network access device and a network.
V24	An ITU-T list of interchange circuits for connecting a terminal device to a modem or one type of communications equipment to another.
VCR	Videocassette recorder—originally a trade name but now a generic term commonly used for a device that plays videotapes.
VDSL	Very-high-speed-digital subscriber line.
vector quantization	A vector is a mathematical representation of a force (in video the force is a frequency, or a series of

483

frequencies). In vector quantization blocks of pixels are analyzed to determine their vectors. Vectors are then used to select predefined equations that describe them in a more efficient way. These equations are expressed as codes, which might be only a few bits long. A decoder interprets these abbreviated codes and uses them to describe a frequency with a given hue and saturation.

vertical blanking	An interval that occurs at the beginning of each video field in which the reproducing spot is turned off during the process of retracing from the bottom right to the top left of the screen. After the vertical blanking period has ended the electron starts shooting electrons at the screen again, starting at the top and working toward the bottom.
vertical resolution	The total number of horizontal scan lines.
vertical sweep signal	The electrical signal that moves each horizontal scan line vertically.
VGA	Video Graphics Array. Developed by IBM, this has become the dominant standard for PC graphics. VGA provides two different graphics modes: 640 by 480 pixels at 4-bits per pixel or 320 by 200 pixels where 8 bits are used to describe a pixel.
VHF	Very High Frequency. The frequency band used to transmit television channels two through 13.
video	A sequence of still images that, when presented at a sufficiently high frame rate, give the illusion of fluid motion. In the US, full-motion video (TV) is sent at 30 frames per second. In Europe and most of the rest of the world the frame rate is 25 per second. Motion pictures send video at 24 frames per second.
video capture board	A PC circuit board that can capture the two fields that comprise a single video frame (included in most modern PCs). See also frame grabber.
video mail	A multimedia version of electronic mail that includes moving or still images that are embedded into the message. Also known as v-mail.
video server	A specialized file server with enormous hard disc capacities (often measured in terabytes or trillions of

bytes). A video server stores compressed audio and video images for access. Often called video jukeboxes, these servers provide service to end-users over high-speed networks.

video wall Multi-screen video system in which a large number of video monitors or back projection modules together produce one big, bright image. This is achieved by splitting the incoming video signal across many monitors.

Video-On-Demand VOD. The ability for subscribers to control when and how they view a movie. VOD delivers features that mimic VCRs (e.g., rewind, pause, fast-forward(.

videoconferencing A collection of technologies that integrate video with audio, data, or both and convey the aggregate signal, Real-Time, over distance for the purposes of a meeting between dispersed sites.

videophones Telephones with video capability.

virtual circuit A seemingly dedicated connection between two points in a packet-switched network. In a virtual circuit, packetized data from multiple devices is placed on the same circuit.

virtual private network Virtual private network technology allows an organization to use the Internet as a private network by employing tunneling or data encryption techniques in a firewall.

VLSI Very Large Scale Integration.

voice-activated microphones Microphones that automatically capture and transmit audio at the trigger of a voice. Transmission ceases when the triggering sound stops. Videoconferencing system manufacturers effectively incorporate voice-activated microphones with cameras that capture and transmit images based on who is speaking.

voltage Electrical pressure caused by electrons repelling other electrons.

VSAT Very Small Aperture Terminals. Small transportable satellite earth stations used in videoconferencing BTV applications.

VTS 1.5 Compression Lab's first codec that was introduced in

1982. It had an operating bandwidth requirement of 1.544 Mbps (T-1), remarkable for the time.

— W—

watts

A measure of the power or electrical energy of a signal or waveform.

wavelength division multiplexing

A technology that uses lasers to transmit multiple light signals concurrently over a single optical fiber. With WDM, each signal travels within its own color band and is modulated by transmitted data.

waveform

A presentation of the varying amplitude of a signal in relation to time.

WDM

See *Wavelength Division Multiplexing.*

white balance

A camera feature that adjusts the balance between the RGB components that yields white in the video signal. Cool-temperature lighting casts a different tint on a room than does warm-temperature lighting. White balance produces a uniform white regardless of room lighting.

white noise

Noise, containing energy, distributed uniformly over the frequency spectrum.

whiteboarding

The ability for multiple users to share a drawing space, generally a bit-mapped image that all conferees can modify.

wide area network

WAN, a collection of circuits that make up the public switched network and over which organizations communicate.

— X—

X.21

Used primarily in Europe, X.21 is a standard that operates at bit rates between 56 Kbps and 384 Kbps to control network dialing. A valuable feature of X.21 is its inherent dialing functions, including the provision for reporting why a call did not complete. X.21 can be used to connect to both switched and dedicated networks.

— Y—

Y
: The common nomenclature for the luminance signal.

Y/C
: In component video, the Y or luminance signal is kept separate from the C (hue and color saturation signal) to allow greater control and to enable enhanced quality images. The luminance is recorded at a higher frequency and therefore more resolution lines are available. Super-VHS and Hi8 systems use Y/C video.

YCbCr
: This term refers to the three different components that make up component video. Y represents the luminance portion of the signal. Cb and Cr represent the two different chroma components. CCIR-601 specifies 8-bit encoding for component video. White is luma code 235. Black is luma code 16.

YIQ
: A trichromatic color system used in NTSC color television systems. The luminance signal is the Y signal. There are two elements of color, hue and saturation. The color information is modulated onto the subcarrier with the phase of the sine wave describing the color itself and the amplitude describing the level of color saturation. Y uses 4.5 MHz, I uses 1.5 MHz and 1 uses 0.5 MHz.

YPbPr
: Part of the CCIR Rec. 709 HDTV standard that refers to three components (luma and two color components) that are conveyed in three separate channels with identical unity excursions. YPbPr is employed by component analog video equipment such as M-II and Betacam. In YPbPr, Pb equals (0.5/0.886) multiplied by Bgamma minus Y. Pr equals (0.5/0.701) multiplied by Rgamma minus Y.

YUV
: The complete set of component signals, that comprise luminance (Y) and the two color difference signals, U (B-Y) and V (R-Y).

— Z—

Zigzag scanning
: Used in the Discrete Cosine Transform, zigzag scanning results in the reordering of DCT coefficients from the lowest spatial frequency to the highest. Since

the highest spatial frequencies have values that tend toward zero, reordering them through zigzag scanning results in long strings of zeros that can be abbreviated using run-length encoding.

INDEX

494